0/04

Beasts of Eden

The publisher gratefully acknowledges the generous contribution to this book provided by the General Endowment Fund of the University of California Press Associates.

Beasts of Eden

WALKING WHALES, DAWN HORSES, AND OTHER
ENIGMAS OF MAMMAL EVOLUTION

David Rains Wallace

UNIVERSITY OF CALIFORNIA PRESS

BERKELEY   LOS ANGELES   LONDON

University of California Press
Berkeley and Los Angeles, California

University of California Press, Ltd.
London, England

© 2004 by the Regents of the University of California

Library of Congress Cataloging-in-Publication Data
Wallace, David Rains, 1945–.
  Beasts of Eden : walking whales, dawn horses, and
other enigmas of mammal evolution / David Rains
Wallace.
      p.   cm.
  Includes bibliographical references and index.
  ISBN 0-520-23731-5 (alk. paper).
      1. Mammals—Evolution.   2. Mammals, Fossil.
  I. Title.
  QL708.5.W25   2004
  599.138—dc22

                                              2003022857

Manufactured in the United States of America
13   12   11   10   09   08   07   06   05   04
10   9   8   7   6   5   4   3   2   1

Printed on Ecobook 50 containing a minimum 50%
post-consumer waste, processed chlorine free. The balance
contains virgin pulp, including 25% Forest Stewardship
Council Certified for no old growth tree cutting,
processed either TCF or ECF. The sheet is acid-free
and meets the minimum requirements of
ANSI/NISO Z39.48-1992 (R 1997) (Permanence of Paper). ∞

TO STEVE FISHER, WHO BRINGS ART TO SCIENCE

Actually most scientific problems are far better understood by studying their history than their logic.

ERNST MAYR
quoted in Leo F. Laporte,
George Gaylord Simpson:
Paleontologist and Evolutionist

CONTENTS

ILLUSTRATIONS

ACKNOWLEDGMENTS

A number of individuals and institutions helped me write this book. Back in 1993, before I even got the idea, David Webb and Bruce MacFadden of the Florida State Museum of Natural History talked to me about mammal evolution. Spencer Lucas of the New Mexico Museum of Natural History, Brent Breithaupt of the University of Wyoming Museum of Geology, and Greg McDonald, then of Hagerman Fossil Beds National Monument in Idaho, showed me around fossil sites and answered questions in 1997. Alan Rabinowitz gave me advice about a trip to Thailand in 1998, and Mark Graham provided advice and hospitality when I was in that country.

David Wake of the U.C. Berkeley Museum of Vertebrate Zoology advised me about sources. Chris Beard, Mary Dawson, and Zhexi Luo of the Carnegie Museum of Natural History answered questions and provided invaluable reference material when I visited there in 2000. Mary Ann Turner of the Yale Peabody Museum spent most of a morning showing me the Zallinger murals, and she, Joyce Gherlone, and other staff members subsequently provided much invaluable information and other help. Andrew Petryn helped me with questions about Zallinger's artistic techniques.

David Archibald of the San Diego State University talked to me at length about a relatively little known subject, and generously provided reference material and constructive criticism. Lowell Dingus of the American Museum of Natural History answered questions, and Matthew Pavlick helped me with illustrations. William A. Clemens of the Museum of Paleontology at U.C. Berkeley spent a morning showing me some of their collection, and Patricia Holroyd of the same institution provided much useful information. Ruthann Knudson of Agate Fossil Beds National Monument helped me with information and illustrative materials. Joan Burns of Williams College generously provided me with a photo of her father, George Gaylord Simpson.

This book would have been impossible without the work of the many authors, living and dead, who've written about mammal evolution. I'm particularly grateful to writers on evolutionary history, including Martin J. S. Rudwick, Adrian Desmond, Peter J. Bowler, and Ronald Rainger, among others, for the inspiration their work has provided. Vincent Morgan and Spencer Lucas were especially helpful in letting me see prepublication copies of their bulletins on Walter Granger and the Osborn Fayum expedition, and by answering questions that arose from that. And, of course, I'm grateful for the writings of the paleontologists themselves, from Cuvier to the amazingly prolific Simpson.

This book also would have been impossible without the services of the University of California at Berkeley, particularly the Biosciences and Earth Sciences libraries, as well as the Doe, Bancroft, and Anthropology libraries. The Geology Library at the Ohio State University, Columbus, was a valuable resource as well.

Finally, I wish to thank my agent, Sandy Taylor, and my editor at the University of California Press, Blake Edgar, for their indispensable help in transforming this project from an ambitious proposal to a manuscript.

For misinterpretations, misquotations, factual errors, or other flaws, of course, I take full responsibility.

PROLOGUE

The Fresco and the Fossil

If there is a Sistine Chapel of evolution, it is Yale University's Peabody Museum. Of course, the Peabody's neo-Gothic brick edifice is less august than the Vatican's papal shrine. Past a dim corridor off the museum's vestibule, however, a painting as breathtaking as the Sistine murals covers the wall of a soaring room with an immense landscape of rosy cliffs, exotic vegetation, and life-sized dinosaurs. This is Rudolph Zallinger's *Age of Reptiles,* and, like Michelangelo's murals, it is a fresco, brushed on plaster day after day over a period of years. Like the Sistine's biblical vision of human creation and judgment, it shows the beginning and end of a world, that of the great saurians that left their bones in western North America.

Comparing a big dinosaur picture to Renaissance painting's supreme achievement seems presumptuous, to be sure, and Rudolph Zallinger is not considered a titan even of modern art. Born to Siberian refugees in 1919, he attended Yale's School of Fine Arts on scholarships during the Depression, when it was training illustrators, and taught there after graduation. The Peabody hired him to paint the mural in 1943 at $40 a week, because the director felt its Great Hall "resembled a dismal barren cavern devoid of color." Zallinger took a crash course in paleontology and finished the job in 1947. Three years later, abstractionists purged illustrators from Yale's art school, and Zallinger might have spent his career doing ads in his home town of Seattle if the Peabody had not appointed him "artist in residence," a position he held—teaching elsewhere—until his death in 1995.

The art historian Vincent Scully was speaking outside the mainstream when he complained in 1990 that Zallinger's work is not "valued as it ought to be by modern critics." Most critics aren't aware it has artistic value. In a way, however, Zallinger had more in common with Michelangelo than do the modernists who have prospered from critical and commercial patronage. Great noncommercial institutions—religion, science—patronized both muralists, with similar parsimony, and both murals have become icons, endlessly reproduced in popular media. In one stamp issue alone, the U.S. Postal Service printed six million copies of scenes from the *Age of Reptiles*. And although the modernist canon excludes Zallinger's mural, its status is more than popular. Scully called it "abundantly entitled" to a "distinguished position among contemporary mural paintings." Soon after its completion, the art historian Daniel Varney Thompson, a fresco specialist, said "that wall is the most important one since the fifteenth century." It won the Pulitzer Prize in 1949. W. J. T. Mitchell, a University of Chicago art critic and cultural historian, recently described it as "a modern monument" that displays prehistoric life's evolution as "a single, unified landscape panorama, a symmetrical tableau of stately reptilian demigods in a peaceable arcadian kingdom."

"I was moved nearly to tears by the Zallinger fresco in the Great Hall when I visited there as a callow college senior," one dinosaur scientist, Peter Dodson, wrote in 1999. "This portrayal of the history of 350 million years of life on land is familiar to every paleontologist and to every reader of natural history books, one of the high-water marks of natural history illustration in the 20th century." Another dinosaur scientist, Robert Bakker, traced his vocation to seeing the picture in a *Life* magazine article at his grandfather's house in 1955.

Such enthusiasm is justified when one sees the "arcadian kingdom" across its hall of dinosaur skeletons. Yet the eminence of the *Age of Reptiles* illuminates something strange about it, an oddity so ingrained in our attitude toward evolution that it seems normal. What if, rather than painting the biblical story in the Sistine Chapel, Michelangelo had painted the Hellenic one—with Zeus, not Jehovah, presiding over the earth? It would still be great art, but it would not be about *us* in the way that it is. A Hellenic Sistine Chapel would have left out the mythology that mainly formed Western civilization. Zallinger's *Age of Reptiles* does something similar. It presents a core vision of evolution that is not really about *us*, and in doing so, it reflects a blind spot in our view of life that seems to have grown, oddly, the more we have learned.

In the past century, as we have understood life's history and functions better, we increasingly have told ourselves, in part of our minds, that evolution is something that happened to creatures so unlike us—dinosaurs—that they not only are long extinct, but have left no heirs. (And even if they have left heirs in the form of birds, as is now widely believed, comparing a tyrannosaur and a chicken seems to mock the very notion of evolutionary descent.) A glance in any bookstore or library will demonstrate this. Dinosaurs prevail overwhelmingly in evolution sections. They are so prevalent that W. J. T. Mitchell called them the "totem" of industrial civilization, "*the* animal image that has, by a complex process of cultural selection, emerged as the global symbol of modern humanity's relation to nature." The media's obsession with dinosaur size and strength supports this. Movie dinosaurs have as much in common with machines as organisms, and the sense they convey of human *non*relationship to nature's evolutionary past is part of their appeal.

Of course, many people believe that evolution has been happening to creatures like us since dinosaurs first evolved. The Great Hall's mural gives a nod to the existence of such warm-blooded, hairy animals during the reptile age. At its end, next to Zallinger's signature, a tiny, long-snouted mammal named *Cimolestes* crouches underfoot of tyrannosaurs and hadrosaurs. The fact largely seems relegated to a smaller, dimmer part of our minds than dinosaurs occupy, however, and the Peabody Museum also demonstrates this. Mitchell wrote that Zallinger "never produced any work that came remotely close to his masterpiece either in scale or ambition or in cultural impact," and this is true in the sense that he never painted a bigger or more famous picture. Yet there is another mural full of marvelous animals in evolution's Sistine, past the Great Hall in a smaller, dimmer room at the back.

I first saw this other picture in a Time/Life book, *The World We Live In,* when I was ten years old. It was called the *Age of Mammals* and showed the life that inhabited western North America in the sixty-five million years after the dinosaurs' demise. Zallinger's *Age of Reptiles* occupied the preceding pages, and his dinosaurs impressed me, of course. But the *Age of Mammals* fascinated me, although its attractions were less obvious. The mammals' earth-colored pelts blended with the background instead of standing out like the celestially colored saurians. Many were confusingly small, and even the big ones seemed fuzzy compared to the crested, armored dinosaurs. They and their forest and prairie settings had a familiarity, however, that appealed to me more than the otherworldly tree ferns and pinnacles of the dinosaurs' world. If the *Age of Reptiles* was an Olympus of "reptilian demigods," then the *Age of Mammals* seemed an Eden—not a monument to a dead world, but the embryo of the living one. At the picture's lush beginning, a giant serpent dangled from a bough as though offering something—perhaps the fruit in its tree, perhaps the secret that would eventually transform prehistoric mammals like *Barylambda*—a flat-skulled, thick-tailed, almost dinosaurian beast—into horses and elephants.

The two pictures seemed so unlike that I had assumed that they were by different artists until I read Mitchell's book in 1998. In fact, Zallinger painted both murals, although he had to wait to do the smaller one until Yale could raise funds in the 1960s. (The reproductions in *The World We Live In* were not of the murals themselves but of "cartoons," preliminary studies with some differences in content and execution.) It is as though Michelangelo had finished covering the Sistine Chapel with Hellenic mythology, and then, years later, stepped into an antechamber to paint the biblical story.

The book's dinosaurs and mammals were on the same scale, so the contrast in the murals' size and location surprised me when I first saw them in October 2000. "This room is a kind of annex to the reptile hall," Mary Ann Turner, the collections manager, said as we stood in the windowless gloom before the mammal mural, recessed above display cases in its low-ceilinged chamber. "If there's a reception or other event that involves food, they have it in here." I had hoped that someone could explicate the painting, but no early mammal expert was available, and although Turner was helpful and got a technician to floodlight it, she couldn't tell me much. She wasn't even sure whether it was a fresco, painted on plaster like the *Age of Reptiles*. She thought it might be on canvas. My artist wife suspected other technical differences. Once floodlit, the richness of the mural's colors struck us—its scarlet and gold Ice Age foliage echoed the glorious Indian summer morning outside—and we wondered if Zallinger had used a new kind of paint. But Turner wasn't sure about that either.

More than physical obscurity surrounded the *Age of Mammals*. Unlike the reptile hall's newly remounted dinosaur fossils, the annex's few mammal skeletons had a forgotten air. A wolf-sized, massive-skulled one under the mural looked as though it belonged to one of the painted early mammals, but it was labeled differently from any of them, *Synoplotherium,* and that was all the label said. Turner said nobody paid much attention to *Syn-*

Figure 1. O. C. Marsh's *Dromocyon* skeleton, now named *Synoplotherium*. Courtesy American Museum of Natural History Library.

oplotherium—it was "not one of the more popular animals, research-wise." It seemed a kind of paleontological unclaimed body, although she added that it had been the first skeleton mounted in the museum after its founder, Othniel C. Marsh, America's first paleontology professor, died in 1899. A man of legendary possessiveness, Marsh disapproved of fossil reconstructions, except in drawings or papier-mâché, because they took bones out of his collection drawers.

This apathy toward one of the first fossils to be reconstructed seemed typical of present popular attitudes to our own evolutionary branch. "The class entered the museum's great hall," began a 1999 description of a visit by kindergartners to the Peabody. "Here was . . . *Triceratops* . . . *Chasmosaurus* . . . *Mosasaurus*. . . . The children moved from fossil to fossil mesmerized by the massive skulls with their spooky, vacant eye sockets and menacing sharp teeth. . . . The next room was filled with mammal fossils. . . . But the children walked past these hard-won treasures quickly. After all, they were not dinosaurs."

This hasn't always been so. A lavishly illustrated 1910 book by Henry Fairfield Osborn, Marsh's successor as America's reigning paleontologist, prominently displays a photograph of the Peabody's *Synoplotherium* skeleton, resoundingly—if confusingly—labeled "*Dromocyon vorax*, a mesonychid creodont from the Upper Eocene epoch." The same page shows a dramatic reconstruction by a famed scientific artist, Charles R. Knight, of a "similar form," snarling over the carcass of a uintathere, a primitive herbi-

Figure 2. Charles Knight's restoration of Cope's *Mesonyx*. Courtesy American Museum of Natural History Library.

vore. A few pages later, in describing the Eocene epoch, Osborn says, even more confusingly: "Here also the skulking and swift-footed *Mesonyx* (or *Dromocyon,* Mesonychidae) is represented."

Osborn's 1910 book was a scientific tome, but the public knew about "the skulking and swift-footed *Mesonyx.*" Knight's reconstruction had originally illustrated a vivid 1896 Osborn article on early mammals in *The Century,* a popular magazine. "The next animal one sees is among a grove of young sequoias, standing over the skull of a uintathere," Osborn says of *Mesonyx.* "He has a very long, low body, somewhat like that of a Tasmanian wolf, terminating in a powerful tail, short limbs, and flattened nails. . . . The wide gape of his mouth exposes a full set of very much blunted teeth, which proves that this huge flesh-eater could hardly have killed the uintathere, but has driven away another beast from the carcass. Perhaps, like the bear, he had a taste for all kinds of food."

Once upon a time, people paid attention to the fossil beneath the fresco. It began with O. C. Marsh's archenemy, Edward D. Cope, who spent his career fighting Marsh's bid to monopolize paleontology. Their feud became nineteenth-century America's great scientific scandal, culminating in 1890, when they spent a month accusing each other of theft and plagiarism in the *New York Herald,* then the nation's leading newspaper. Famous for fighting over dinosaurs, they fought longer and harder over mammals, starting in 1872 with a Cope expedition that Marsh tried, not unsuccessfully, to wreck. Among Cope's finds were two fragmentary skeletons of blunt-toothed, flat-nailed creatures that he named *Synoplotherium* ("joined hoof beast") and *Mesonyx* ("half claw"), although he later dropped *Synoplotherium* after deciding both fossils belonged to the genus *Mesonyx.* When Marsh acquired a similar skeleton in 1875, he ignored Cope's names, and called it *Dromocyon* ("swift running dog"). Neither man accepted his hated rival's names if he could help it. Naming fossils has a magical side, because it is the first step in "resurrecting" extinct organisms, and Cope and Marsh jealously clung to their necromantic incantations. When they died, their three mesonychid genera were among hundreds of fossils, similar or identical, to which they had given different names.

Henry Fairfield Osborn took Cope's side in the feud, but later found it a "painful duty" to devote thirty years "to trying to straighten out this nomenclatural chaos." His confusing use of *Mesonyx* and *Dromocyon* in his 1910 book shows his dilemma. For a while, it seemed that they were the same beast, whose name then would have been *Mesonyx,* because Cope's discovery had preceded Marsh's. But then Osborn's colleagues J. L. Wortman and W. B. Scott decided that *Mesonyx* and *Dromocyon* really were different genera, in which case the Peabody skeleton would have kept its Marsh name. But then another Osborn colleague, W. D. Matthew, decided that *Dromocyon* and Cope's long-discarded *Synoplotherium* were the same, and, since *Synoplotherium* preceded *Dromocyon,* it became the valid name of the Peabody's reconstructed skeleton. Cope's name for Marsh's mesonychid fossil finally prevailed in Marsh's museum.

That would not have amused the professor, and neither would another historical happenstance. It was no accident that *The Century* almost always captioned Charles Knight's illustrations for Osborn's articles with Cope names—and it infuriated Marsh. Knight's *Mesonyx* in the 1896 *Century* seemed familiar to me, and I realized why the next time I looked at the *Age of Mammals* cartoon. Zallinger adapted his own snarling *Mesonyx,* right down to the long, banded tail, from Knight's illustration, simply reversing the image and skewing the angle to make it seem more three-dimensional. *Mesonyx* also appears in the Peabody mural, almost directly over the *Synoplotherium* skeleton, although Zallinger further modified the beast, which would have had trouble standing with the hind legs Knight gave it. So

Cope's mesonychid names ended by trumping Marsh's not once, but twice, in the Peabody Museum.

"Gad! Gad! *Gad! Godamnit!*" an eavesdropper once heard Marsh cry when comparing a Cope paper with some of his own fossils. "*I wish the Lord would take him!*"

Mesonyx's background is typical of Zallinger's *Age of Mammals.* Controversy involved most of the mural's central figures, so it is implicitly a historic as well as a prehistoric tableau. Whereas Charles Knight had his *Mesonyx* snarling over a uintathere carcass, Zallinger showed his snarling at a live uintathere, which brandishes saberlike tusks in defiance. As it happened, uintatheres were the flashpoint of Cope's and Marsh's first pitched battle, fought just after Cope's beleaguered 1872 expedition. Zallinger's confrontation implies a prehistoric reincarnation of their feud. A little earlier in the mural, two hippolike beasts roaring at an enemy belong to a genus, *Coryphodon,* discovered in the 1840s by the pioneer English paleontologist Richard Owen, who spent five decades fighting over mammals with early evolutionists. Farther along in the painting roars an even bigger herbivore, a brontothere, which fueled a twentieth-century debate between Henry Fairfield Osborn and younger paleontologists, including one of the greatest and most contentious, George Gaylord Simpson.

Zallinger's oblique references to the old bone hunters remind me of the figures of saints and prophets that lend brawny individuality to the Sistine Chapel's cosmic dramas. Not that men like Marsh and Cope were saintly—they were indeed like snarling beasts sometimes. Yet there was much in them of Michelangelo's larger-than-life visionaries. They saw things that normal humans did not, things that many still find incredible, but that nevertheless, on the evidence, seem true. They were hard on each other and on their assistants and colleagues, but they were hardest on themselves, spending health and wealth on work that brought them little more than the peculiarly human power of seeing deep into time.

Zallinger's mammals also resonate with recent controversies. When Cope described *Mesonyx* in his *Vertebrates of the Tertiary Formations of the West* (called "Cope's Bible," because it is nearly a foot thick and weighs over ten pounds), he noted its strangeness. "The flat claws are a unique peculiarity, and suggest affinity to the seals, and an aquatic habit," he wrote. "The teeth, moreover, show a tendency in the same direction, in the simplicity of their crowns." He guessed that *Mesonyx* had fed largely on the freshwater turtles common in Eocene sediments. But the seallike features puzzled Cope, because, as he wrote: "The structure of the ankle forbids the supposition that these animals were exclusively aquatic, as it is the type of the most perfect terrestrial animals."

Mesonychids continued to puzzle scientists. In the 1960s, a paleontologist, Leigh Van Valen, noted similarities between their teeth and those of primitive whales, but the connection seemed far-fetched. Then, in 1979, Philip Gingerich, a University of Michigan researcher hunting early mammals in Pakistan, found a 50-million-year-old fossil with teeth resembling those of *Mesonyx*, but whalelike ear bones. He named it *Pakicetus* ("Pakistani whale"), but as more of its coyote-sized skeleton emerged, it became clear that it had looked and acted more like a mesonychid than a whale. Able to run fast, but also to swim well, it had probably spent its life wandering beside rivers and estuaries, eating whatever land or water prey came its way. In 1992, a 48-million-year-old Pakistani fossil turned up that had similarities to *Pakicetus* but had clearly lived a more aquatic life, since it had short, sprawling legs, webbed feet, and an otterlike tail. Its discoverer, Gingerich's student Hans Thewissen, named it *Ambulocetus* ("walking whale"), and guessed that it had behaved like a crocodile, feeding on fish, but also lurking in the shallows to catch drinking animals. It even had an elongated, crocodilelike snout, and it had grown to four hundred pounds. The discovery of other, successively whalelike genera implied that the outlandish creatures had been part of a sequence that led from mesonychids to whales as relatives of Cope's "perfect terrestrial animals" gradually traded legs for flippers.

Doubts later arose as to whether mesonychids were whale ancestors. Some researchers proposed a closer relationship between whales and artiodactyls, "even-toed" beasts, the largest group of hoofed animals, or ungulates. (Camels, pigs, deer, sheep, goats, antelope, and cattle are artiodactyls—the other living ungulates are the "odd-toed" perissodactyls, including horses, rhinos, and tapirs.) Most modern artiodactyls are specialized herbivores, of course, but many earlier ones had more omnivorous habits, which might have extended to fish-eating. *Pakicetus*'s teeth and ankle bones were different enough from those of mesonychids to suggest that the genus had another ancestor, one "not inconsistent" with an artiodactyl connection. In particular, researchers saw a link to one semi-aquatic artiodactyl, the hippopotamus.

Still, the ongoing controversy doesn't vitiate a fundamental insight. "Evidence suggests," Gingerich wrote in 1983, "that *Pakicetus* and other early Eocene cetaceans represent an amphibious stage in the gradual evolutionary transition of primitive whales from land to sea." Such a transition is an example of macroevolution, the evolution of one group of organisms—whether mesonychids or artiodactyls—into another, whales. Anti-evolutionists have been even less willing to acknowledge this than microevolution, the transition from one species to another, since it gives evolution an even greater role in life's history. But *Pakicetus* and *Ambulocetus* allowed Gingerich to say: "Fossils contradict the notion that whales suddenly appeared full-blown, without intermediate forms. Intermediates, missing links, are everywhere."

Such insights permeate Zallinger's mammal mural, because fights over mammal fossils have probably played a greater part in the growth of evolutionary ideas than any other paleontological phenomenon. It is one thing to think that dragonlike beings such as dinosaurs lived in the faraway past. Humans have had such ideas since the first recorded myths, and no evolutionary explanations are necessary. It is another thing entirely to think that beasts similar to, but also different from, present ones lived long ago. Although they have vanished, too, their similarities to the living raise deep implications.

Cope's and Marsh's squabbles may have been unseemly, but they also epitomized the chief opposing evolutionary camps of their time. Marsh's work on the mammals in the Peabody murals provided the best support then known for Charles Darwin's evolutionary ideas, which came, of course, to be generally accepted. But Cope's work on the same mammals provided support for challenges to Darwin that seemed compelling to many scientists at the time, and that still remain partly unanswered. Mammals have continued to provide major support for, and challenges to, the various paradigms of how life has changed through time. In this sense, the mammal mural may represent a stronger evolutionary vision than the reptile one.

The mammal mural may also be a more accurate vision. Zallinger's plodding dinosaurs look stiff compared to the galloping ones in recent reconstructions, because when he painted them in the 1940s, science rejected the idea—first posed by pioneers like Cope and Marsh—that dinosaurs' similarities to birds meant that they also had high body temperatures and led active lives. Robert Bakker traced his conversion to the revival of this "hot-blooded dinosaur" paradigm to standing in the Great Hall and thinking: "There's something very wrong with our dinosaurs." The mammal mural also has anachronisms, such as *Planetetherium,* a lemurlike beast that glides like a frisbee through the forest canopy. Before my Peabody visit, I'd stopped at Pittsburgh's Carnegie Museum, where Mary Dawson and Chris Beard, early mammal authorities, told me research has shown that *Planetetherium* was not a glider, but a hedgehoglike creeper. Yet we need not wonder whether *Planetetherium* was "warm-blooded" and gliding mammals like the living Asian colugos did inhabit North America. This difference between the murals may seem paradoxical, since Zallinger drew on Charles Knight's paintings for both, but he may have drawn more on Knight's mammals, because his dinosaurs look too active for 1940s ideas of sluggish saurians.

It is hard to draw conclusions about the murals' creation, however, because their past is murky. Documentation is sparse even on the reptile one, largely consisting of Zallinger's technical account of its execution. Of the other, he said only that the Museum's director, Carl O. Dunbar, "had always wanted the mammal mural and was primarily responsible for making it happen." Dunbar did shed some light on the mammal cartoon's origin shortly before his death in 1979, writing that he had chosen "the richly fossiliferous formations of the Rocky Mountain region and the adjacent Great Plains" for the subject because it contained "the finest record of evolution of the mammals." Dunbar "drew a linear profile to be developed into an idealized landscape" and called in Joseph Gregory, the Peabody's curator of vertebrate paleontology, and Roland Brown, curator of paleobotany at the Smithsonian, as advisers. "Then as Rudy blocked out the landscape, we spent many hours searching the literature for precise data (size, special structure, etc.) for the animals to be shown. For the next eight months I looked over Rudy's shoulder almost daily as we conferred about details."

Zallinger worked so hard on the cartoon that he once fell asleep on his feet while they conferred. This was about the only detail of its execution that Dunbar mentioned, however, and he had nothing to say about the mural, painted after his 1959 retirement. When I asked Joseph Gregory and his successor as Zallinger's advisor, Elwyn Simons, about it, they were so reticent that I wondered whether the mural's painted confrontations reflected more than historical squabbles. But then, none of the mammal specialists I asked about it had much to say, in contrast to Dodson's and Bakker's eloquence about the *Age of Reptiles*. One said only that he preferred the reptile painting's colors.

This raises the question of whether, as W. J. T. Mitchell implied, the *Age of Mammals* is simply not as good a picture as its famous neighbor. There certainly is a unique sense of discovery in the reptile painting. Zallinger was said to have known so little about dinosaurs when offered the job that he went home and looked them up in the encyclopedia. An excited naïveté synergizes with a huge space to generate its Olympian plenitude.

The mammal mural lacks this serendipity, yet Zallinger's increased knowledge of anatomy and botany—and skill in drawing, modeling, and color—give it a supple liveliness. (The Peabody hired him again in 1951 to do the mammal cartoon, at the instigation of *Life* magazine, so he had had over a decade to develop the mural.) There is a sense of animal movement in habitat that compares favorably with the reptile mural's stateliness. It is a little as though my hypothetical Michelangelo, in the years between painting his Hellenic murals and his biblical ones, had developed a style closer to Raphael's illustrative brilliance than his original stony grandeur. The liveliness has worked against the mammal mural's artistic reputation. As Mitchell observed, illustration is not considered fine art today, and the reptile mural's stiffness, which prompts comparisons with medieval muralists like Giotto, has enhanced its artistic cachet. But there are stirring rhythms in the mammals' deftly rendered musculature. It is not just a painting of warm-blooded animals—it is a warm-blooded painting.

Anyway, I find the Peabody murals more alike than different. A colleague of Zallinger's told me that he had painted both on plaster, with the same pigments, and simply used more cadmium, magenta, and vermilion to "jazz up" the mammal one. A characteristic quality imbues both, an unforgettable sense of animal life "assembling" in landscape, as Mitchell put it. "Zallinger was essentially not allowed to 'invent' at all," wrote Vincent Scully. "But he had to find cunning ways 'to discover things not seen' and to present 'to plain sight what does not actually exist.'" I think he did so in both paintings. Again, there are Sistine parallels. Michelangelo had advisers (not least two popes), and was certainly not expected to "invent." He also made artistic borrowings, drawing from Hellenic as well as biblical iconography. Both painters used old stories and pictures in new ways, however, integrating them into narratives that allow contemplation, not simply of framed events, but of life's continuous flow. This sets them apart from contemporaries. Botticelli's and Perugino's murals in the Sistine, and Charles Knight's in the American and Field museums, are blown-up easel paintings by comparison. The other painters give us pieces of mythic deep time. Michelangelo and Zallinger guide us into its current.

Zallinger did leave one clear expression of his artistic aims. "In natural history museums," he wrote, "the traditional convention for painted restorations of ancient animals made use of a single animal or a group of one or perhaps a few species, which strictly observed a geological time frame and locations . . . I ultimately proposed a different convention, that of using the entire available wall . . . for a 'panorama of time,' effecting a symbolic reference to the evolutionary history of the earth's life."

This book will invoke his panorama as I try to do for prehistoric mammals what dozens of books have done for dinosaurs—tell the story of their discovery and evolution. In some ways, it is a better story than that of the dinosaurs. It is a Cinderella story. Although mammals first evolved at about the same time as dinosaurs, during the Triassic Period, which ended over 200 million years ago, they remained small and inconspicuous during the next 150 million years, as dinosaurs ascended to spectacular gigantism. Only after the great Mesozoic dinosaurs vanished did mammals evolve the strikingly large forms of Zallinger's mural, not only ungulates, like horses and deer, but more exotic living beasts, like proboscideans, and even stranger extinct ones, like mesonychids and uintatheres. And then some mammals—whether mesonychids or artiodactyls—evolved into the largest animals ever—the whales, which surpass the greatest dinosaurs in size.

It is also a mystery story. The origin of mammals was one of the nineteenth century's great enigmas. Henry Fairfield Osborn's last conversation with Cope, on his friend's deathbed, was an "animated" argument about "this most coveted of all relationships." And although we now know what kind of animals mammals evolved from, other details of their origins remain unclear, and much of their later evolution is even less well understood.

The reason for the mammals' Mesozoic eons of apparent "arrested development" is still one of paleontology's problems, provoking much argument. The reason for their Cinderellalike transformation at the Mesozoic's end is even more mysterious. Dinosaur-age mammals were not only small and inconspicuous. They were so unlike modern ones that it is unclear how and why animals like Zallinger's tiny *Cimolestes,* much less even older, stranger creatures with names like "triconodonts" and "pantotheres," evolved into the amazing diversity of living beasts. Early mammal fossils, especially Mesozoic ones, are usually teeth and bits of bone, seldom whole skeletons, and very seldom traces of hair, internal organs, or other soft parts, so we may never be sure exactly how the transformation occurred. But the enormously accelerated research of the past few decades is beginning to shed light, sometimes in surprising ways, on the origins of living beasts: the monotremes that lay eggs—platypuses and echidnas; the marsupials that give birth to fetal young and nourish them externally—possums, kangaroos, and wombats; the placentals that nourish their fetuses internally—elephants and manatees . . . anteaters and armadillos . . . rabbits, squirrels, and humans . . . shrews, bats, lions, rhinos, giraffes, and dolphins.

Of course, dinosaurs are wonderful, and I don't mean to slight them. Paleontology is wonderful. Fossils indeed seem, as our ancestors saw them, a kind of divine revelation, the sculpture of eternity. But mammal evolution has a special claim on our attention because, as Zhe-xi Luo, another early mammal specialist at the Carnegie Institute, pointed out during my visit there, it resonates in our daily lives. "We all go to the dentist because mammals alone, of living vertebrates, have permanent teeth," Luo told me, sitting in an office stacked with exquisitely preserved, newly discovered fossils of dinosaur-age mammals. "The reason is that mammals have a determinative growth pattern."

Other living toothed vertebrates, from fish to crocodiles, keep growing throughout their lives and keep replacing their teeth as old ones wear out, a pattern that allows for a steady supply of healthy teeth. But it has a drawback. The disposable teeth are not very specialized and don't allow very efficient food-processing. Crocodiles can't chew each bite even once, let alone the nine times of nursery admonitions. They *have* to gulp their food, because they lack molars to chew with. On the other hand, mammals' permanent teeth have evolved into a complex set of shears, slicers, piercers, crushers, and grinders that processes food with unique thoroughness—until they break, decay, or wear out. (Some mammals have continually growing permanent teeth, but these are secondary adaptations that humans, anatomically primitive in many ways, unfortunately lack.)

Fillings, crowns, and worse can seem a high price to pay for fancy teeth, but the first mammals' dental gamble also led to some of the things we value most. The jaw and palate adaptations that evolved with them helped our ancestors to taste, smell, and even hear better, and to process their food efficiently enough to become increasingly active and perceptive. Although we still don't understand fully how they evolved, those unique qualities of our small early relatives are why we are here today.

ONE

Pachyderms in the Catacombs

THE MOST STRIKING FIGURE IN the Peabody's *Age of Mammals* comes toward the end, among the Ice Age's brilliant foliage. It is a wooly mammoth, and it takes up most of the wall's height with its rufous bulk, curling tusks, and high-domed cranium. It is the only figure, except for a soaring bird of prey, that extends above the horizon. Unlike the mural's coryphodonts and uintatheres, it is not engaged in a confrontation but gazes forward serenely as though confident of its preeminence. Even the naked pink nostrils at the end of its trunk have a confident air. The entire 60-foot-long painting, with its grandly shifting scenery and dozens of figures, might have been laboring to produce this magnificent and intelligent beast.

Yet if the mammoth implies the culmination of certain valued mammalian qualities, there is another giant beast even nearer to the mural's end that does not. It stands on its hind legs, head slightly cocked, to look back toward the mammoth in a way that is hard to read. It might be challenging, fearful, or curious. Or it might not be looking at the mammoth at all, but simply gazing vacantly into the haze of time. It is hard to imagine what's in the creature's mind, because it is so strange. Unlike the elephantine mammoth, *Megatherium* resembles no familiar living creature. Indeed, there is something of Zallinger's slightly toylike dinosaurs about it. Standing propped on its massive tail, it is nearly as tall as the mammoth, but its small head and pigeon-toed feet make it seem clumsy rather than

Figure 3. *Megatherium* and *Glyptodon* (Pleistocene) from
Zallinger's *Age of Mammals* mural. Courtesy Peabody Museum of Natural History, Yale University, New Haven, Conn.

majestic, despite its gigantic claws. Compared to the mammoth's, its eyes are tiny and dull.

The two figures facing each other like heraldic beasts, the one familiar and alert, the other bizarre and sluggish, lend a certain ambiguity to all the activity that precedes them, and this seems more than accidental. While Zallinger places them prominently at the end of mammal evolution's prehistoric story (both vanished less than ten thousand years ago), they figured with equal prominence at the beginning of its historical one, a beginning that also mingled confidence with obscurities. They embody a basic question about evolution—whether it is going "somewhere"—progressing toward "higher" traits like a mammoth's intelligence—or "nowhere"—producing smart and stupid creatures with aimless impartiality. Since the question involves human as well as mammoth intelligence, paleontologists have debated it vigorously since the science began.

The idea that mammals have progressed anywhere over time is a recent one, although the concept of mammals is ancient, reflected in the Greek word for a hairy animal, *therion,* as opposed to a scaly one, *herpeton.* Aristotle recognized a distinct group of air-breathing, live-bearing creatures. "Some animals are viviparous, some oviparous," he wrote. "The viviparous are such as man, and the horse, and those animals which have hair; and of the aquatic animals, the whale kind as the dolphin." He also noted that all viviparous quadrupeds then known had teeth, and described them according to their various dentitions. Aristotle's observations decayed into hearsay and fantasy in Roman writers and medieval bestiaries, but post-Renaissance taxonomists like John Ray reaffirmed them. When Linnaeus created the class Mammalia, based on the feeding of young with milk, in the 1759 edition of his *Systema Naturae,* he included the animals now recognized as such. Except for mythic creatures like griffons, however, the idea that beasts might have been very different in the past—that the earth's living fauna might have changed in major, indeed startling, ways—did not occur to naturalists even during the Enlightenment. They would have to dig into the obscurities beneath living fauna to encounter it.

That period's intellectual capital, still known as the city of light, actually overlies greater areas of darkness than most. One of my more vivid Parisian memories is of a walk through the catacombs below the boulevard Raspail, where grave diggers stacked the bones of six million people when cemeteries overflowed in the late eighteenth century. I don't know of a larger or tidier ossuary. Chamber after chamber was piled to the ceiling with carefully sorted femurs, crania, tibias, and pelvises, and I saw only a few of the man-made caverns, which extend for over 200 miles. They weren't dug to house the bones, but because Paris happens to be located on one of the world's best deposits of calcium sulfate, also known as gypsum and, when powdered, as plaster of Paris. Malleable, durable, and snowy white, gypsum is a first-rate building material, and miners began quarrying it when the Romans founded the town of Lutetia on the site of Paris two millennia ago.

Authorities have used the gypsum mines to dispose of human remains at least since Romans threw the beheaded corpses of St. Denis and other missionaries into one on "martyr hill"—Montmartre. But more than human bones rest in the gypsum, a sedimentary rock that forms in shallow, coastal ponds where the climate is warm. Such places supported rich prehistoric faunas, as when the Paris gypsum originated some fifty million years ago, and shallow ponds preserve bones well. Fossils were common finds in the quarries as building boomed in the Enlightenment, and miners who had previously discarded bones found that they could sell them to the "savants" who thronged the capital. Displayed in "cabinets" along with crystals and other curiosities, the gypsum fossils included turtle and crocodile skeletons, but most were mammalian, as their owners would have perceived. Most other toothed animals have rows of identical spikes or pegs in their jaws; most mammals have a Swiss Army knife set of incisors, canines, premolars, and molars. Since teeth are the most durable vertebrate fossils, collectors could recognize even fragmented beasts.

Enlightenment collectors perceived little else about the plaster-of-Paris bones, however. Georges-Louis Leclerc, comte de Buffon, director of the royal natural history collections, did not encourage such perceptions, ignoring the gypsum fossils in his voluminous writings, and declaring that "the bones, horns, claws, etc. of land animals are seldom found in a petrified state." Like other eighteenth-century naturalists, Buffon was interested mainly in constructing a general "theory of the earth" to match the seventeenth century's Cartesian and Newtonian cosmologies. Finding and classifying bones played no great part in the undertaking.

If Buffon had remarked on the Paris gypsum fauna, its crocodiles and other tropical aspects probably would have pleased him. They coincided with his theory that the earth, gradually cooling from a molten state, had been warmer in the past, and that tropical animals had then inhabited the north, as apparent elephant and rhino bones in Europe and North America seemed to show. Buffon was vague as to how such animals had originally been "born," as he put it, but he assumed that, aside from their emigration to the present-day tropics, they had not changed much since. An elephantine fossil from Ohio seemed to have nonelephantine teeth, but Buffon's colleague Louis Daubenton thought hippo teeth had gotten mixed with the skull. Buffon did not dwell on such confusions anyway. Intellectual consistency was not required of Bourbon courtiers, and the fact that the Paris gypsum did *not* contain elephants or rhinos might not have bothered him even if he had noticed it.

After Buffon's death in 1788, aristocrats had more to worry about than classifying fossil mammals. A revolutionary mob stripped the count's sarcophagus of its lead lining, and his son went to the guillotine. The fossil trade picked up again after the Terror, however, and a 26-year-old newcomer to Paris was prepared to regard the gypsum bones in a new light. Georges Cuvier, who arrived in 1795 for an interview at the Musée d'histoire naturelle, was not an aristocrat, or even a savant in quite Buffon's sense. He'd grown up in Montbéliard, then attached to the duchy of Württemberg, and although it was a French-speaking town, it was Lutheran. He'd attended a German academy, the Karlschule in Stuttgart, where he'd studied natural history in the firsthand way that was developing north of the Rhine. Naturalists such as Abraham Werner, a professor at the mining school in Saxony, were more concerned with describing phenomena accurately than with system-building on the Buffon model. They thought the old theorists superficial.

When Cuvier graduated, he had gone to work tutoring the heir of a Norman noble family, the d'Hericys. Although he was at first enthusiastic about the Revolution, he had seen the Terror at work in the city of Caen, and he was relieved when the d'Hericys retreated to the safety of their estate on the coast near Fécamp. There he had improved his spare time by collecting the creatures of tidepools and mudflats, perceptively comparing their diverse structures. He had walked the coastal hills and valleys, observing in Wernerian fashion how local rocks were arranged with fossil-bearing, evidently younger strata toward the top and fossil-barren older rocks at the bottom. Naturalists called the lifeless strata "Primary" and the fossil-bearing ones "Secondary" and "Tertiary." (Although it wasn't quite that simple, because scattered, primitive fossils turned up in upper Primary strata, requiring the addition of a "Transition" category.)

Cuvier was prepared for a museum job after several years of this, and the upheavals that emptied Buffon's tomb had left openings. He was a presentable young man. A portrait from that time, possibly by himself, shows long hair, soulful large eyes, and a sensitive but firm mouth. The clothes are slightly *en déshabille,* giving a rustic air, which must have appealed to the admirers of Jean-Jacques Rousseau, the philosopher of natural harmony, in the museum's older generation. Within the year, young Georges was substituting for an elderly superior as a lecturer in comparative anatomy, and he soon became a member of National Institute, the successor to Buffon's Royal Academy. He didn't let this go to his head, and he kept working diligently. While continuing his invertebrate research, he branched out into vertebrates with the collections his predecessors had accumulated.

Although admiring Buffon's prose and erudition, Cuvier had little respect for his theorizing. He saw from teeth and other features that the fossil European "elephant," called a *mamut* by Siberians, who sometimes encountered its carcass in permafrost, was a different species than the living Indian and African ones. He saw, furthermore, that it was a species that had probably ceased to exist, although not, he thought, because the climate had become too cold. It differed from the elephant the way "the dog differs from the jackal and hyena," he wrote, and "since the dog tolerates the cold of the north while the other two only live in the south, it could be the same with these animals, of which only the fossil remains are known." Elephants had not fled south to escape a cooling planet's chill. Another species, adapted to the cold, had disappeared from some other cause, perhaps a catastrophic incursion of the sea.

Zallinger's mammoth seems to commemorate Cuvier's confident insight, and an earlier historical encomium by Henry Fairfield Osborn might caption the great beast's symbolic role in the mural. "The wooly mammoth is the classic of paleontology; it is the first extinct animal to be found by man; it is the first to be used as proof of a universal deluge; it is the first to be used as proof of the existence of a long extinct world of mammalian life antecedent to the deluge; it is the first to receive a scientific description in the Latin language; it is the first to receive a scientific name." Looming against glaciated peaks, furred to its enormous toes, *Mammuthus* leaves no doubt that it "differs from the elephant."

Cuvier further identified vanished kinds of bears, crocodiles, rhinos, and deer, speculating that "some kind of catastrophe" might have extinguished them, too. So much for Buffon's magniloquent "theory of the earth" and its cooling planet. Yet such animals were still enough like living ones to make the idea of complete disappearance, extinction, seem tentative. Cuvier fixed that in his next paper, however. A colossal skeleton from South America had arrived in Madrid, and he acquired drawings of it. The twelve-foot-long beast had walked on massively clawed feet, and the American savant

Thomas Jefferson thought a similar one was a giant carnivore, a reasonable conjecture from such massive claws. Cuvier knew enough mammal anatomy, however, to see strong similarities between the skeleton and those of the much smaller herbivorous tree sloths still living in South American forests. The ancient beast's few teeth were peglike, hardly a carnivore's, and he concluded that it was a giant, ground-dwelling sloth. He named it *Megatherium*, "great beast," and it was indeed one of the largest that ever lived, as the specimen that Zallinger painted gazing dimly toward his mammoth demonstrates. *Megatherium* also was one of the strangest mammals ever, as the mural shows, and there was no record in 1792 of any such monster living. If it had been alive in the unknown American interior, Cuvier reasoned, word of it would have reached the coast, and no such reports existed. It was almost certainly extinct.

Extinction was a fairly new idea, transgressing assumptions common since Aristotle about a stable natural order. Cuvier was not the first to have it. Buffon had toyed with it, and Jean-Baptiste Lamarck, Cuvier's senior colleague at the museum, believed that prehistoric species had disappeared. Lamarck had a very different interpretation of the phenomenon than Cuvier, however. He thought species had disappeared not by dying out but by transmuting into new species, a feat accomplished partly by passing newly acquired characteristics to their offspring. Habitually reaching up to browse in trees (to give a popular, if oversimplified, example of Lamarck's thinking), the giraffe gradually might have been transformed from a vanished, short-necked species into a living, long-necked one. Life was a process of ever-ascending change, with "animalcules" continually generating spontaneously in water and soil, then transmuting progressively from worms to fishes to lizards to beasts, and eventually to savants.

Rousseau had helped launch Lamarck's career, and the aging naturalist's theory was optimistic in keeping with prerevolutionary assumptions about nature's goodness and change's benign possibilities. That Lamarck had developed it during the 1790s may seem surprising in an impecunious member of the minor aristocracy who stayed in Paris through the Terror, but a half century of Enlightenment evidently had influenced him more than mass executions. His vision also reflected his professional specialty. Originally a botanist, he had been named the museum's curator of invertebrate animals when the republic reorganized it, and had become an expert on fossil shells, more abundant in the strata around Paris even than mammal bones. The fossil shells largely were different from living ones, he found, but not all of them, and shells were abundant and diverse from the lowest strata in which they occurred right up to the highest. This seemed good evidence of the dynamic and continuous process he envisioned.

Cuvier conceived a darker vision, particularly after 1796, when he began studying Paris gypsum bones that the museum had acquired from a defunct collector. They were diverse as well as abundant, so much so, in fact, that confident young Georges was daunted. "I found myself as if placed in a charnel house," he wrote, "surrounded by mutilated fragments of many hundred skeletons of more than twenty kinds of animals, piled confusedly around me." Professionalism overcame bewilderment, however. "The task assigned me was to restore them all to their original positions," he continued. "At the voice of comparative anatomy every bone and fragment of a bone resumed its place." Cuvier developed a technique he called "correlation of parts" to reconstruct animals from incomplete fossils. Even if only the teeth and feet of an animal remained, he could tell if it was a carnivore or an herbivore because carnivores had shearing teeth and claws; herbivores grinding teeth and hooves. The technique would prove to have limitations, but it worked so well for Cuvier that the novelist Honoré de Balzac hailed him as a magician.

As he studied them through the next decade, Cuvier realized that the gypsum fossils were much more unusual than anyone had thought. In the first place, they were embedded deep in the sedimentary rock, unlike most fossil bones then known, which came from loose surface deposits of sand or gravel. This meant, according to Wernerian stratification, that they were much older than fossils such as mammoths. Cuvier thought that the gypsum had formed "many thousands of centuries" before the present. In the second place, some of the species he restored were much less like living French mammals even than mammoths were like elephants.

The bones included eight species—in two genera—of hoofed mammals, which Cuvier called "pachyderms"—thick-skinned beasts. The first genus he reconstructed had a head and teeth resembling a South American tapir's but feet more like a camel's. He named it *Palaeotherium,* ancient beast, and identified five species, ranging from horse to sheep size. The second genus, which he called *Anoplotherium,* "unguarded beast," because it lacked canine teeth, was even stranger. Although they had similar teeth and feet, the three species he placed in it were very different in outward appearance. *A. commun,* the commonest, had had a long tail and "much the same stature as an otter." He thought it had probably lived in water, although its teeth showed that it had eaten plants rather than fish. A short-tailed one, *A. medium,* was "light like a gazelle or roe deer" and had prob-

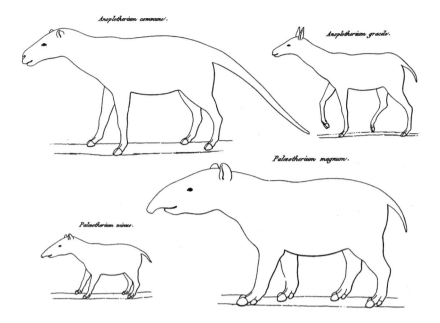

Figure 4. Cuvier's restorations of Paris gypsum mammals. From Georges Cuvier, *Recherches sur les ossemens fossiles* (2d rev. ed., Paris, 1822), 3: 38, pl. 66.

ably looked like one, although only distantly related to antelope or deer. The third species, *A. minus,* had the size and proportions of a hare, but, again, only a distant relationship to rabbits.

Another fossil, the skeleton of a small clawed mammal, was an even greater departure from living European ones, and it permitted a spectacular demonstration of Cuvier's anatomical powers. He quickly discerned a similarity to American opossums, but the skeleton's gypsum matrix hid one proof of its marsupial nature—a pair of pelvic bones thought to support the animal's pouch. Cuvier surmised that the bones lay under some of the vertebrae, and took a calculated gamble with the precious fossil. "So I sacrificed the remains of these vertebrae," he wrote.

> I excavated carefully with a sharp, steel point, and had the satisfaction of exposing . . . the two supernumerary or marsupial bones I was looking for. This operation was done in the presence of some persons to whom I had announced the results in advance with the intention of proving to them—by the act—the justice of my zoological theories . . . from then on, nothing was left to be desired for a complete demonstration of

the proposition already so singular and indeed important, that there are in the plaster quarries that surround Paris, at great depth and under various beds filled with marine shells, the remains of animals that can only be of a genus now confined entirely to America.

Cuvier realized, also, that this world of tapirlike and opossumlike mammals differed from anything else in the Paris basin rocks. Below the gypsum, the basin's older stratum was chalk, a soft limestone that contained seashells and marine reptiles, like a giant "crocodile" he had mentioned in his elephant paper (he later realized that it was a seagoing lizard, eventually named *Mosasaurus*), but no land animals. Above the gypsum, there were still more beds of marine and freshwater sediments, full of shells, but also empty of land vertebrates. Only in the loose, pebbly deposits at the top did significant land mammal bones again appear, and they were of species closely related to living ones, like deer and cattle. Unlike shellfish, land mammals had inhabited the Paris region only at widely separated intervals.

It was all very well for Lamarck to go on about an ever-ascending continuum of life, with his shells in every stratum below the city. Cuvier saw a brilliant mammalian world that had risen, Edenlike, from the ocean, but then had sunk again, leaving no connections with living France, and only tenuous ones with continents half a world away. The gypsum contained no evidence that *Palaeotherium* or *Anoplotherium* had transmuted into some extant genus in accordance with Lamarck's theory. Indeed, one of the species, *Palaeotherium crassum,* "stout ancient beast," had a skeleton so *like* a living tapir's that Cuvier was "persuaded that most travelers would have confused the two animals if they had existed in the same epoch." If transmutation was the rule of life, why would an extinct animal remain so much like a living one despite their separation by time and space? Tapirlike mammals might have migrated from France to America across a land bridge after the Paris gypsum had sunk, but that would not have required that they be transformed. Mummified animals that the museum zoologist, Étienne Geoffroy Saint-Hilaire, had brought back from Napoleon's Egyptian expedition seemed the same as living ones, although four thousand years older.

Cuvier's was a vision of change, but of change without progressive continuity, and thus without the Enlightenment's worldly optimism. He came to see animal life as divided into four *embranchements*—vertebrata, articulata, mollusca, and radiata—which, like prerevolutionary France's social *états,* were too essentially different to transmute from one into the other.

An 1805 portrait suggests a different mood from the decade before, with hair in crisp Empire style, eyes shrunk and hardened, and lips with a slight, disdainful curl. To Lamarck, who sank into blindness and poverty in old age, Cuvier's vision must have seemed a grim retreat for natural philosophy, closing barely glimpsed vistas of progress. If lower animals such as reptiles had not given birth to mammals, how had mammals appeared? If extinct species had not given birth to living ones by hereditary transmutation, how had living animals appeared? It must have seemed to leave life's history in darkness, illuminated only by the catastrophic interventions of an inscrutable Providence.

Cuvier's vision has been interpreted as a reactionary return to biblical fundamentalism. An 1813 English translation of his *Discours sur les révolutions de la surface du globe* with notes by Robert Jameson, a conservative Scottish naturalist, assumed that his fauna-swallowing catastrophes provided evidence for Genesis. Recent historians of science like Martin Rudwick, however, find no real indication of biblical creationism in the original French. Indeed, Cuvier was quite prepared to debunk supposed evidence of biblical events. Since 1725, scholars had regarded a human-sized fossil skull and ribcage from a German limestone quarry as a victim of Noah's flood, a *Homo diluvii testis*. On an 1811 visit to Amsterdam, where the fossil resided, the anatomist chipped away the stone beside the ribcage to reveal stubby legs, demonstrating that it was the skeleton of a giant Tertiary salamander, not a biblical man.

Cuvier was made a baron and peer of France under King Louis-Philippe, and later portraits of him depict a fat, bemedaled member of the elite. Liberals like the American historian George Bancroft cordially disliked him. Yet he did not participate actively in the post-Napoleonic period's religious and intellectual reaction. France's leading Protestant, he married a Catholic and ignored the evangelical fervors that swept Europe before his death in 1832. His devoutly Protestant daughter failed to evangelicize him, and he spent much of his later career modernizing the French school system.

Cuvier probably saw nothing backward in his vision. As one of a new generation of empiricists, he prided himself on not answering questions without evidence. He was prepared to say that mammals came after reptiles because he found no mammal bones among the giant lizards of the sub-gypsum chalk. He likewise was prepared to say that mammoths and megatheres came after palaeotheres, and that humans came after them all. Indeed, he was the first to say such things from fossil evidence instead of speculation—the first, as he said, to have "burst the limits of time." The concrete evidence of his gypsum mammals raised the novel possibility that the strange life of the past might be reconstructed. "[S]uch reconstructions . . . were the most vivid expressions of his ambition to demonstrate that reliable human knowledge of the prehuman world was not unattainable," Rudwick observes. "The best guarantee of such knowledge was his demonstration of the sheer 'otherness' of the animal world he had discovered; it was not a mere variant of the present, but a truly different 'ancient world.' A *real* history of life on earth was within grasp."

Yet for all his precise empiricism, Cuvier was enigmatic. Although he envisioned mammal faunas changing through migration, he showed little curiosity about the migrations. While debunking Buffon's assumption that mammothlike beasts had failed to reach South America, he insisted that the Spanish had not found a single quadruped exactly like the Old World's on that continent. Although he suspected that causes other than present-day natural forces might have destroyed fossil worlds, he refused to speculate at any length as to what they might have been. To have devoted so much of his life to retrieving the past but shown so little public interest in the question of origins and ends suggests a deliberate reticence. And Cuvier did not simply ignore the theorizing of others, he attacked it. His eulogy of Lamarck—so mocking that the Académie des sciences refused to have it printed as Cuvier's envoy read it to them—dismissed savants who "laboriously constructed vast edifices upon wholly imaginary bases, resembling those enchanted palaces of our old romances, that one may cause to vanish in thin air by shattering the talismans upon which their very existence depends."

Lamarck's remains went from a rented grave into the catacombs, uniting him forever with the Paris gypsum. Cuvier's higher status entitled him to tomb, but the gypsum's compound of suave lime and caustic sulfur seems to have entered his bones while he lived. Historians often quote Balzac's praise: "Is not Cuvier the greatest poet of our century? . . . our immortal natural historian has reconstructed worlds from bleached bones." The lengthy passage that contains the praise is equivocal, however, as if questioning the value of knowing that yesterday's beasts were so different from today's without knowing why. The novel in which it appears, *La Peau de chagrin* (translated as *The Wild Ass's Skin*), is a fable on life's incalculability, the story of a magic horsehide that shrinks every time its owner makes a wish on it. Balzac invokes Cuvier after describing the junk shop where his doomed hero gets the skin:

Have you ever plunged into the immensity of space and time by reading the geological treatises of Cuvier? . . . As one penetrates from seam to seam, from stratum to stratum and discovers, under the quarries of Montmartre or the schists of the Urals those animals whose fossilized remains belong to antediluvian civilizations, the mind is startled to catch a vista of the milliards of years and the millions of people which the feeble memory of man and an indestructible divine tradition have forgotten . . . he digs out fragments of gypsum, decries a footprint, and cries out: "Behold!" And suddenly marble turns into animals, dead things live anew, and lost worlds are unfolded before us! . . . In the presence of this awesome resurrection due to the voice of a single man, that tiny grain granted to our use in this nameless infinity, which is common to all spheres and which we have baptized as TIME, that minute of life seems pitiable. We wonder, crushed as we are under so many worlds in ruin, what can our glories avail, our hatreds and our loves, and if it is worth living at all.

TWO

Doctor Jekyll and
the Stonesfield Jaws

THE PEABODY MURAL DOESN'T SHOW Cuvier's Paris gypsum mammals, since they didn't live in western North America, but some beasts in its early sections resemble them. *Palaeosyops*, "old piglike animal," a tapir relative that stands behind snarling *Mesonyx*, inhabited America at about the time that Cuvier's *Palaeotherium* lived in France. Another example is *Peradectes*, "pouched biter," an opossumlike creature perched on a mossy log near the dinosauresque *Barylambda*. It lived even earlier than the French marsupial, but the group has changed so little in fifty million years that Zallinger painted *Peradectes* to resemble the extant Virginia opossum, with whitish fur, pink skin, and a goofy look. It already seems out of its depth beside larger placentals like *Barylambda*.

No living mammals in North America appear more primitive than opossums as they waddle myopically about their business. I once encountered a pair on an Ohio golf course one gray November day. One of them climbed a pin oak, but the other simply flopped on its side, jaws gaping, tongue lolling, clearly overcome by my presence. It stank, but its unconditional surrender seemed to require some action from me, as a "higher mammal," so I took it home and put it in the garage to see when it would revive. Marsupials' artless airs can be deceptive. I had left the garage doors closed tight, but when I went out the next morning, the opossum was gone. I have never figured out how it escaped.

Cuvier got a surprise from an opossumlike fossil later in his career, when he had become not only France's premier naturalist, but Europe's. It didn't disappear unexpectedly, *au contraire,* its surprise appearance when he was far from home must have made the baron wonder if reconstructing the past was as *raisonnable* as the Paris gypsum had made it seem, although he didn't let on publicly to any such bewilderment.

The English admired Cuvier because of his Protestant empiricism, and because he was an exalted professional naturalist at a time when no Englishman was. Britain had no such prestigious jobs before 1830; naturalists were clergymen, physicians, or simply educated gentlemen, like a young rock-fancier named Charles Lyell, who paid the baron a deferential visit in 1823. He marveled that Cuvier's study contained no fewer than eleven desks, each fully supplied with writing materials. The great man proceeded from desk to desk according to his current project. Of course, even Cuvier was not infallible. When Lyell showed him a tooth that his friend Dr. Gideon Mantell had found in southern England, Cuvier said it was a rhino incisor from a surface deposit. But Mantell eventually proved that it came from the much older, Secondary strata, and belonged to a giant saurian. Cuvier admitted his mistake in 1825 after Mantell presented more evidence.

Cuvier did better at identifying another English fossil, but it may have cost him some sleep. British savants had been collecting old bones as long as French ones, and a good place to do so was at the town of Stonesfield, near Oxford. The fossil-bearing stratum there was limestone, but it was hard enough to use for roofing slates, and quarrymen had to dig deep underground shafts to reach it. These features suggested that it was much older than the Paris gypsum, and in fact it overlaid a stratum, called the oolite or "egg stone" because its texture resembles fish roe, that was known to be of mid-Secondary age. The Stonesfield "slate" yielded large saurian bones, and William Buckland, a geologist and custodian of the University's Ashmolean Museum, had acquired many of those.

Buckland, also canon of Christchurch College, was a vintage clergyman-naturalist, a jocular eccentric who liked to serve dinner guests crocodile or ostrich meat and kept a menagerie that included a bear, which his son would dress in academic robes for garden parties. Another young rock-fancier, John Ruskin, marveled at the Buckland establishment's "Aladdin's Cave" ambience: "not a chair fit to sit down upon—all covered with dust . . . frogs cut out of serpentine—broken models of fallen temples, torn papers—old manuscripts—stuffed reptiles." Such antics distressed more serious-minded investigators like Lyell, and Buckland's eccentricity later declined into dementia. But he was an good geologist, and he became famous in 1821 by finding a prehistoric hyena den in a Yorkshire cave. To prove that hyenas had inhabited it, he acquired a living one named Billy and compared the way it cracked and gnawed bones with the cave's fossils. "So wonderfully alike were these bones in their fracture," he wrote a friend, "that it was impossible to say which bone had been cracked by Billy and which by the hyenas of Kirkdale!" A lithograph of Buckland creeping into the cave by torchlight to observe resurrected hyenas gnawing pachyderm bones was one of the earliest full-blown prehistoric mammal reconstructions.

As a cleric, eventually dean of Westminster Abbey, Buckland valued Cuvier's reputation for piety as well as his anatomical knowledge. When he published a paper on a Stonesfield giant saurian that he had named *Megalosaurus,* he sent Cuvier a copy and received his blessing, predictably, since it upheld Cuvier's belief that giant saurians occurred only in Secondary strata. Yet the Stonesfield slate held more problematic things. For decades, the quarries had been yielding inch-long jaws that seemed more like a shrew's than a lizard's. Buckland had acquired one from a student in 1812, but he had done little with it. The idea of shrews living with giant saurians would have seemed bizarre. The largely marine nature of known Secondary strata suggested that the planet had been too soggy to support higher land animals, and most giant saurians had apparently been aquatic, like the recently discovered plesiosaurs and ichthyosaurs. Buckland suspected that the atmosphere had been unbreathable by warm-blooded mammals.

The tiny jaws came to prominence in 1818, however, when Cuvier visited London to see the collection of skeletons and pickled specimens at the Royal College of Surgeons' Hunterian Museum in Lincoln's Inn Fields. Buckland showed Cuvier around the museum, and the baron then accompanied him to Oxford to see the university's fossils. This trotting off to a provincial town is a little surprising, given Cuvier's grandeur and rudimentary English, but, while there, he saw a tiny Stonesfield jaw and identified it as an opossum's. The identification is not surprising—the jaw looks not unlike a miniature of a modern opossum's. But it must have been a shock to Cuvier, apparently coming from deep in the same rock that had produced giant saurians. Its implications for his clear-cut scheme of Secondary saurians and Tertiary mammals were disturbing. A Secondary occurrence of marsupials, which seem more primitive than placentals because they give birth to fetal young instead of gestating them, would have hinted at transmutation and progress, the Lamarckian process.

Cuvier kept his sangfroid. He made no mention of oolitic opossums in the revised second edition of his *Recherches sur les ossemens fossiles* (1821–24). When Buckland mentioned the jaw in his *Megalosaurus* paper in 1824, Cuvier seems to have passed the problem to a junior associate named Constant Prévost, who wrote a paper maintaining that the Stonesfield slate was really a Tertiary rock, like the Paris gypsum. Cuvier did concede in the third edition of his *Recherches* (1825) that the fossil would be "a remarkable exception" to his rule of mammal occurrence if it really was from Secondary strata, and he reconfirmed his identification, naming it *Didelphis prevostii,* the same genus as the living Virginia opossum. He apparently ignored the matter afterward. There is no record that the jaw came up when he visited England again in 1830. Still, it may have. The 26-year-old anatomist who showed the old baron around the Hunterian Museum on that occasion would eventually trample the jaw's Lamarckian implications as vigorously as a younger Cuvier might have.

Richard Owen, the Hunterian's newly appointed conservator, was a tall, bug-eyed, strangely big-handed man who already had been practicing anatomy for ten years. His zoology and his French (learned from his mother) so charmed Cuvier that the baron invited him to return the visit, and Owen spent the next July in Paris, frequenting the museum and zoo, attending lectures, and taking part in Cuvier's Saturday night salon. He met both Cuvier's allies and his adversaries, who were increasing. Lamarck had died in 1829, but visions of progressive biological change persisted. Cuvier's old museum colleague, Geoffroy Saint-Hilaire, was not so sure that the animal mummies he had brought back from Egypt were entirely the same as modern species. In direct opposition to Cuvier, he thought comparative anatomy demonstrated that all animal phyla were related— worms to mollusks, mollusks to fish, and so forth—an arrangement he called "unity of plan." Geoffroy was skeptical of Lamarck's inheritance of acquired characters, but thought that transmutation between species had occurred, perhaps through changes during embryonic development. Such ideas were increasingly popular. Balzac preferred them to the baron's cold empiricism, calling Geoffroy "the victor over Cuvier in this point of higher science."

There were British transmutationists, like Robert Grant, an Edinburgh anatomist transplanted to London who shared not only Lamarck's ideas but his specialty of invertebrate biology. Grant was a rising figure in 1831, recently appointed professor of zoology at London University, and active in the prestigious Zoological Society, which ran the new London Zoo.

Paradoxically, admirers were calling him "the English Cuvier." A free-thinking francophile with republican sympathies, he summered in Paris, and he was staying in Owen's hotel. They became acquainted, sometimes sharing meals, and Grant showed Owen around the lecture circuit and recounted his continental travels. He didn't know what he was getting into by opening his mind to the apparently affable younger man, however. Owen was an orthodox Anglican and probably had reservations about his urbane guide's cosmopolitanism, bachelorhood, and other unconventional traits. He certainly had reservations about Lamarckism and its variants. They implied that humans had been transmuted from beasts, and Owen hated the idea with a virulence that must have shocked the rationalist Grant when he perceived it.

Transmutationism was perhaps too vivid a reminder of something beastly in the young anatomist's own experience. Owen's youth had been troubled. His merchant father had died bankrupt when he was five, and his sister remembered him as an "exceedingly mischievous" child. At sixteen, he had apprenticed to a surgeon, and, when his master served at the Lancaster gaol, he had assisted at autopsies, which first horrified, then fascinated him. Following the autopsy of a "negro patient in the gaol hospital," he had fallen prey to such an "anatomical passion" that "all other resolves and scruples were forgotten." He had bribed the guard to let him return that night and stolen the man's head to dissect at home, but had slipped and dropped it descending an icy hill. "As soon as I recovered my legs," he recalled, "I raced desperately after it but it was too late to arrest its progress. I saw it bounce against the door of a cottage facing the descent, which flew open and received me at the same time, as I was unable to stop my downward career. I heard shrieks, and saw the whisk of the garment of a female, who had rushed through an inner door; the room was empty. The ghastly head was at my feet. I seized it, and retreated wrapped in my cloak."

Owen had managed to pull back from the brink of such career-threatening criminality. After his apprenticeship, he had moved on to the highly regarded Edinburgh medical school, where he had expanded his anatomical studies into nonhuman subjects, although he never became a field naturalist like Lyell and Buckland. He had read the great anatomists, Lamarck as well as Cuvier, and in 1825, he had moved to London to study at the College of Surgeons. His skill there so impressed John Abernethy, president of the college and the London Hunterian Museum, that he ap-

pointed him his prosector—his dissecting assistant in anatomy lectures—
and later offered him a museum job under the conservator, William Clift.

The museum was just the place for a licensed practice of the "anatomical passion." Its founder, John Hunter, was probably a model for Mary
Shelley's archetypal mad scientist, Victor Frankenstein. His physiological
experiments had included repeated attempts—at least once successful—
to graft newly extracted human teeth onto living rooster's heads, and his
enthusiasm for more complete human specimens was as persistent, if more
circumspect. Coveting the eight-foot skeleton of a circus giant named
Charles O'Brien, Hunter had awaited the man's natural death before bribing the corpse's custodians to relinquish it, then personally boiled it down
in a huge kettle and set the bones up as his museum centerpiece. Such
grotesqueries, typical of early anatomical museums, have strange effects on
the psyche. (I shall never forget a 1964 visit to a remarkably well-preserved
nineteenth-century one in Montpellier, France, whose specimens included
the genitalia of a woman who had killed herself with a knitting needle and
the toadfishlike head of a tertiary syphilis victim.)

Owen made himself so useful cataloguing Dr. Hunter's treasures and acquiring new ones that he became not only Clift's successor but his son-in-
law. He blossomed into an eminent Victorian, and his wife knew him as
the soul of kindness. Yet youthful sins had apparently left their mark.
Anatomy students often brag callously about their gruesome adventures in
dissection, but the mature Owen remained so attached to his that his
grandson-biographer referred to his prison hospital burglary as "the Negro's head story, which the professor told so well." Owen even added a
melodramatic preamble, describing the "female" in the cottage as the
daughter of a slave trade seaman. Ambivalently mourning her father, killed
in a barroom brawl, she hears a "heavy blow" to the door, and sees "the
phantom of a negro slave lying on the floor; which turned its ghastly head
and glared for a moment upon her with white protruding eyeballs. A figure
in black entered as she fled screaming." Owen then would identify his own
"craniological longings" as the source of "the mystery."

Such fantasies imply wayward impulses, and Owen's contemporaries
saw a "mischievous" side in him that does suggest the vile Mr. Hyde into
whom, in a converted dissection room lab, Robert Louis Stevenson's
Dr. Jekyll transforms himself. One acquaintance, the naturalist Edward
Forbes, called him "one of the oddest beings I ever came across & seems as
if he was constantly attended by two spiritual policemen, the one from the

upper regions & the other from the lower—the one pulling him towards good impulses and the other towards evil." Owen's general rectitude and goodness to his wife were doubtless real, and he was charming to aristocrats and others of high status, although Thomas Carlyle's wife likened his smile to "sugar of lead." He could be hard on inferiors and dependents, however—his son drowned himself—and he was known for deliberate and devious viciousness to potential rivals like Robert Grant. "It is astonishing with what an intense feeling of hatred Owen is regarded by the majority of his contemporaries," observed one of the chief haters, a rival named Thomas H. Huxley.

A year after visiting Cuvier in Paris, and probably in reaction to the transmutationist ferment he had encountered there, Owen stepped smartly into the late baron's role as upholder of anatomical class distinctions. One of the Lamarckians' favorite examples of a link between two animal classes was the strange little duck-billed platypus discovered in New South Wales the century before. Named for the sensitive "bill" with which it probes for food in stream bottoms, the platypus is furry and web-footed like a muskrat, and a nimble swimmer, as I learned at age ten from trying to get a look at one in the Bronx Zoo. Despite its fur and warm-blooded behavior, it was said to lay eggs, and Lamarck had placed it in a new vertebrate class between birds and mammals—the monotremes, "one-holed" animals. Geoffroy and Grant thought its rumored oviparity, single excretory-genital orifice, and apparent lack of mammary glands demonstrated a nature transitional between mammals and reptiles. Cuvier, however, classed it with sloths as a true mammal.

Owen agreed with Cuvier, and he set out energetically to prove the platypus's true mammalhood, taking advantage of numerous reports and specimens that observers were sending back from New South Wales. They had found baby platypuses with stomachs full of milk, and Owen determined that females did indeed lactate, not from teats, but from tubular glands in the skin of the belly. He even succeeded in throwing doubt on platypus egg-laying, although naturalists had found apparent platypus nests, and he himself had seen specimens with eggs in the ovaries. He suggested that the oviducts were so narrow that the eggs had to be hatched internally. Geoffroy and Grant continued to believe that platypuses laid eggs, but Owen had forced them to acknowledge that they feed their young with milk and thus are mammals of a sort. In 1834, the Zoological Society elected Owen a fellow in recognition of his platypus work.

Owen's growing reputation brought him into a renewed controversy over the Stonesfield jaws. In 1838, Cuvier's successor at the Paris museum, an anatomist named Henri de Blainville, published two papers maintaining that the jaws really were reptilian and changing their genus from *Didelphis,* "two womb," to *Amphitherium,* "double beast." Blainville acknowledged that the teeth, including incisors, canines, and molars, were more complex than most reptiles', but he cited a recently discovered, apparently reptilian, marine genus named *Basilosaurus,* "ruling lizard," that did have cusped molars like the Stonesfield teeth. Blainville was not a transmutationist, and his papers appealed to creationists because they denied a mammalian presence in the Secondary. Yet Geoffroy and Grant also liked *Amphitherium* because it implied a progressive link between reptiles and mammals.

All this was vexing to William Buckland, who started calling the Stonesfield fossil *Botheratiotherium,* "bothersome beast," but still considered it a marsupial. He had tried to buttress the case for Secondary marsupials by describing some even older footprints, recently found in Germany, as similar to a kangaroo's. The presence of such an "advanced" marsupial even before the Stonesfield beasts would have been a blow against Larmarckist progress. Grant examined similar, equally old tracks near Liverpool, however, and concluded they belonged to a crocodilelike reptile, a diagnosis with which Owen had to agree. Buckland finally had to agree, too, and compensated for his embarrassment by writing Owen that "Dr. Grant seemed disappointed that he could not differ from me" on the matter at a Geological Society meeting.

Buckland retaliated by going after *Amphitherium.* Blainville had laid his classification open to question by writing his papers without the jaws in hand. Buckland took two of them to Paris on a continental tour that summer, and gave casts to the members of the Académie des sciences. Examining them, a zoologist named Achille Valenciennes decided that the jaws belonged to marsupials after all, although he changed the genus from *Didelphis* to *Thylacotherium,* "pouched beast," because they seemed more like a South American mouse opossum's than a Virginia opossum's. Even Geoffroy acknowledged that the jaws were a marsupial's, although, as with monotremes, he put marsupials in a transitional class.

Blainville and Grant were not convinced, so Buckland turned the Stonesfield jaws over to Owen. As with the platypus, Owen was able to bring the ever-growing resources of empire to the problem. He had a variety of Australian specimens to compare with the jaws, including the recently discovered numbat, *Myrmecobius* ("lives with ants"), a shrewlike marsupial whose long, narrow jaw contains nine molars, the same as in Buckland's original Stonesfield jaw. In an 1838 paper, Owen said that this similarity definitely put the Stonesfield jaws in the marsupials. Grant argued forcefully against this conclusion, however, at the November and December meetings of the Geological Society where Owen read his paper. Owen had not, after all, addressed Blainville's original point that the Stonesfield jaws were like those of *Basilosaurus,* supposedly a marine reptile. *Basilosaurus*'s owner was due to arrive in London with the fossil in January, and Grant expected this to support his views. But he was walking into some Owen mischief.

Basilosaurus's owner was a Philadelphia doctor named Richard Harlan, a lecturer in anatomy at the American Philosophical Society. Fossil bones increasingly were arriving from the frontier, and the society had received some giant vertebrae from Louisiana in 1832. Harlan had decided they belonged to a marine reptile, but he had had doubts when similar fossils showed up from Alabama in 1835, because the teeth seemed more complex than known reptile teeth. Still, the jaws were long and hollow like a marine reptile's, and he had concluded in a lengthy paper that *Basilosaurus* was one. An enterprising man, Harlan sent copies to European naturalists, and he must have been gratified by the attention it received. His January 1839 trip was an attempt to get in on the fun, although it is unclear exactly where he stood in the controversy. He was not a transmutationist, although it was Grant who had invited him to speak at a meeting of the Geological Society.

Coached by Buckland, however, Owen got to Harlan before the meeting. He examined *Basilosaurus* and soon convinced its owner that it was, in fact, a mammal, an archaic whale. Its long hollow jaws and concave vertebrae were not unlike a sperm whale's, and its teeth were undoubtedly mammalian. They consisted of an outer ring of cement and interior tubular structures, a mammalian characteristic, and the molars were double-rooted, as no reptile tooth is. When Harlan appeared at the Geological Society, he recanted his reptilian *Basilosaurus,* then gave the floor to Owen, who read a detailed paper demonstrating that the fossil was a whale and changing its name to *Zeuglodon,* "yoked tooth," a reference to its double-rooted molars. He was quite right about the creature being an archaic whale—it is another link in the chain connecting Philip Gingerich's *Pakicetus* with living cetaceans—although taxonomy has rejected his arrogant name change, and retains the genus *Basilosaurus,* despite its literal meaning of "ruling lizard."

Robert Grant's reaction to this paleontological ambush is unrecorded, but can be imagined. The reptilian *Basilosaurus* could hardly have been debunked in a more underhanded and humiliating fashion. "From now on there was open hostility between Owen and the radicals," Adrian Desmond writes. Owen's wife Caroline commented in her diary that "Dr. Grant was forced to admit, in spite of his teeth, that they were mammalia and not saurians," but Grant never accepted that the Stonesfield jaws were marsupial, and he was still describing the creatures as a semi-reptilian *Amphitherium* ten years later. His influence was waning as Owen's grew, however. Owen pushed him out of the Zoological Society leadership, which meant that Grant no longer had access to exotic animals to dissect. He stopped publishing significantly after 1840 for want of time and funds, and dwindling income from his reduced lecturing eventually forced him to live in the slums. An 1830s social fossil in a swallowtail coat, he continued embarrassing Victorian medical students with deist witticisms until his death in 1874.

Most naturalists accepted the Stonesfield jaws as Secondary mammals after the *Basilosaurus* coup, and Owen must have felt that he had vindicated Cuvier despite disproving his rule that mammals lived only in the Tertiary. He had upheld the baron's deeper rule that neither anatomy nor fossils showed reptiles transmuting into mammals. Owen instead of Grant had become the official "English Cuvier," a gratifyingly high status for the thirty-five-year-old son of a failed West Indies merchant. Indeed, he had become the first truly professional English naturalist, and in 1841 he coined the most famous paleontological term when he named Buckland's and Mantell's saurians "terrible lizards"—dinosaurs. Like his monotreme and marsupial work, this was a blow at the Lamarckians, since it held that reptiles had been more developed, more mammal-like, in the past, and had degenerated instead of progressing.

Owen's attainments never made him rich, like the highborn Lyell, but his social status rose steadily. He became a favorite of Queen Victoria, who knighted him, the director of natural history programs at the British Museum, and recipient of most available medals, honorary degrees, pensions, and other distinctions.

As time passed, however, the Stonesfield jaws began to belie a hidden decay within Owen's courtly façade. Despite his crushing of Grant, transmutationists increasingly saw them as transitional creatures, ancestral mammals, if not semi-reptiles. "It is an interesting circumstance that the first mammifers should have belonged to the marsupialia, when the place of that order in the scale of creation is considered," declared the first widely popular English book on such ideas, *Vestiges of the Natural History of Creation,* in 1844. "In the imperfect structure of their brain, deficient in the organs connecting the two hemispheres—and in the mode of gestation, which is only in small part uterine—this family is clearly a link between the oviparous vertebrata (birds, reptiles, fishes) and the higher mammifers." Soon afterward, Owen would make discoveries that undermined his own 1830s denial of a link between reptiles and mammals. But by that time, he would be too set in his antitransmutationist role, and too hated by colleagues, to turn them to professional advantage.

THREE

The Origin of Mammals

ONE REASON THE PUBLIC PREFERS dinosaurs to prehistoric mammals may be that they are less threatening. There's a certain aloof impartiality about a tyrannosaur's aggressions. Despite baleful red eyes, the one in Zallinger's *Age of Reptiles* seems bucolically oblivious of the hulking potential prey all around it. The computer-enhanced dinosaurs that rampage at Mack Truck speed through movies lack a certain discernment. We tend to think of large, predatory mammals as more deliberate and perceptive, and they still sometimes skillfully stalk, kill, and eat real humans, as no real nonavian dinosaur ever did. Hoofed mammals trample and gore more than predators eat. Cow elk are the most malevolent ungulates I've en-countered—they've gone after me twice and once nearly got me.

There's something to be said for Richard Owen's discomfort with beastly ancestry. If life is "going somewhere," the question must arise of whether increasing intelligence combined with animal aggression will reach a pleas-ant destination. A figure in Zallinger's mammal mural seems to epito-mize the mythically malevolent potential of our class. The painting has no savager-looking beast than a yellowish, wolverine-sized one named *Oxy-aena,* which arches its back and bares fangs and claws under a palm tree. There is nothing aloof or impartial about its attitude.

Oxyaena (which means something like "sharp baleful thing," not one of the more enlightening genus labels) belonged to an early Tertiary group that Edward Cope named creodonts, "flesh-toothed," because their cheek teeth were so clearly adapted for shearing meat. "Their variety was greater than that presented by their carnivorous successors," he wrote, with proprietary pride, "and their numbers were proportional to the general luxuriance of the life which furnished their subsistence. There were species whose size and powers of destruction equaled those of the bears, lions and tigers of modern times." According to Cope's description of it, *Oxyaena* might have been especially scary to two elfin creatures named *Pelycodus* that huddle near it in Zallinger's *Age of Mammals*. "I have been studying the skeleton of a fossil carnivorous beast from the Eocene period lately," he wrote his daughter.

> The jaws show a head as large as a wolf, and the teeth are dangerous cutters. But there is something queer about the hind foot. It looks as though it had a thumb-like big toe, as an opossum has, and also the little toe opposable in the same way. The hind foot may have been split, two toes one way and then the others. I haven't fully made it out yet. What if it climbed trees for its prey?

Whether or not *Oxyaena* specialized in leaping fiendishly through the treetops after early primates like *Pelycodus* (Zallinger's version looks more terrestrial), it might have been an archetypically fearsome brute. Creodonts were as well armed for killing as modern carnivores but had smaller, smoother brains and, perhaps, less of the sense of self-preservation that diverts living superpredators like wolverines from unbridled aggressions. The last ones died millions of years ago, so we'll never know, but creodonts may truly have been inexorable killers like the ones that haunt our nightmares. Two large herbivores that confront *Oxyaena* appear to sense, and reciprocate, a basic viciousness. The creodont seems too small to threaten them, but they have it cornered on a rock, as if determined to trample it.

Richard Owen would have bristled at any linkage of creodonts and humans. Yet Owen's professional position after 1840 would become a bit like *Oxyaena*'s in the Peabody mural. He, too, would find himself cornered by enemies, somewhat unjustly, because of his predatory nature. So it is ironic that the large herbivores cornering *Oxyaena* in the mural were among Owen's greatest discoveries. From a jawbone containing a single tooth found off the Essex Coast in 1844, Owen inferred that a fossil creature he called *Coryphodon*, "pointed tooth," had been a large, hoofed herbivore, a particularly daring display of Cuvier's "correlation of parts."

Figure 5. *Coryphodon* and *Oxyaena,* with *Pelycodus* (Eocene) from Zallinger's *Age of Mammals* mural. Courtesy Peabody Museum of Natural History, Yale University, New Haven, Conn.

O. C. Marsh would vindicate it in 1876 when he announced the discovery of a whole *Coryphodon* skeleton in the Rocky Mountains. The genus, named for its tusklike canine teeth, was a hippo-sized, five-toed herbivore, "older in time, earlier in date, than *Palaeotherium.*" It was the most primitive large mammal then known.

Of course, Owen didn't seem like a cornered predator from the outside. He spent the half century after his 1830s triumphs over Grant and other early transmutationists filling in the mammalian history that Cuvier's Paris basin studies had begun. He pushed back the vast darkness that had disturbed Balzac, exercising a great deal of disciplined imagination in the process. Beside adding to Cuvier's early Tertiary menagerie, he saw that creatures similar to *Palaeotherium* had inhabited Europe in later times as well, and that they also had similarities to modern creatures such as horses. The more light he shed on the subject, however, the harder it was for him to follow Cuvier's example in evading the issue of why and how later mammals came to differ from earlier ones.

To Owen's credit, he didn't want to evade the issue. "It is to be presumed that no true researcher after truth can have a prejudiced dislike to conclusions based upon adequate evidence," he wrote to Robert Chambers, an Edinburgh publisher who happened to be the anonymous author of the transmutationist best-seller *Vestiges of the Natural History of Creation.* Yet Owen was unable to be straightforward, and he ridiculed *Vestiges* elsewhere, particularly after he discovered that its anonymous author was a commoner and not, as he had imagined, the Tory landowner-politician Sir Richard Rawlinson Vyvyan. He did more than just ridicule the book. In 1848, Owen led Chambers through the Hunterian Museum and showed him material that he secretly thought possible evidence of how species are transformed, then smugly wrote a friend that Chambers "saw nothing of their bearing." Such mischief fostered an atmosphere in which Owen would *seem* to evade the issue of transmutation.

Ways of thinking about mammal prehistory were proliferating. Charles Lyell, the rock-fancier who had admired Cuvier's lab, had developed a theory that was both novel and respectable. Impressed, on a trip to Sicily's Mount Etna, by the sheer mass of rock strata, Lyell had mused that the earth might be so old that life could be virtually eternal, making the question of transmutation moot. This was a nonbiblical idea, but it had venerable antecedents. Aristotle had thought similarly, as had the eighteenth-century savant James Hutton, who had been more successful than Buffon in building a terrestrial system. Hutton's influential *Theory of the Earth* had seen "no vestige of a beginning,—no prospect of an end" in the cycles of geological events. The earth's internal heat elevated mountains and volcanoes, rain and wind eroded them into sedimentary rock, and internal heat pushed the rock into mountains again, ad infinitum. Why shouldn't life's cycles be equally indeterminate?

Lyell, whose patrician birth detached him from the struggle to rise in society, suspected, not without reason, that fossil evidence of progress from simple to complex organisms was an illusion. The existing evidence was scanty. "Although the bones of mammalia in the tertiary, and those of reptiles in the secondary, afford us instruction of the most interesting kind," he wrote, "yet the species are too few, and confined to too small a number of localities, to be of great importance."

The small fossil record was growing so bewilderingly in the mid nineteenth century, anyway, that any conclusions seemed premature. No longer could prehistory be divided into Primary, Transition, Secondary, and Tertiary strata. The Transition and Secondary opened like a conjurer's box into a welter of new periods commemorating fossil locations—Silurian, Devonian, Carboniferous—Triassic, Jurassic, Cretaceous. Lyell cut the Tertiary Period into Greek cognate epochs—Eocene, "dawn recent"; Miocene, "less recent"; and Pliocene, "more recent"; he also fabricated a Pleistocene, "most recent," epoch that came after the Tertiary. In 1841, another English geologist, John Phillips, pushed the Transition's and Secondary's burgeoning formations into Paleozoic and Mesozoic eras, "old" and "middle life," and the Tertiary's into a Cenozoic Era, "recent life."

Amid such chronological sawing and hammering, it seemed quite possible that fossils like the Stonesfield jaws would continue to turn up in earlier strata. Indeed, there might have been another "age of mammals" like the Cenozoic during some dry, cool period in the distant past. On the other hand, future climate might revert to the Mesozoic's warm, wet conditions. "Then might those genera of animals return, of which the memorials are preserved in the ancient rocks of our continents," Lyell wrote in his *Principles of Geology.* "The huge iguanadon might reappear in the woods, and the ichthyosaur in the sea, while the pterodactyl might flit again through the umbrageous groves of tree ferns." This purple passage provoked a Cuvierian geologist, Henry de la Beche, to caricature a professorial ichthyosaur in a gilt-trimmed academician's uniform lecturing on a fossil human skull to a grinning saurian audience. "You will at once perceive," it declaims, "that the skull before us belonged to some of the lower order of animals. The teeth are very insignificant, the power of the jaws trifling, and altogether it seems wonderful how the creature could have procured food."

Other readers found Lyell's vision less fanciful. It acquired a hard-headed adherent in Thomas Henry Huxley, a struggling young anatomist in the 1850s. Lyell's social status accorded with his complacent denial of change, but Huxley's embrace of the "uniformitarian" doctrine was more complex. His youth had been as sordid as Owen's, and their early photographs have a similar harried intensity, although Huxley's sharp, snub-nosed features are the opposite of Owen's heavy ones. Raised in a turbulent lower-middle-class family (a secret disgrace forced an older sister to emigrate to Tennessee), he, too, had apprenticed to a surgeon, and he had apparently also committed youthful beastliness. "I confess to my shame," he recalled, "that few men have drunk deeper of all kinds of sin than I." Huxley only hinted at his depravity instead of bragging about it like Owen, so its nature is unknown, but he also sought redemption in work and love. He had studied anatomy and shipped on a four-year naval expedition to New Guinea and Australia, where he had met a respectable young woman and channeled wayward impulses into Victorian romance.

After his return in 1850, England's social stability was a barrier to Huxley as he struggled to find a job. It was five years before he could bring his fiancée out from Australia and marry her. Yet he disdained the transmutationist's progressive notions. His time in the South Pacific confronting nature's brute abundance probably made such theorizing seem effete, and he was in any case more interested in ascending the social pyramid than rearranging it. Lyell's uniformitarianism, as opposed to radical Lamarckism, was a respectable way of pushing against the Cuverian establishment, and Huxley became a spokesman for it. In an 1862 talk to the Geological Society, he maintained that the Stonesfield beasts were in no way "more embryonic, or of a more generalized type" than modern opossums. He even speculated in an 1859 letter to Lyell that a *homo ooliticus* might have hunted them the way Australian aborigines hunted possums.

Although Owen was more socially conservative than Lyell and Huxley, he was intellectually progressive enough to reject divine intervention as unscientific. Moreover, his superior grasp of anatomy made him more open to the concept of biological change. "To what natural laws or secondary causes the orderly succession and progression of such organic phenomena may have been committed we are as yet ignorant," he acknowledged in an 1849 lecture, "The Nature of Limbs." Owen thought, with some justice, that Lyell's eternal cycles merely evaded the issue of change.

For him, the problem was to understand how change had occurred naturally without reverting to Lamarckian materialism, so he sought the "secondary causes" in transcendentalist ideas that had grown up in the vacuum of Cuvier's enigmatic empiricism. Owen decided that anatomical patterns must reflect a higher, eternal reality, such as that conceived of by the Platonists. Mammals differed from reptiles in so many distinct, predictable ways that there had to be an ideal mammal, an archetype, behind the physical organism. Change thus somehow involved the archetype, raising it above the materialist transmutation that saw physical links between men and beasts.

Such thinking was widespread. Stevenson's Dr. Jekyll recalls that his "scientific studies . . . led wholly towards the mystic and transcendental." But Owen also believed that this natural but transcendent process must have left traces in the fossil record and living organisms, and he was prepared to seek them. The archetype might seem mystical, but it also provided a gauge by which change could be measured. The archetypal mammal had five digits on each limb, for example, so mammals with fewer digits must have changed from that primal state. There also must be archetypal qualities by which a change from reptile to mammal could be understood. His demonstration that the platypus and the Stonesfield jaws weren't links between reptiles and mammals did not mean that a link would never be found. It *did* mean that Richard Owen, with his spiritual wisdom, would be the one to find it.

Once again, the expanding British empire served Owen well. An Army road builder, Andrew Bain, began finding strange fossils in the South African Cape Colony's Karoo Basin. The Karoo was good, semi-desert bone-hunting country, although the sandstone was so hard that Bain had trouble distinguishing bone from rock. But the hardness was part of the rocks' interest. They were very old, from the Triassic Period near the Mesozoic's beginning—much older even than the Stonesfield slate. In 1844, Bain began sending bones to the London Geological Society, which paid him well for them. Owen encouraged this, and described a stream of Karoo bones over the next four decades. He took them with him when he became the British Museum's natural history director in 1856, and put the museum's stonemasons to work chiseling them out. Even the royal family got into the act. The Prince Consort had asked Owen to tutor his children, and when Alfred, the third son, visited South Africa in 1860, he returned with two Karoo skulls.

Prince Alfred's skulls belonged to a genus Owen had named *Dicynodon* ("two-dog-tooth"). The name expressed their strangeness. Their skulls—low and flat—seemed saurian, yet their teeth were not the pegs or snags of most reptiles. They had long canines—dog teeth—which Owen likened to manatee tusks. Other skulls, also reptilian in shape, were even more anomalous, and Owen gave them names like *Galesaurus* ("weasel lizard"), *Tigrisuchus* ("tiger crocodile"), and *Cynodracon* ("dog dragon"). They had not only had canines, but incisors and molars, suggesting that they had used their teeth and jaws not just to seize and swallow food but to shear and masticate it like mammals. "In no other saurian," Owen wrote, "are incisors so divided from molars by a single canine, in none is the definition of the three kinds of teeth so plain and unequivocal." He placed these "mammal-like reptiles" in a new reptile order, the Theriodontia ("beast-toothed"), and described a half dozen genera in an 1876 catalogue. One of them, a pinkish gray creature named *Cynognathus* ("dog jaw"), crouches underfoot of high-stepping early dinosaurs in the Peabody reptile mural's Triassic section.

In a catalogue summary, Owen noted the theriodonts' similarities to marsupials, even to placental mammals. *Cynodracon major,* for example, was a lion-sized predator with a flexible forepaw for grasping victims and canines like a saber-toothed cat's, features "utterly unknown and unsuspected as reptilian ones." These were "great gains in organization," and they had continued in early mammals, although not, Owen was sure, through any Lamarckian process. It was unclear from Owen's descriptions just how mammalian features *had* appeared on reptilian skeletons. Still, here was an evident link between two classes that younger anatomists like Thomas Huxley should have found instructive.

While Owen discovered mammal-like reptiles, however, Huxley was discovering yet another model of biological change that Lyell's friend Charles Darwin had come up with since returning from the five-year voyage of H.M.S. *Beagle* in 1836. In a sense, Darwin was a born transmutationist. His grandfather, Erasmus Darwin, a freethinking physician and poet, had conceived of the inheritance of acquired characteristics decades before Lamarck. But young Darwin had disliked Lamarckism, partly because he had associated it with the unconventional Robert Grant, who had spouted it at him in 1827 when he was an Edinburgh medical student. First befriending Darwin by leading him on seaside zoology rambles, Grant later had alienated him by taking credit for an observation of marine invertebrates that the younger man considered his own. Disliking cadavers and surgery, Darwin had dropped medicine in favor of Cambridge divinity studies, although he put off a clerical career in favor of circumnavigating the globe as a gentleman naturalist companion to the *Beagle*'s captain, James Fitzroy.

Darwin's background was as genteel as Lyell's, but the voyage had shaken his Anglican faith in life's stability. As the expedition coasted Argentina, they had found skeletons of "great quadrupeds" eroding out of sea cliffs and river bluffs, bones of mastodons and Cuvier's *Megatherium,* even a horse tooth, although no horses had lived in South America when Europeans arrived. Other bones were of entirely unknown creatures—bizarre skulls, weird hooves, and huge bony plates, which he had tried to link to *Megatherium.* This graveyard welter of the known and unknown had troubled Darwin. "It is impossible to reflect on the changed state of the American continent without the deepest astonishment," he marveled in his journal. "Formerly it must have swarmed with great monsters; now we find mere pygmies, compared with the antecedent, allied races." When the *Beagle* reached Australia, he had watched platypuses in a river and examined one a companion had shot. "Earlier in the evening," he recalled, "I had been lying on a sunny bank and reflecting on the strange character of the animals of the country as compared to the rest of the world. A disbeliever in everything beyond his own reason, might exclaim, 'surely two distinct Creators must have been at work.'"

Back in England, Darwin had shown the South American bones to the then 32-year-old Richard Owen, whose interpretations of the strange skeletons had intensified his bemusement. During a morning's discussion at Owen's museum quarters, Darwin wrote Lyell, they had made out the remains of eleven, or perhaps twelve, great beasts. Owen had identified the bones of mastodons, ground sloths, and other known genera, and he had also seen affinities between the unfamiliar skeletons and known mammals. He had thought that one he named *Toxodon* ("bow tooth") because of its curved, flinty molars might be related to the capybara, a giant South American rodent. Another, which he named *Macrauchenia* ("long neck") seemed similar to the American relatives of the camel, llamas and guanacos. The mysterious bony plates, he decided, belonged not to *Megatherium,* as Darwin had guessed, but to a kind of giant, herbivorous armadillo, which Owen named *Glyptodon* ("grooved tooth").

Reconstructions show why Owen thought this. Zallinger painted Darwin's "great quadrupeds" for a 1960 sequel to *The World We Live In* based on the *Beagle* journal, and his *Toxodon* and *Macrauchenia* do look like bigger, bonier versions of capybaras and guanacos, although *Macrauchenia* has a small trunk, and *Toxodon* seems as much pachyderm as a giant rodent. Owen had noted such ambiguities, but the extinct giants' resemblance to the living mammals is still striking. Zallinger also illustrated a succession of carapaced mammals, from cat-sized Eocene ones, to boulderlike *Glyptodon,* back to cat-sized living ones.

Such successions had profound implications for Darwin. They showed that change *had* occurred, and although it was not necessarily progressive, à la Lamarck, it did seem continuous. As far as he had been able to tell during the voyage, *Glyptodon* had inhabited virtually the same South American landscapes as living armadillos. No floods or other catastrophes seemed to divide past from present. This implied a genealogical relationship, and Darwin had spent the decade after his return seeking a non-Lamarckian mechanism for change. He had read Thomas Malthus's observation in his *Essay on the Principle of Population* that the vast majority of organisms die young because reproduction far exceeds food supply, and he had begun to see life as a struggle for existence, not a divinely ordained hierarchy. But how did some organisms prevail over others?

Observing how livestock breeders produce new strains by selecting desired variations in size, color, or other features, Darwin had envisioned a process of "natural selection" in which wild organisms with variations favoring their survival, their "fitness" in life's struggle, would leave the most offspring. As environments changed, new variations might increase an organism's ability to adapt, and enhance its fitness. Natural selection would modify wild populations over time as the fittest organisms spread their traits through populations, eventually creating new species. Despite their paltry size, then, living armadillos and tree sloths had survived mighty *Glyptodon* and *Megatherium* because they had been better able to leave descendants. Even the outlandish platypus had survived for that reason, perhaps "from having inhabited a confined area, and from having thus been exposed to less severe competition."

By mid-century, Darwin's theory was well developed, although he had grown increasingly apprehensive about its implications. He concealed it from conservatives like Owen and Buckland but discussed it with Lyell and younger naturalists. In 1856, he invited the 31-year-old Huxley to his modest estate in the Kent countryside, Downe House, and showed him the experimental evidence he had amassed, including cage after cage of domestic pigeons transformed by artificial selection from ancestral rock doves into fantastic shapes and sizes. Then he likened the contrast between his tame pigeons and their wild ancestors to that between extinct mammals and living ones. His Argentinean *Macrauchenia* had many similarities to llamas and camels, for example, as Owen had demonstrated.

Huxley was skeptical. Squire Darwin might revel comfortably in his clever "struggle for existence," but the idea made life seem dangerously like the London slums from which the young anatomist had narrowly escaped. Huxley clung to uniformitarian gentility. Still, he found Darwin's brand of transmutation more congenial than Lamarck's progress or Owen's archetypes. It didn't require that change be revolutionary or transcendent, since natural selection worked only to adapt species to their surroundings, not to realize a goal or an ideal. Darwin's well-supported arguments for what he called "descent with modification" shook even Lyell, and they were not altogether incompatible with a uniformitarian world of eternal cycles. If evidence of "modification" was inconclusive in known geological strata, admittedly a problem with Darwin's ideas, it might mean that environments had not changed enough to cause it. Huxley speculated that even mammals might have originated in the unknown time before the Paleozoic era.

Huxley had to swallow his misgivings about "descent with modification" after 1858, when Darwin got a letter from a young naturalist in Malaysia, Alfred Russel Wallace, outlining a strikingly similar but independently conceived idea. The letter forced Darwin to go public with his theory, and he and Wallace presented a joint paper on their ideas to the Linnean Society that year. It made little stir, but the publication in 1859 of Darwin's book *On the Origin of Species by Means of Natural Selection; or, The Preservation of Favoured Races in the Struggle for Life* caused a sensation, which nudged Huxley into the role of "Darwin's bulldog," natural selection's public defender against the various antitransmutationist pieties. Although craving respectability, Huxley disliked sanctimony. In a February 1860 Royal Institute lecture, he used Darwin's parallel between wild and domestic pigeons and extinct and living mammals as a major selling point for the theory.

Owen had won a Geological Society medal for his work on Darwin's "great quadrupeds" in 1838, but he probably came to regret his service to the younger man's reputation. He found the newly revealed Darwinism, if anything, more deplorable than Lamarckism. At least Lamarckian materialism acknowledged a progression toward higher beings of a sort. Darwin's variety seemed to acknowledge little except aimless adaptation, a "law of higgledy-piggledy" as the eminent astronomer Sir John Herschel put it. It wasn't that Owen himself was unreasonably wedded to transcendental uplift. In 1860, he published a paleontology text that recognized that the main trend of Tertiary mammals seemed to be toward more specialized species rather than "higher" ones. If Cuvier's Paris gypsum *Anoplotherium* had lacked the complex stomach that adapts modern ruminants for digesting tough plant fibers, deer and cattle were not necessarily closer to transcendence for having it. Yet Owen still saw no support for transmutational explanations of such adaptations: "observation of the actual change of any species into another through . . . hypothetical transmuting influences, has not yet been recorded," he wrote. "And past experience of the chance aims of human fancy, unchecked and unguided by observed facts, shows how widely they have ever glanced away from the gold centre of truth."

Owen wrote a spiteful review of *The Origin of Species,* singling out for ridicule its assertion that "certain facts in the distribution of the inhabitants of South America, and in the geological relations of the present to the past inhabitants" might "throw some light on the origin of species—that mystery of mysteries." He scoffed that "what the 'certain facts' were, and what may be the nature of the light which they threw on the mysterious beginning of species, is not mentioned or further alluded to in the present work." The review placed him conspicuously in opposition to Darwin's increasingly influential circle, although covert antagonism had been growing for a decade. In 1848, an American visitor, Ralph Waldo Emerson, had noted that Owen "indemnified himself in the good opinion of his countrymen, by fixing a certain fierce limitation to his progress, & abusing without mercy all such as ventured a little farther; these poor transmutationists, for example."

Darwin had benefited greatly from Owen's help, but he came to deplore the professor's devious egotism and high-flown airs. "It is consolatory to me that others find Professor Owen's controversial writings as difficult to understand and reconcile with each other, as I do," he wrote in reply to Owen's review of *Origin.* His enmity was mild compared to Huxley's, who snarled that Owen was "not referable to any 'archetype' of the human mind with which I am acquainted." It was as though a common touch of beastliness made the two great dissectors hate each other as mirror images. The prickly younger man had felt slighted even when Owen helped him in his early 1850s job search. Asked for a recommendation, Owen at first had made "no answer whatever," putting Huxley in "a considerable rage." When they had met by chance a few days later, Owen had offered to "grant" the request so condescendingly, Huxley recalled, that "if I had

Figure 6. Caricature by Frederick Waddy of Richard Owen astride
a skeletal *Megatherium,* from *Cartoon Portraits and Biographical
Sketches of Men of the Day: The Drawings of Frederick Waddy*
(London: Tinsley Brothers, 1873; 2d ed., 1874), opposite p. 36.

stopped a moment longer, I must knock him into the gutter." And that, in a sense, was what Huxley did in the next four decades.

The milder-tempered Robert Grant had been easy prey for Owen, because his radicalism made him disreputable. By the 1860s, the more conservative, if militantly agnostic, Huxley had acquired the status to counter Owen's intrigues with impunity, and he did so with dispatch. When invited to give a lecture series at Huxley's School of Mines, Owen styled himself its professor of paleontology, and Huxley exploited that presumption as a pretext to sever relations and start an Owen-dissecting campaign that lasted until his rival's death decades later. He methodically pushed Owen out of the Zoological Society leadership and other positions of power and took every chance to publicly criticize his work, ridiculing his claim that human brains include a structure, the hippocampus minor, lacking in other mammals. Huxley did this, indeed, at an 1862 British Association for the Advancement of Science lecture where Owen was present, reprising Grant's public humiliation over *Basilosaurus* two decades before.

Owen's painstaking work on his South African theriodonts sank into this professional miasma. Darwin simply ignored it in a tentative vertebrate genealogy of 1871. "[T]he monotremes now connect mammals with reptiles in some slight degree," he wrote, "But no one can at present say by what line of descent the three higher and related classes, namely, mammal, birds, and reptiles, were derived from the two lower vertebrate classes, namely, amphibians and fishes." Huxley's conversion to Darwinism meant he no longer could ignore the possibility of links between classes, but he still could push them into Paleozoic mists. Instead of acknowledging Owen's evidence of a Triassic reptile-mammal link, Huxley elaborated a theory that mammals had originated directly from an amphibian ancestor, arguing in an 1879 paper that the pelvic bones of Owen's theriodont, *Dicynodon,* were too unlike a mammal's for there to be any relationship. Huxley thought the earliest mammals had a platypuslike pelvis, and maintained that such a pelvis resembled a salamander's more than a lizard's. "[T]hese facts appear to me to point out the conclusion that the Mammalia have been connected with the amphibia by some unknown pre-mammalian group," he wrote, "and not by the known forms of Sauropsida."

Owen's beastly side facilitated such dismissals of his work. The English Cuvier was a notorious dog in the manger with his bones, occasionally snapping even at his British Museum assistant Henry Woodward, who complained that Owen was "eager to conceal his treasure from the curious and inquiring eyes of youthful aspirants." Huxley let Owen snap away, and "though he often described the Museum's fossils, generally left the South African theriodonts alone." Yet Owen would not have been "eager to conceal" his 1876 theriodont catalogue from "youthful aspirants"—not if he wanted his scientific reputation to last. Huxley probably looked at it, but he felt he had ample cause for ignoring it and may have relished the poetic justice of his own ideas about mammalian origins. Forty years before, Owen had used details of platypus anatomy to bury Grant and Geoffroy's claim that the platypus was a link between reptiles and mammals. Huxley now used details of platypus anatomy to obscure Owen's claim that theriodonts were a link between reptiles and mammals.

In fact, other studies tended to vindicate Owen. In the 1870s, taxonomists decided that the Stonesfield jaws were not in fact marsupial, but from a more primitive mammal group, refuting Huxley's claim that they were unchanged from modern opossums. Then, in 1883, a young Cambridge biologist, W. H. Caldwell, went to Australia, and, motivating aboriginal hunters by charging them extortionate prices for flour, tea, and sugar, got enough platypuses and echidnas to demonstrate that monotremes lay eggs with reptilian characteristics. These developments would not have pleased Owen entirely after his campaigns to refute the Stonesfield jaws' transitional nature and platypus egg-laying, but they were even less convivial with Huxley's amphibian mammal ancestry.

A more basic vindication came in the early 1900s. Dr. Robert Broom, later famous for hominid discoveries, studied theriodonts in South Africa and concluded that they were indeed transitional between reptiles and mammals. They had "well-developed limbs which enabled the animals to walk with a mammal-like gait with the body well supported off the ground," suggesting a more active life than a reptile's, and, by implication, the possibility of warm-bloodedness, hair, and other mammalian traits. Imaginative modern reconstructions show them living in burrows, forming pair bonds to hatch eggs, and even yapping and wagging their tails, like a pair of "cynodonts" that nimbly dodge Triassic dinosaurs in a 1999 BBC computer animation series, *Walking with Dinosaurs*.

Broom's work came too late for Owen. Tripped by his wayward impulses, Owen slipped down the dark side of the Cuvierian pedestal he had ascended in 1840, and his luster had faded even before his death in 1892. When the Reverend Richard Owen wrote his grandfather's official biography two years later, he asked Huxley to assess Sir Richard's character and professional stature. The request astonished Darwin's aged bulldog, although Huxley had magnanimously proposed a commemorative Owen statue now that the original was out of the way. He declined to address Owen's character, protesting that he had never been inside his house, nor Owen in his. He did write a guarded scientific assessment, half about anatomical history, half about Owen's philosophical limitations. A few pages in the middle summed up his defunct rival's paleontological work, with a bare mention of "the wonderful extinct faunae of South Africa."

Owen's scientific reputation has recovered somewhat since, but the perennially popular Darwinian saga still casts him as a Dr. Jekyll. According to his only recent biographer, Nicolaas Rupke, he was "systematically written out of history by Darwin and his followers. By blackening the behavior of their adversary, they undercut his scientific credibility and succeeded in turning his stance vis-à-vis Darwin into the touchstone of his historical worth." Even Owen's statue is black, Rupke notes, whereas Darwin's is white.

Owen may have felt blackened. His portraits seem a reversal of Huxley's progress from youthful dishevelment to aging dignity. The young Richard Owen looks dapper, if tense. Later photos seem hangdog, with straggling whiskers and poached egg eyes. "With age, his raw-boned looks became more pronounced," Rupke observes, "and as if by sympathetic magic he began to resemble his vertebrate fossils." To increasingly suspect oneself the descendant, not only of an ape, but of a "weasel lizard" and a "tiger crocodile" may have seemed secret proof of primal guilt. "He was wild when he was young," Stevenson's narrator concludes of the fallen Jekyll. "Aye, it must be that; the ghost of some old sin, the cancer of some concealed disgrace; punishment coming, *pede claudo,* years after memory has forgotten and self-love condoned the fault."

FOUR

The Noblest Conquest

DARWIN COULD SNUB OWEN'S KAROO THERIODONTS easily enough. They sound like something out of a Lewis Carroll parody. If he had ignored another group of animals in which Owen had been the first to see progressive biological change, however, he would have had a much harder time convincing even friendly skeptics like Huxley and Lyell of "descent with modification." Indeed, if Darwin had believed in a divine providence for transmutationists, he might have seen that other group's existence as evidence of it. Asked to name the Creator's salient trait, he might have replied, not "an inordinate fondness for beetles," as did twentieth-century geneticist J. B. S. Haldane, but "an inordinate fondness for dead horses."

Although not up to beetle abundance and diversity, horses and their forebears have perhaps left more bones in more places over a longer time than any other large mammals. Today, every fossil collection contains drawer after drawer of them, and the hardest parts, the curved, fluted teeth, are often simply boxed up in heaps. I once volunteered briefly at an Idaho horse quarry, Hagerman Fossil Beds National Monument, which contains so many skeletons of a zebralike Pliocene species named *Equus simplicidens,* "simple-toothed horse," that it has supplied most of the world's museums since its discovery in 1928. A paleontologist estimated in 1930 that the bone-bearing stratum was 5,000 square feet in extent. In 1997, skeletons remained so abundant that my job mainly was brushing dust away from a pavement of articulated bones. Greg McDonald, the Hagerman Monument paleontologist at the time, thought a herd had starved to death there during a drought.

Early paleontologists quickly became aware of the abundance of horse fossils, and Darwin had every reason to be thankful for it. As Owen's nasty review pointed out, *The Origin of Species* was unable to invoke a fossil sequence demonstrating that "descent with modification" had occurred, much less that natural selection had caused it. Darwin explained that the "extremely imperfect" geological record was the reason "why we do not find interminable varieties, connecting together all the extinct and existing forms of life by the finest graduated steps," but the excuse was weak. Some vestiges of "interminable varieties" would have to be found. Darwin's Argentinean fossils were a step in the right direction, and the remains of giant Tertiary kangaroos and wombats found in Australia's Wellington Caves in 1836 were another. As Lyell wrote Darwin, the latter showed that "the peculiar type of organization which now characterizes the marsupial tribes has prevailed from a remote period in Australia." Still, likening the Volkswagen-sized *Glyptodon* to the modern armadillo, or ten-foot prehistoric kangaroos to modern wallabies, hardly demonstrated "the finest graduated steps," and it was lucky that the ancestry of the horse was starting to become apparent when the *Origin* was published. Buffon had said more than he knew when he wrote in his *Histoire naturelle:* "The noblest conquest man has ever made is that of this proud and fiery animal who . . . even dies the better to obey."

Dead horses emerged to transmutational notice on a Greek streambank in the late 1830s, when a Bavarian soldier serving with Greece's German-born King Otto stumbled on bones at a place called Pikermi on the Megalorhevna River, northeast of Athens. They appeared to be encrusted with gemstones, and included a skull that looked human, so he took them home to Munich and tried to sell them, but the "gems" were worthless calcite crystals, and the skull raised suspicions of grave-robbing. A zoologist named Andreas Wagner identified the skull as that of a baboonlike monkey, however, and he realized that the bones were very old because calcite crystals grow on fossils in the ground. After Wagner published a paper on the bones in 1839, naturalists flocked to Pikermi.

They found a bone deposit so rich that a single cubic yard once yielded eight monkey skulls. From the bones, they reconstructed a world almost as strange as Cuvier's Paris gypsum one. As the monkeys showed, its fossil creatures were more like African than European animals. There were leopardlike cats, hyenas, giraffes, rhinos, and antelopes. Yet the Pikermi fauna was not quite like living African ones. Instead of elephants, there were four-tusked mastodons, and there were saber-toothed cats, bearlike beasts, and deer. There also were horselike beasts, but unlike any living in Europe *or* Africa today. The Pikermi fauna seemed a link between Cuvier's strange Eocene Paris basin fauna and the more familiar Ice Age deposits, and were dated to Lyell's Miocene epoch

Pikermi was the first fossil site to provide a comprehensive look at a vanished world. Artists, including Cuvier himself, had done reconstructions of the Paris gypsum fauna, showing *Palaeotherium* and *Anoplotherium* browsing by swampy subtropical rivers, but the "pachyderm" fauna had seemed unrealistically depauperate in its absence of other herbivores or predators. Pikermi provided enough skeletal information about its fauna to allow full-scale reconstruction. Indeed, Zallinger evoked its half-prehistoric, half-modern world in a series of book illustrations soon after he finished the Peabody's *Age of Mammals.* In them, deer congregate with mastodons, antelope, and a rhino in a Floridian landscape of palms and oaks, and a saber-tooth springs on one of the horselike beasts.

When a French paleontologist named Albert Gaudry made two expeditions to Pikermi in the 1850s and 1860s to study its "intermediate forms," the commonest fossils he found were those of the horselike beasts. Any bone extracted at random from the matrix was likely to belong to a genus named *Hipparion* ("lesser horse"), which had been found before, but never in such abundance. It was unlike living horses in having three toes, a large functional one and two small vestigial ones. Its molar teeth had high crowns and convoluted cusps like a living horse's, but again, less so. Altogether, it was a less specialized creature than living horses, more like a tapir or rhino, which have multiple toes and low-crowned teeth. Zoologists had long believed that horses are related to those ungulates, which was why they classed them as "odd-toed" perissodactyls. The Pikermi fossils hinted that the relationship was ancestral as well as anatomical.

Gaudry was not a Darwinian. Like many naturalists, he thought natural selection emphasized competition and chance too much to reflect the harmony and order he saw in living landscapes. The *Origin's* popularity helped make transmutationism a respectable approach to paleontology, however, and Gaudry took advantage of this. In the 1860s, he drew a genealogy showing a small Pikermi species, *H. gracile,* giving rise to a larger one, *H. crassum,* in the subsequent Pliocene epoch, and *H. crassum* then giving rise to a single-hoofed true horse, *Equus robustus,* in more recent

Pleistocene deposits. *Hipparion* might not have transmuted into *Equus* through a "struggle for existence," but "descent with modification" did seem to have occurred.

The trouble with Gaudry's diagram was that it gave no idea of how *Hipparion* itself had originated. A brilliant young Russian was working on that by 1870, however. Disliking tsarist autocracy, Vladimir Kowalevsky had left Leningrad law school to study in Germany with Ernst Haeckel, a zoologist and enthusiastic Darwinian. He had visited Darwin at least twice as he traveled around Europe, had attended Huxley's lectures, and was eager to address the issue of "intermediate forms." As a poor émigré, he lacked the funds and status to mount a bone-hunting expedition, so he explored the fossils already piled in museum drawers, particularly the abundant ones in Cuvier's Paris collections. There he found an even earlier genus of horselike creature that Cuvier had named *Anchitherium,* "near beast."

Like *Hipparion, Anchitherium* had three-toed feet and high-crowned teeth, but the toes were all functional, like a tapir's, and the molars also were tapirlike. In many respects, *Anchitherium* seemed transitional between *Hipparion* and a long-legged species of Cuvier's *Palaeotherium.* If so, here was a plausible lineage spanning many "hundreds of thousands" of years indeed, from Montmartre's gypsum swamps to Pikermi's savannas and Ice Age prairies. As a Darwinian, moreover, Kowalevsky saw a plausible reason for the fossils' changes—the animals were adapting to shifting environments. *Anchitherium's* three-toed feet and relatively low-crowned teeth suggested a very different way of life from a horse's. Like a tapir or a palaeothere, it had lived in woodland, and browsed on soft vegetation. *Hipparion's* toes and teeth indicated a somewhat harsher, but still warm and relatively lush environment, like savanna. *Equus's* hard hooves and flinty teeth spoke of the cool, dry grasslands where wild horses had lived during the Ice Age.

Publishing his findings in French and German as well as Russian, Kowalevsky reached a wide audience and made a big impression among fellow Darwinians and younger naturalists. His activities seem to have been less welcome to the older generation of European paleontologists, who ran the university departments and museums. Justifiably or not, Kowalevsky developed a reputation for playing fast and loose with fossil collections, even for outright theft. His leftist politics probably contributed to this—he was involved in the Paris Commune in 1871, when radicals took over the city after the catastrophic defeat of the Franco-Prussian War. Unable to find a scientific position in paleontology, he started a career as a chemist and taught at Moscow University, but he suffered from depression and financial problems, and he finally chloroformed himself to death in 1883 at the age of forty-one.

Kowalevsky's work fared better than its author. Encouraged by Darwin (who had cited Owen's *Palaeotherium-Equus* link when pitching his theory back in 1856), Huxley drew on the Russian's work in constructing his own horse lineage, which he used in lectures advocating what he had begun to call "evolution." (The word, from the Latin *evolutio,* "unrolling," has several meanings—its primary definition in *Webster's* is simply "a process of change in a certain direction." The philosopher Herbert Spencer substituted it for Darwin's unwieldy "descent with modification" in the 1860s.) The lectures were popular. "With his publicist's flair, Huxley fixed on one creature, meaningful in an equestrian age," his biographer, Adrian Desmond, writes; "a model of faithful subservience, the horse showed Nature in yoke." Huxley's borrowed lineage was a jump, of course, from his Lyellian ideas of the 1850s, when he had scorned Owen's rudimentary notions of horse ancestry. But he admitted in an 1870 talk that work like Gaudry's had provided "much ground for softening the somewhat Brutus-like severity with which, in 1862, I dealt with a doctrine, for the truth of which I should have been glad to find a good foundation."

As it happened, there was considerably more "ground for softening" on evolutionary doctrine than even Huxley knew of in 1870. Despite some rich fossil beds like Pikermi, paleontologists were realizing that Europe had major drawbacks from a bone-hunting standpoint. The land area west of the Urals was relatively small, and mountains and inland seas had punched frequent gaps in its fossil record. Forests and human activities made even existing fossils hard to reach. Around 1863, one of the first professionals, Professor Ferdinand Roemer, told his class at Breslau University in Germany that it was "not worthwhile to spend time in the thickly settled regions." The "most inviting fields for Paleontology" were unsettled places, like the American West. An ambitious young New Englander named Othniel Charles Marsh happened to be present when Roemer said this, and it was not lost on him.

The West, of course, was one of the planet's most unsettled places at the time, as Roemer had found while hunting Texas trilobites in the 1840s. It had all the bone-hunting advantages that Europe and eastern America lacked. It was vast, and huge fossil beds existed there, many of them unaffected by human or geological changes. It was dry, with only grass or brush growing in many places. In others, wind and water erosion had removed all vegetation, and fossil bones littered the ground for square miles. Hundreds of such places, badlands, extended over the Great Plains, Rocky Mountains, and Great Basin, and most of the periods and epochs that Lyell and his confreres had invented were much better represented there than in Europe. It was a bone hunter's candy store, albeit expensive and dangerous to patronize. Hazards included not only angry Indians, storms, and other frontier dangers, but deep, thinly covered mudstone crevasses, into which the unwary might fall.

The Great Plains' premier boneyard was in the White River region of the Nebraska territory. "From the uniform monotonous prairie," wrote a geologist, John Evans, "the traveler suddenly descends one or two hundred feet into a valley that looks as though it had sunk away from the surrounding world. . . . Embedded in the debris lie strewn, in the greatest profusion, organic relics of extinct animals." Now part of South Dakota's Badlands National Park, the White River beds still yield bones to casual observers after nearly two centuries of collecting. When I took a stroll there in 1997, I tripped on a legbone, mineralized a beautiful creamy yellow, embedded in the dirt road, although I wasn't able to learn whether it came from a horse ancestor or some other beast, since park ethics require fossils be left in situ.

In 1846, a fur trader had sent part of a jaw from the White River to a St. Louis physician named Hiram Prout. Prout was familiar with Cuvier's work, and he thought the jaw similar to the Paris gypsum *Palaeotherium*'s and identified it as such. The jaw, and many other badlands fossils, eventually found their way to Philadelphia, where Joseph Leidy, who had succeeded *Basilosaurus*'s discoverer, Richard Harlan, as the American Philosophical Society's reigning paleontologist, classified them. Born in Philadelphia in 1923 and trained as a doctor, although shyness kept him from practicing, Joseph Leidy was heir to a New World variant of the Cuvier and Owen tradition. The best American anatomist of his day, he was less of a specialist and systematizer than the two Europeans. Cuvier and Owen had studied mainly vertebrates, with excursions into sea creatures. Facing a largely unknown continent, early American naturalists tended to emphasize observation over organization. Like the freewheeling Philadelphia naturalists who preceded him, John and William Bartram, Leidy studied everything, even the lowly life of mud puddles.

Leidy lacked the time and funds to follow the frontier tradition by collecting western bones himself, but he was prepared to study quantities that would have intimidated a less eclectic worker. The bones he got from geological explorers like Ferdinand Hayden were even stranger than the Pikermi ones. Vast numbers of ungulate and carnivore skeletons emerged from the White River badlands, and most were unknown in the Old World. Pikermi's elephantlike creatures were absent, as were its apes, lions, and other African animals. Hiram Proutt's supposed *Palaeotherium* turned out to be another creature entirely, although it was indeed an odd-toed perissodactyl like *Palaeotherium.*

A section of Zallinger's Peabody mural, largely based on White River fossils, shows how strange the time was. On a lush plain studded with pines and cottonwoods roam beasts tantalizingly similar to living ones, but they almost always turn out to be different. A beast named *Mesohippus* appears vaguely horselike, but is three feet tall. What looks like an outsized wild boar rooting under a sycamore is *Archaeotherium,* an animal with only distant links to pigs, while a hyenalike creature is actually a creodont, *Hyaenodon,* a relative of snarling Eocene *Oxyaena.* A rhinolike *Subhyracodon* and a tapirlike *Protapirus* browsing by a pond are more closely related to modern beasts, but a *Bothriodon* wallowing in the pond's center lacks living counterparts, as do the honey-colored beasts that graze in the distance. These might be taken for sheep, deer, or antelope but they are, in fact, oreodonts ("mountain tooths"), which were artiodactyls, "even-toed" ungulates, like sheep and deer, belonging to a long-vanished group.

The fossils' strangeness obsessed Leidy. "You have no idea how much my mind has become inflamed on this subject," he wrote to Spencer Baird, a Smithsonian naturalist, in 1851. "Night after night I dream of strange forms: Eocene crania with recent eyes in them." He published lengthy compendiums of badlands fossils in the 1860s and 1870s, but seems to have held them in a kind of egalitarian awe, refusing to theorize about them like his European counterparts. Although sympathetic to Darwinism, which he likened to an intellectual meteor that "flashed upon the skies," Leidy refrained even from Gaudry's mild genealogizing. The White River fauna clearly was transitional between Eocene and Miocene ones (a new epoch eventually was created for it, the Oligocene, "less recent"), but he didn't try to fit it into the hierarchical edifices of Old World museums and universities. He called his work "a record of facts . . . as the author has been able to view them," disavowing any attempt "at generalizations or theories which might attract the momentary attention and admiration of the scientific community."

It remained to Leidy's young colleague, Marsh, to take charge of bringing American paleontology up to European standards, and he did it in a peculiarly Yankee way. Marsh's life was a kind of Horatio Alger story. The son of a ne'er-do-well New York farmer, his mother dead in his childhood, he had left home at sixteen and had seemed bound for nonentity well into early manhood. But he had a rich uncle, indeed, the richest. George Peabody, his mother's brother, had parlayed door-to-door peddling in Washington, D.C., into an international financial empire and a net worth of twelve million dollars by the 1850s. He eventually paid for his straggling 23-year-old nephew to attend Andover, and when Othniel was in his second year there, something made him decide to "take hold." He started excelling at schoolwork and also showed a flair for political maneuvering, which included enlisting his uncle's support for further education. Uncle George put him through Yale's new Sheffield Scientific School, then sent him to Europe for further study.

Another trait Marsh shared with Horatio Alger heroes was an apparent freedom from some of the wayward emotions that beset men like Owen and Huxley. A preference for hunting and fishing over farmwork seems to have been his only youthful vice. Aside from decorous, rejected suits to two wealthy and well-connected young ladies, he showed little interest in the opposite sex and remained a bachelor, focused entirely on his career. This raised suspicions of what Victorians discreetly called "inversion," but there is no evidence of sexual activity on his part, nor of any intense emotional involvement except professional rivalry, which led to a secretiveness that grew obsessive. "My belief is that he always lived apart from those about him," wrote Yale's president, Timothy Dwight, complaining that Marsh's taciturnity made a meeting with him into a kind of diplomatic conference.

Marsh had wanted to teach mineralogy at Yale, but the only faculty opening was in paleontology, so he had pragmatically abandoned his first ambition and followed Professor Benjamin Silliman's advice to pursue bones. Convincing George Peabody of the obscure subject's worth, he had arranged for his uncle to give Yale a large sum to build a paleontology museum, with himself as its director. He had then gone to Germany to learn his new subject from the likes of Breslau's trilobite-hunting Professor Roemer, where he had heard something else that was not lost on him. "When a student in Germany some twelve years ago," he wrote in 1877, "I heard a world-renowned professor of zoology gravely inform his pupils that the horse was a gift from the Old World to the New, and was entirely unknown to America until introduced by the Spaniards. After the lecture, I asked whether no earlier remains of horses had been found on this continent and was told in reply that the reports to that effect were too unsatisfactory to be presented as facts of science. These remarks led me, on my return, to examine the subject myself."

Marsh wasted little time in going west and finding horse remains after taking up his Yale paleontology chair, the first in America. On an 1868 jaunt to the end of the unfinished transcontinental railroad with an American Association for the Advancement of Science field trip, he applied Uncle George's money to investigating rumors of human and tiger bones dug from a well at a Nebraska site called Antelope Station. He tipped the conductor to hold the train while he examined the well diggings, and "soon found many fragments and a number of entire bones, not of man, but of horses, diminutive indeed, but true equine ancestors." When the conductor grew impatient, he tipped the station agent to hold the bones. Picking them up on the way back, Marsh regaled his colleagues with an impromptu lecture that described "evidence of no less than eleven inferior animals, all extinct." In particular, a "small horse was strongly in evidence, and interested all observers. He was then and there christened *Equus parvulus*. During life, he was scarcely a yard in height, and each of his slender legs was terminated in three toes."

Marsh's memoir went on to say that "later research" proved the little horse "to be a veritable missing link in the genealogy of the modern horse." The "later research" also proved Marsh to be a particularly resourceful and acquisitive naturalist, or scientist, as they were beginning to be called. Owen might have a far-flung British empire to send him fossils. Marsh had an exploding American one to send him *to* the fossils. Well-connected in government as well as academe, he mounted four big expeditions from 1870 to 1873, using Yale students and U.S. Army detachments to do the work. This provided publicity as well as fossils, since the patrician students wrote up their adventures for eastern newspapers and magazines.

Marsh already had become a media figure by debunking a notorious hoax in 1869. A freethinker named George Hull had fabricated a giant human figure from gypsum and buried it in the hamlet of Cardiff in upper New York in an attempt to hoodwink fundamentalists who believed in biblical "giants in the earth." Probably to Hull's surprise, some eminent scientists had hailed the 10-foot figure as a genuine relic of prehistory, and James Hall, the New York State geologist, had called it "the most remarkable object brought to light in this country." The "Cardiff Giant" became a sensation and made thousands for the hoaxers as they displayed it in ensuing weeks. Marsh had taken one look at the figure, however, and had pointed out that groundwater would have dissolved the conspicuous tool marks on the soft gypsum if the figure were more than a few years old. "I am surprised that any scientific observers should not have at once detected the unmistakable evidence against its antiquity," he scoffed in a letter widely distributed to newspapers.

Marsh became the first celebrity bone hunter, a kind of scientific Sherlock Holmes who penetrated knavery with a gimlet eye and esoteric knowledge. He didn't entirely look the part. Although lanky in early portraits, he soon developed a taste for luxuries, particularly the era's gala banquets, and he was getting portly by his early forties. Still, he hunted bison, parlayed with Indians, and hobnobbed with the likes of Buffalo Bill Cody until such adventures palled after a particularly rugged autumn expedition to the White River badlands during an 1874 Indian uprising, when the weather was so bad that he had to break icicles off his moustache to eat. Marsh then developed an efficient machine of hired field collectors, a motley, sometimes thuggish, crew, also called "bone sharps," to send him fossils, which a staff of graduate assistants and museum technicians then processed.

Marsh's fossil machine pumped a cascade of bones into New Haven from all over the West, and horse bones flowed in fastest of all. Even his hasty 1868 bone haul had included four equine species. From then until 1892, Marsh added nineteen new species and eight new genera to the seventeen American horse species previously named—mostly by Leidy. Some of these turned out to be the same as European genera that were already known. *Equus parvulus,* which Marsh renamed *Protohippus,* "before the horse," was the same as *Hipparion.* A genus that Marsh named *Miohippus* was the same as *Anchitherium.* Regardless, Marsh used his piles of horse fossils to construct a much more detailed genealogy than Kowalevsky's. From Wyoming Eocene deposits came a fox-sized but horselike creature he named *Orohippus,* "mountain horse." The four digits on each front foot showed that it was even more primitive than the three-toed *Palaeotherium.* A later White River badlands creature that he named *Mesohippus* had three toes on its front feet, but the rudiment of a fourth. Then came *Miohippus-Anchitherium* and *Protohippus-Hipparion* in the Miocene. In the Pliocene, a one-toed *Pliohippus* was a smaller forerunner of *Equus,* which emerged in the Pleistocene.

"The line of descent appears to have been direct," Marsh concluded in 1874, "and the remains now known supply every important intermediate form." When the revered Huxley brought his horse genealogy on a lecture tour to the United States in 1876, Marsh chivalrously but firmly buried him in western horse bones. Sumptuously housed in George Peabody's

New Haven apartments, Huxley wrote his wife that the Peabody Museum's fossil collection was the "most wonderful thing I ever saw." Marsh trotted out genus after genus showing the gradual transition from four-toed *Orohippus* to one-toed *Equus.* If Huxley questioned a link, Marsh simply told his assistant to bring more bones, until the Englishman "turned upon him and said: 'I believe you are a magician; whatever I want, you just conjure it up.'"

"One of Huxley's lectures in New York was to be on the genealogy of the horse," Marsh recalled. "My own explorations had led me to conclusions quite different from his, and my specimens seemed to me to prove conclusively that the horse originated in the New World and not in the Old, and that its genealogy must be worked out here. With some hesitation, I laid the whole matter frankly before Huxley, and he spent nearly two days going over my specimens with me, and testing each point I made. He then informed me that all this was new to him, and that my facts demonstrated the evolution of the horse beyond question, and for the first time indicated the direct line of descent of an existing animal."

Huxley had swallowed Marsh's genealogy whole by the time he gave his horse lecture at New York's Chickering Hall on September 22. He declared that the Yale professor had worked the western fossil fields "in an incomparable fashion, so that nothing too much could be said of the care taken by him, the extent of the discoveries that had been made, or their scientific importance." His enthusiasm impressed even the *New York Herald,* then at its peak as America's most popular newspaper and not partial to English highbrows. "[T]he lecture was so strictly confined to the horse, in so far as it elucidated the theory of evolution, that a wag might have been tempted to observe that the professor was certainly riding a hobby," its report began skeptically, but continued: "It must be acknowledged, however, that he rode those antique, ossified chargers up to the outerworks of the tacitly designated enemy with much spirit, and with a perfect command of hand and knee." Evidently preferring a Yankee Darwinian to a foreign one, the *Herald* concluded that "the audience listened throughout with marked attention, and loudly applauded Professor Huxley's remarks touching the eminent services rendered by Professor Marsh of New Haven."

Huxley prophesied that Marsh soon would find even earlier horse ancestors, "completing the victory of *equus* over *fides,*" as the *Herald* put it. When they had discussed the possibility at the Peabody, Huxley had playfully drawn a five-toed caricature of a "dawn horse," or *Eohippus,* then, with residual Lyellism, had placed an ape-man, *Eohomo,* on its back. His lecture less fancifully predicted the discovery of a four-toed *Eohippus* with a rudimentary fifth toe on its front feet. "Seldom has a prophecy been sooner fulfilled," recalled Huxley's son. "Within two months, Professor Marsh had described a new genus of equine animals, *Eohippus,* from the lowest Eocene deposits of the West, which corresponds very nearly to the description given above." Marsh had simply pulled it out of his collection. "I had him 'corralled' in the basement of our Museum when you were there," he wrote Huxley, "but he was so covered with Eocene mud that I did not know him from *Orohippus.* I promise you his grandfather in time for your next horse lecture if you will give me proper notice."

Marsh failed to find the creature's "grandfather," although he predicted that it would be five-toed and "not larger than a rabbit, perhaps much smaller." And his *Eohippus* proved to be very like *Hyracotherium,* a genus that Richard Owen had described from England in the 1830s. (Owen's *Hyracotherium* fossil consisted only of some teeth and a skull, however, so its relationship to horses was unclear. Owen thought it resembled the hyrax, a small living African elephant relative.) Otherwise, Marsh's triumphant ascension from celebrity bone hunter to "American Darwin" was complete, although photographs make him seem more superconfident tycoon than cloistered savant. In a famous keynote speech to the American Association for the Advancement of Science's 1877 Annual Meeting in Nashville, he was able to proclaim: "To doubt evolution today is to doubt science, and science is only another name for truth."

Marsh's horses helped to dispel Huxley's lingering doubts about Darwinian evolution. "[N]o collection which has hitherto been formed approaches that made by Prof. Marsh in the completeness of the chain of evidence by which certain existing Mammals are connected with their older Tertiary ancestors," he wrote the American geologist Clarence King. When Marsh visited England in 1878, Huxley told him: "When I was in America, you showed me every extinct animal that I had read about, or even dreamt of. Now if there is a single living lion in all Great Britain that you wish to see, I will show him to you." Darwin showed his own appreciation for Marsh's help in bringing Huxley fully into the fold. He wrote the Yale professor in 1880 that his work had "afforded the best support to the theory of evolution which has appeared in the last 20 years."

Marsh's triumph resonates in the Peabody mural. Horses are its most numerous and persistent figures, featured in almost every epoch, and they tie Zallinger's panorama together. The Eocene section shows two specimens of *Eohippus-Hyracotherium,* although only one is alive, scuttling away from the *Oxyaenea* versus *Coryphodon* brouhaha. (The other, interestingly, is the carcass over which *Mesonyx* snarls at *Uintatherium.*) The little dawn horses have distinctly equine heads, although the rest of them is less horselike, and they have short legs and long, canine tails. They wouldn't have been much of a mount for Huxley's *Eohomo,* unless he was a foot tall. The group of bay-colored *Mesohippus* in the Oligocene glade have longer legs and a more equine stance, but still seem too small and slender to be rideable. A glossy chestnut *Merychippus* herd that gallops through the grassy Miocene is markedly horsey, however, and *Pliohippus,* browsing among Pliocene aspens, is hard to distinguish from a pair of furry *Equus* galloping under Ice Age peaks.

FIVE

Terrible Horns and Heavy Feet

MARSH'S DARWINIAN APPLE CONTAINED A WORM, however, and the Peabody mural shows it along with his horse genealogy. It is a very large "worm"—*Uintatherium,* the bizarre, knob-skulled giant that confronts *Mesonyx* in the Eocene section. A similar but even larger beast rampages through the forest a few hundred feet back. The first has six paired knobs on the top of its head, while the second has four, prolonged to near-antler size. Both beasts have elongated canine teeth that gleam like sabers against earth and foliage. Knobs and sabers clearly leave little room for brains, and it is hard to imagine the lives of brutes so unlike anything living. Perhaps, as with creodonts like *Oxyaena,* they manifested a primitive version of later mammal behavior and were even more willing than rhinos to attack anything that troubled their dim perceptions.

But such speculations are unrewarding, because, as the mural also shows, the giants left no descendants and hardly even seem to have had ancestors. Compared to Zallinger's frieze of horses, they might not have evolved but have dropped from the sky. What was known about them in the 1870s, at least, was the antithesis of the confident equine genealogy that Marsh served up for Huxley's delectation, and that made the Yale professor the apple of Darwin's eye. And even that "noblest conquest" wasn't as complete as it seemed.

Marsh's horse genealogy may have been the definitive Victorian version

Figure 7. *Uintatherium* and *Eobasileus* (Eocene) from Zallinger's *Age of Mammals* mural. Courtesy Peabody Museum of Natural History, Yale University, New Haven, Conn.

of evolution, but it wasn't the only one. Another, developed from almost as rich a fossil collection, was startlingly different. Like Marsh's, it began with four-toed beasts and ended with single-toed ones, but it ignored his classifications. Its author was Edward D. Cope, who had, in fact, found a jaw fragment of a "dawn horse" in 1872, well before Marsh discovered *Eohippus* in the Peabody basement. Like Owen with *Hyracotherium,* Cope had not understood the significance of his initial find, but in 1880, his field collector, J. L. Wortman, dug up most of the creature's skeleton in Wyoming. Four years later, when *Eohippus* remained disarticulated bones in Peabody Museum drawers, Cope used Wortman's skeleton, which he named *Protorohippus,* to make the first dawn horse reconstruction, and that became the model, via Charles Knight's paintings, for Zallinger's Peabody versions of *Eohippus-Hyracotherium.* Indeed, *Protorohippus* may become the American dawn horse's ultimate name, since recent research suggests that the currently accepted name, *Hyracotherium,* applies only to the Eurasian dawn horse.

Cope's hard-won discoveries, unannounced in the popular press, led him to feel that Marsh was occupying the new scientific limelight unjustly. He scorned his rival's "American Darwin" status, and, although his animus arose largely from jealousy at Marsh's eminence, he had a point. Darwin drew on all natural history to support his theory. Marsh simply drew on a lot of fossil vertebrates to support Darwin.

To be sure, Marsh's limitations were part of his strength. About theory, at least, he felt few of Darwin's complex doubts. "As a cause for many changes of structure in mammals during the Tertiary and post-Tertiary," he said in his 1877 Nashville speech, "I regard as the most potent Natural Selection. . . . Under this head I include not merely a Malthusian struggle for life among the animals themselves, but the equally important contest with the elements, and all surrounding nature. By changes in the environment, migrations are enforced . . . and with change of location must come adaptation to new conditions, or extinction. The life history of Tertiary mammals illuminates this principle at every stage, and no other explanation meets the facts."

The confident Yankee also assumed that evolution was progressive, a surety that eluded Darwin. Marsh's huge skull collection had allowed him to trace changes in brain size and shape using gelatin casts of craniums, and he made another tidy exhibit of them. "The real progress of mammalian life is well illustrated by the Brain-growth, in which we have the key to many other changes," he proclaimed in Nashville. "The earliest known

Tertiary mammals all had very small brains, and in some forms this organ was less than in certain reptiles. There was a gradual increase in the size of the brain during this period, and it is interesting to find that the growth was mainly confined to the cerebral hemispheres, or higher portion of the brain. In most groups of animals, the brain has gradually become more convoluted, and thus increased in quality, as well as quantity . . . in the long struggle for existence during Tertiary time, the big brains won; and the increased power thus gained rendered useless many structures inherited from primitive ancestors."

Marsh's choice of the brain as natural selection's master organ exposed a weakness, however. It raised the subject of consciousness, a stumbling block for Darwinist materialism. Changes in horse teeth and hooves might correlate directly with a drying environment, but did brain growth? Mental growth suggested not just mechanical adaptation but an element of volition. Did the horses' improving brains imply that they were somehow willing their adaptations? Such questions pointed back toward Lamarck, a direction that Darwin and Huxley disdained. But if willed improvement didn't influence progressive evolution, how *did* the physical variations that natural selection favored come into existence? Darwin didn't address the question confidently, and many scientists noticed.

One of the quickest to notice was Cope, who presented a paper at the 1877 American Association meeting that directly addressed the questions the Yale professor skated over. "The origin of variation of animal structures is, *par excellence,* the object of the doctrine of evolution to explain," Cope wrote. He maintained that "sensibility to impressions" had led to variation of structures by encouraging the "acquisition of new movements." Memory then fixed the new movements in the individual acquiring them, and an embryonic process called "recapitulation" passed on the variation to the next generation. "The ascending development of the bodily structure in higher animals has thus been, in all probability, a concomitant of the evolution of mind," Cope concluded, "and the progress of the one has been dependent in an alternating way on the progress of the other."

Although the public largely ignored Cope's ideas, they impressed scientists, particularly Europeans like Gaudry and the German paleontologist Karl von Zittel. The ill-fated Russian horse genealogist Kowalevsky spent two weeks studying Cope's collection when he visited the United States. And Cope had already been treading on Marsh's heels for over a decade in 1877. Marsh had looked askance at him when they had first met as students in Germany in 1863 and recalled having doubted Cope's sanity. Cope had been emotionally troubled at the time, but Marsh would have to revise his contemptuous first impression. Cope had a powerful, if erratic, brain, and he would devote much of his career to perceptively undermining Marsh's overall scientific status, as well as his simplified Darwinism.

Like Joseph Leidy, Cope was heir to the colonial tradition of eclectic natural history. While Leidy was the naturalist-tortoise, however, quietly plodding through his unspeculative papers, Cope became the scientist-hare. Born to a rich Philadelphia Quaker family in 1840, he had begun studying animals systematically in grammar school, and had published his first scientific paper, on salamander classification, at nineteen. Although he never bothered to attend a college, he would publish over 1,400 more before his death in 1897. He rivaled Leidy in the breadth of his professional interests, which included seminal contributions to herpetology and ichthyology as well as paleontology. Fueled by family wealth and an energy that verged on the hyperthyroid (his eyes bulge in photographs), Cope far surpassed Leidy in the scope of his active research. After returning from Europe in 1864, he had spent his summers traveling remote regions in search of specimens. Although he held no academic position during most of his career, and made his expeditions with small hired crews or alone, he amassed a vertebrate fossil collection second only to Marsh's.

Cope's insight into animals, fossil or living, bordered on the uncanny. "Under his expert guidance I felt I had stepped back into an ancient world—filled with all sorts of bizarre and curious things," Charles Knight wrote, recalling their work together on fossil reconstructions in the 1890s, "and in imagination I could picture quite distinctly just what these mighty beasts looked like as they walked or swam in search of food." Knight's description of Cope's Philadelphia lab recalls Balzac's Cuvierian junk shop: "Inside everything was unique and completely dust covered. Never have I seen such a curious place," Knight recalled. "I looked into the front parlor . . . the floor was completely hidden by the massive bones of some vast creature."

Cope's brilliance spawned many stories. An eminent herpetologist told me that when Cope once visited a colleague who held a newly discovered lizard in his hand, he was able to work up a description and classification merely from looking at it as they chatted. "I can't understand to this day how he could have been so prescient," the herpetologist said, "his reputation among his peers was as having been absolutely brilliant and *extremely* perceptive." The story also touches on a darker side of Cope's character, however, because he telegraphed the lizard's description to a scientific publication after the encounter, thereby grabbing authorship of his colleague's specimen. He could be vain and greedy, sometimes unscrupulous. Charles Knight found his lab a "most sinister domicile."

Indeed, Cope reversed Owen's and Huxley's Victorian progress from youthful waywardness to middle-aged respectability. His family was devout, and he himself had been fanatically so in youth. His friend Henry Fairfield Osborn thought that religious doubts arising from his European studies, along with an unhappy love affair, had caused the emotional turmoil Marsh noticed. "If I know myself I need every possible aid to distract myself from myself," Cope wrote his father in 1864; "what it would result in if my various outlets for my activities were not to my hand I cannot tell, but I do not much doubt, in insanity." Back from Europe, he had staidly married a Quaker cousin, "an amiable woman, not over sensitive," and his 1870s expedition letters to his wife were sanctimonious about frontier life. Yet Cope had a roving eye that contributed to marital separation in later life. Charles Knight told Osborn that no woman was safe "within five miles" of him—that his mind was "the most animal," and his tongue "the filthiest" he had encountered. Knight perhaps exaggerated—Osborn thought Cope retained "a conviction that God made the world, and that right and wrong existed very tangibly in it." But Cope certainly abandoned piety, adopting Unitarianism after his father's death, and living a raffish semi-bachelor life in his last years. Osborn described him puffing cigars in cheap restaurants.

Cope never even considered adopting Darwinism, although student letters from Europe show that he had quickly understood evolution. He had approvingly mentioned *Archaeopteryx,* the reptile-bird link that had done much to support the *Origin* after its discovery in 1861. He flirted with creationist ideas into the mid 1860s, however, writing in one article that a "great change in temperature" had destroyed all animal life at the end of the Cretaceous Period, "requiring the introduction of entirely new forms." He had dropped such catastrophism by 1869, but simultaneously started attacking Darwinism. In an 1871 pamphlet, he acknowledged that "the law of natural selection [is] the cause of modification of descent" but pointed out that it does not explain how organisms vary, as they must "in order that materials for exercise of natural selection should exist." Darwinism was thus "only restrictive, directive, conservative, or destructive of something already created." Cope proposed "to seek for the originative laws by which these subjects are furnished."

Marsh's reaction to such modest proposals from a clever amateur nine years his junior can be imagined, and their relations, which were superficially cordial at first, had become covertly hostile by 1870. Marsh suspected Cope of fossil-stealing, and pointed out gleefully that the young Philadelphian had mistakenly reconstructed a marine reptile at his city's Academy of Sciences with the head on the tail. Cope suspected the more affluent Marsh of bribing stone quarries to let him monopolize fossils, and thought that Marsh had reneged on a promise to deliver herpetological specimens from his 1868 western trip. His reaction to Marsh's triumphal 1870 expedition can also be imagined.

Cope went west himself in August 1871, but only reached Kansas. He found Cretaceous reptiles there—plesiosaurs and mosasaurs—but they were not what he really wanted. They were already well known from Europe and did not provide much insight into the evolutionary questions he burned to tackle. He wanted to go farther, to the Fort Bridger region of western Wyoming, where the oldest known Eocene deposits perhaps held the key to the origins of Tertiary mammals, perhaps even to the "origin of the fittest." In a letter to General E. O. C. Ord, the district's military commander, Cope wrote: "The history of the life of the successive ages of earth's history has nowhere a greater prospect of elucidation." He showed his usual insight in this. The deposit would provide many of the subjects for Zallinger's mural.

Western Wyoming's arid badlands contained more fossil mammals than living ones. Settlers called one place Grizzly Buttes for Eocene skulls that they mistook for those of modern bears. A physician at Fort Bridger, James Carter, had been sending Leidy fossils since 1868, and, although fragmentary, they were even more fascinatingly strange than those of the White River mammals. "One of the most curious of the extinct mammals of the Bridger Tertiary fauna," Leidy excitedly wrote for a government geology report, "is an odd-toed pachyderm about the size of a living peccary which, with the usual complement of molar teeth, was apparently devoid of canines, and was provided with a large pair of incisors like those of rodents." Leidy happily named the strange creature *Trogosus castoridens,* "the beaver-toothed gnawing hog."

The locustlike descent of the ambitious younger men on Fort Bridger blighted Leidy's enjoyment of such treasures, however. When Marsh brought his student expeditions to the area in 1870 and 1871, they filled crate after crate with fossils, including Leidy's "gnawing hogs," of which Marsh took possession by naming them tillodonts, "nipping teeth." Even more spectacularly bizarre were the bones of extinct giants, which seemed almost as common in the badlands as the skeletons of newly slaughtered bison were on the prairie. "These animals nearly equaled the elephant in size, but had shorter limbs," Marsh wrote. "The skull was armed with two or three pairs of horn cores, and with enormous canine tusks. The brain was proportionally smaller than in any other land mammal. The feet had five toes. . . . These mammals resemble in some respects the Perissodactyls, and in others the Proboscideans, yet differ so widely from any known Ungulates, recent or fossil, that they must be regarded as forming a distinct order." Echoing Owen's "terrible lizards," Marsh proudly named his discovery the Dinocerata, the "terrible-horns."

Marsh conceived such a proprietary passion for the Bridger fields that when he heard Leidy and Cope planned to collect there in the summer of 1872, he tried to keep them out. Although on friendly terms with Leidy, he rejected a suggestion from their mutual friend Dr. Carter that he leave his Yale students behind and help the aging naturalist on his first western trip. He actively tried to stop his younger rival, demanding that Ferdinand Hayden, the government geologist sponsoring Cope's trip, promise to send him elsewhere. Failing in those maneuvers, he tried to make Leidy and Cope agree to send him prepublication copies of any papers they wrote on their discoveries. This academic red tape would have been equitable if all three had been out in the field that summer, facing the same hardships. But Marsh stayed in New Haven, working up the material he had already accumulated, and expecting to beat his rivals to the distinction of describing the strange beasts. The methodical Yale professor worked slowly, however, and Leidy scooped him.

The gentle Leidy was overwhelmed when Carter and Dr. Corson, the Fort Bridger post surgeon, took him to the badlands in July. "The utter desolation of the scene, the dried up water courses, the absence of any moving object, and the profound silence which prevailed, produced a feeling that was positively oppressive," he wrote. "When I thought of the Buttes beneath my feet, with their entombed remains of animals forever extinct, and reflected on the time that the country teemed with life, I truly felt that I was standing on the wreck of a former world." Still, although he had reached Fort Bridger two weeks after Cope, Leidy quickly managed to describe and name a skeleton that Carter and Corson hauled into camp from some distant buttes. It was "the remains of the largest animal which had been brought to our notice," and he named it after the local mountains—*Uintatherium*, "beast of the Uintahs"—in a paper dated August 1, 1872. The description lacked his usual thoroughness. (He mistakenly thought the "enormous canine tusks" found near the skeleton belonged to another, carnivorous creature.) But it was the first published scientific description of one of Marsh's Dinocerata.

Cope found the going rougher. Marsh had henchmen at Fort Bridger, and he let them know that the Philadelphia upstart was not to be indulged. They salted sites near the fort with assorted fossil fragments and sniggered as the greenhorn excitedly assembled them into a "new" species. Cope had to waste weeks trying to recruit a crew, sleeping in the government hay yard for lack of other quarters. The teamster he hired turned out to be a drunk, and the only guide he could get, a shady character who called himself "Sam Smith of the Rocky Mountains," was a Marsh spy. Cope finally got into the field in late July, but the teamster went on a binge and stole supplies, and unknown enemies stampeded the mules. By mid August, Smith had deserted, writing Marsh: "My motive in going with Cope was to ceep him off some places as I think is good bone country close hear." Such stresses, combined with bad water and an undiagnosed ailment, flattened Cope in September. He suffered high fever, skin eruptions, and hallucinations, and his wife had to nurse him at Fort Bridger until October before he could return east. News of these torments doubtless pleased Marsh, who finally went west later that month with a small student expedition.

Yet Cope also scooped Marsh in his erratic way. He was as prescient about finding fossils as naming lizards. On August 17, 1872, he sent a telegram to Philadelphia describing a specimen similar to Leidy's "largest animal," but of course he gave it a new name—*Loxolophodon*, "crested tooth." The telegraph operator changed this to *Lefalophodon*, which means nothing, and the name was invalid anyway because Cope had used it earlier on another creature. But Cope already had written another paper naming the same animal *Eobasileus cornutum*, "horned dawn ruler," and that was published on August 28. "In a word, *Eobasileus* is the most extraordinary fossil in North America," he had grandiloquently written his father, perhaps with a touch of incipient delirium. Luckily for him, *Eobasileus* did turn out to be a different genus from *Uintatherium*. It is the four-knobbed beast in the Peabody mural's background.

Marsh named his first genus, *Dinoceras,* on August 19, and soon named another, *Tinoceras*, "punishing horn." His fossils turned out to be the same as Leidy's and Cope's, however, so their prior names prevailed. *Godamnit!* His rivals had usurped the distinction of naming the beasts *he* had discovered, and Marsh bitterly resented it. Although he eventually accepted Leidy's *Uintatherium,* he would ignore *Eobasileus* for the rest of his life.

Wanting to stay friendly with the venerable Leidy, Marsh turned his wrath on Cope, who, after his summer with the Bridger bullyboys, was eager to reciprocate. They exchanged spiteful letters, then spent much of 1873 squabbling over the horned giants in the usually sedate pages of the *American Naturalist.* In the March issue, Cope reconstructed his *Eobasileus* with an elephantine trunk, tusks, and floppy ears, and concluded that it was "the predecessor in turn of the huge Proboscideans now known." Marsh called Cope's conclusion "quite erroneous" and suggested that the dates and references in his uintathere papers were equally so, implying that Cope had falsified them to beat him into print. Cope defended his trunked *Eobasileus* in the May issue, remarking: "[I]t is plain that most of Prof. Marsh's criticisms are misrepresentations, his systematic innovations are untenable, and his statements as to the dates of my papers are either criminally ambiguous or untrue." Marsh's nine-page reply was so vituperative that the editors relegated it to an appendix, "at the expense of the author." He called Cope's work "a sleight of hand performance with names and dates," and its author a forger as well as a blunderer. Cope responded in an August appendix with a paragraph deploring Marsh's "recklessness" and "erroneousness."

In December, Cope wrote his father that Marsh was not "normally constituted," and might have to be put away. Marsh was normal enough mentally, but the uintathere squabble did seem to unhinge him scientifically. Although he bested his rival politically in 1879, when he took over the U.S. Geological Survey's paleontology section and cut off federal funds for Cope's expeditions and publications, Marsh went off the deep end with his campaign to bury *Eobasileus* under his Dinocerata. From 1872 to 1885, he issued thirty-four papers on the huge beasts, culminating in a 237-page quarto monograph illustrated with 56 plates and 200 woodcuts. It cost the U.S. Geological Survey $11,000 to produce 3,050 copies, and Marsh was so enamored of it that he had a special edition of 500 printed for himself at a personal cost of $1,247.50.

Marsh's monograph was typically meticulous. He provided fine reconstructions of uintathere skeletons, boasting in his preface that what paleontology needed was "not long descriptions of fragmentary fossils but accurate illustrations of characteristic type specimens." This was a dig at Cope, who often did not take the time to reconstruct his fossils systematically. Cope retaliated by accusing Marsh of faking parts of his fossils with plaster. Marsh's uintathere reconstructions became the accepted ones, however. Zallinger's Peabody *Uintatherium* and *Eobasileus* are based on them, even though they don't bear Marsh names, and they are indeed unlike elephants or anything else alive.

Wonderful as they are, however, Marsh's Dinocerata reconstructions had limited scientific value compared to his horse studies. The nomenclature was faulty, because Marsh stubbornly excluded Cope's terms. The two men couldn't even agree on the name of the order, which Cope called the Amblypoda, the "heavy-footed," Marsh the Amblydactyla, "heavy-toed." More important, Marsh's Dinocerata offered little insight into evolution. Despite the abundance of their skeletons around Fort Bridger, bone hunters there found no evidence that uintatheres had evolved from or into anything. They seemed to have appeared fully formed in mid-Eocene, swarmed the West for millions of years, then vanished, leaving no descendants. Although Marsh acknowledged that they faintly resembled elephants and rhinos, he doubted whether the similarity indicated more than some very distant common ancestry, although he did think that they were more closely related to Owen's *Coryphodon,* placing both in his Amblydactyla. The best he could do was speculate that his suborders, Dinocerata and Coryphodontia, had branched off in the late Cretaceous from hypothetical ancestors he called the "Protungulata."

"It is evident that the greater part of the work of writing this monograph remains to be done," jeered Cope just after publication of Marsh's *Dinocerata.* "We should have preferred to have seen this magnificent opportunity improved, so that it should have embraced detailed descriptions of those characters of all the species on which alone the derivation theory can be established or refuted. A natural result of this neglect is a failure to appreciate the true generic relationships of the species."

Even the cheerleading Huxley was reticent about Marsh's Dinocerata, passing over them in his New York lectures. If Darwinian evolution had caused the weird beasts' appearance and extinction, there was precious little evidence of it. Their Eocene environment had apparently been warm and lush when they appeared, warm and lush when they disappeared. Their molars and five-toed flat feet seemed well enough adapted to a subtropical forest environment, but not remarkably so. Their bizarre tusks and "horns" seemed adapted to nothing except defense or sexual display, and that was hard to prove. A rhino-relative called *Diplacodon,* "two broad tooth," that replaced them might have caused their extinction through competition, but that too was hard to prove.

The wily Marsh usually avoided subjects for which the evidence was so scanty. Of course, he could not have foreseen the scantiness when he had first launched into the uintathere monopolization business. Still, a lavish horse monograph would have served him much better. But passion is blind. Despite its auspicious start with the deflation of a trunked and floppy-eared *Eobasileus,* his campaign to crush his rival's Amblypoda under his *Dinoceras* and *Tinoceras* reconstructions backfired as, in Cope's busy hands, the Eocene monsters posed a possible negation of Darwinian evolution.

Ignoring Marsh's attempts to co-opt the group, Cope energetically set out to explore the mystery of their origins. He suspected that they had immigrated from the south, along with other mammals that had replaced saurians above the Cretaceous boundary, so he went to New Mexico's San Juan Basin in 1874. There he found the oldest Cenozoic beds hitherto discovered, empty of uintatheres, but full of bones very like Owen's *Coryphodon,* to which, of course, he gave a new Cope name, *Bathmodon,* "deep tooth."

The San Juan Basin shows why vertebrate paleontology's mainstream moved from Europe to America after 1870. Fossils still dripped from the red and yellow mudstone buttes when I was there in 1997. A yard-long gar skeleton hung on one cliff like a museum display, shedding onyx-like scales, and bits of white turtle shell shone like almonds in a cake. Almost as ubiquitous were blue-black objects that Spencer Lucas, curator of geology at the New Mexico Museum of Natural History, identified as scraps of *Coryphodon* bone. Cope was the first bone hunter to work there, and within a few weeks of his arrival, he wrote his wife that he had found over seventy-five vertebrate fossil species, most of them new. There was no need to dig deep as in the Paris gypsum or Stonesfield slate. Every rainstorm washed fossils into sight. Even when Cope revisited badlands a few years after picking them clean, he could find dozens of newly exposed fossil species within days.

Nearby, under the reddish early Eocene mudstones, he found even older, gray-green beds, which he named the Puerco Formation, a find so significant—although Cope saw no mammal fossils at the time—that Marsh sent a collector named David Baldwin to the San Juan Basin the next year. Baldwin, a frontiersman whose property consisted of a mule and a pickaxe, worked for Marsh until 1880, collecting in winter because he could melt snow for water in the parched badlands of the basin's west side. He found hundreds of Puerco Formation fossils, but Marsh was inattentive, strangely, since he presumably had hired Baldwin to look for things like uintathere ancestors. Baldwin eventually tired of his absent-mindedness about payment, anyway, and went to work for Cope, who quickly saw that the Puerco fossils represented dozens of new species more primitive than any Tertiary mammals then known.

The Puerco fossils were so primitive, from soon after the dinosaurs' disappearance, that paleontologists later placed them in a new epoch, the Paleocene, the "old recent," at the Cenozoic's beginning. Cope classified many as marsupials, including small beasts he called "multituberculates" because of bizarre cusps, or tubercules, on their molars. Zallinger painted one of these, which Cope named *Ptilodus,* "feather tooth," behind his mural's giant serpent. It seems to be crouched on a nest, reflecting Cope's further insight that multituberculate teeth were more like a baby duck-billed platypus's than an opossum's, which raised the possibility that the group were not marsupials, but egg-layers like monotremes. "It is the oldest Mammalian fauna of any extent," Cope exulted in 1888, after presenting a paper on Baldwin's fossils, "there are 93 mammalia, and its types are ancestors of those of later ages. It has enabled me to explain how and what the origin and changes in Mammalian anatomy (Osteology) have been."

In particular, the Puerco fauna enabled Cope to explain, or at least to speculate on, the "origin and changes" of Marsh's obsessive monograph subject. He thought his Puerco fossils included two early members of his Amblypoda, and that the larger of these, which he named *Pantolambda,* was the ancestor of *Coryphodon* and *Uintatherium.* Coyote-sized, *Pantolambda bathmodon* had a tiny brain and flat, five-toed feet like them, although the resemblance ended there, at least superficially. With shortish legs and a long, thick tail, a *Pantolambda* that slinks under a sassafras tree in Zallinger's mural looks more like a living tropical mammal called a tayra than a uintathere. Cope considered it partly arboreal, as tayras are, and possibly omnivorous as well as herbivorous, yet he thought he saw how it could have evolved into hulking *Coryphodon* in a relatively short time.

Cope didn't liken *Pantolambda* to tayras, but the comparison gives insight into his thinking about its evolution. Common in neotropical forests, tayras are members of the carnivorous weasel family, but they also eat a lot of fruit. I once watched a pair forage on the grounds of La Selva Biological Station in Costa Rica, galloping noisily around in a way that seemed almost equine. Fallen fruit is commoner on the forest floor now than it was ten thousand years ago, when big herbivores like mastodons and ground

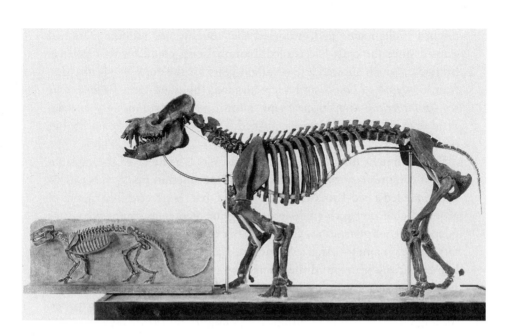

Figure 8. Skeletons of *Pantolambda* and *Coryphodon*. Courtesy American Museum of Natural History Library.

sloths consumed it. This may cause tayras to eat more of it than their ancestors did, and could start a shift toward less carnivorous habits, which, in a continued absence of competitors, might evolve a new group of big herbivores. Cope thought he saw evidence of a similar shift in his Amblypoda's teeth, which evolved from simple cusped molars in *Pantolambda* to yoke-shaped grinders in *Coryphodon* and high-crested ones in uintatheres.

In the Puerco Formation's sudden absence of dinosaurs, a change from arboreal omnivore to "heavy-footed" herbivore made sense. But if *Pantolambda* was *Coryphodon*'s near ancestor, such a transformation was more of a stretch for natural selection than horse evolution's gradual hoof and tooth changes, because it evidently had happened much faster than the *Eohippus* to *Equus* one. This raised a scientific specter that haunted Darwinism in its cradle. Following Lyell's lead, the original 1859 edition of the *Origin of Species* had allotted hundreds of millions, even billions, of years for natural selection to work. In 1866, however, William Thomson, Lord Kelvin, the reigning Victorian physicist and an author of the third law of thermodynamics, computed the earth's age by its rate of cooling from an original molten state and estimated that roughly 100 million years had elapsed since the crust had cooled. Nonphysicists could hardly challenge his figure, which allowed a few million years for the entire Cenozoic Era, from *Eohippus* to *Equus,* and a few hundred thousand for a *Pantolambda* to *Uintatherium* transition. Some more directed evolutionary mechanism than the "law of higgledy-piggledy" seemed not only desirable, but necessary.

By the time he discovered *Pantolambda,* Cope had refined the explanation of evolutionary variation and change outlined in his 1877 Nashville paper. He led a neo-Lamarckian movement that resurrected inheritance of acquired characteristics while jettisoning the eighteenth-century baggage of spontaneous generation. Cope gave his version of transmutation a battery of new names—Archaestheticism, Bathmism, Kinetogenesis—but, like Lamarck's, it involved the interaction of mind and body. "The function of the organism in evolution is to provide variations in its structure as an effect of its motions," he wrote in 1888, reiterating his 1877 paper. "It follows that organic evolution is the result, mediate or immediate, of consciousness."

Cope imagined some small, omnivorous, arboreal ancestor of *Pantolambda* looking down at a world somehow emptied of ground-dominating giant saurians. According to Cope's hostile view of Darwinism, its adaptation to the treetops would cause it to stay there, regardless of the newly available fodder below. Natural selection's mindless favoring of small, random variations would only gradually produce individuals capable of wandering down to benefit from the change. According to Cope's neo-Lamarckism, however, *Pantolambda*'s arboreal ancestor would perceive the lushness going to waste on the ground, would begin climbing down to get some, would develop new muscular structures for ground life, and would pass them on to its offspring by recapitulation, whereby acquired parental characteristics were transferred to the developing embryo. This could lead much sooner than natural selection to semi-arboreal *Pantolambda,* then to *Coryphodon,* then to uintatheres, with a "will to fodder" as a continual force for change.

Cope had noted in his 1871 pamphlet challenging Darwinism that Herbert Spencer had "epitomized" natural selection as the survival of the fittest. "This neat expression no doubt covers the case," he had written, "but it leaves the origin of the fittest entirely untouched." By 1888, he evidently felt that his modest proposal to seek the "originative laws" had been realized, and that his heavy-footed beasts touched the origin of the fittest as Marsh's terrible-horned ones did not. Natural selection might bring about the relatively small changes that result in new species, but it could not cause the more basic changes that led to new genera, families, and orders. Macroevolution, at least of mammals, was "a concomitant of the evolution of mind."

SIX

Mr. Megatherium versus Professor Mylodon

COPE'S ERUDITION AND IMAGINATION gave his *Pantolambda* to *Uinta-therium* lineage a compelling plausibility. The image of a post-dinosaur tree-dweller peering down hungrily at an unprecedentedly vacant forest floor is vivid. But it has to be said that there was then no evidence that *Pantolambda* had such an ancestor. The Peabody mural certainly hints at none. If anything it implies that most early Paleocene mammals spent more time on the ground than in the treetops. The fruit on its Genesis tree looks tempting, but no beasts are up there feeding on it. And although small, arboreal early Paleocene relatives of *Pantolambda* have turned up recently in Asia, that still doesn't prove they evolved into *Pantolamba* through Cope's process.

The less imaginative Marsh evidently considered such fossil lineages sheer moonshine, since he never even complained about them, perhaps the one aspect of Cope's career that he did not complain about. Huxley also ignored them, and snubbed the younger paleontologist when he visited England in 1878. "He took no pains to see me or hear any of my papers," Cope wrote his wife, "a coolness I suspect he would not have shown to friend Marsh." Darwin gave neo-Lamarckism passing mention in the sixth edition of the *Origin,* but complained in an 1872 letter to Cope's fellow theoretician Alpheus Hyatt that he found it obscure. "It has quite annoyed me that I do not clearly understand yours and Professor Cope's views," he wrote, "and the fault lies in some slight degree, I think, with Professor Cope, who does not write very clearly."

There was justice in Darwin's complaint. Cope's brilliance worked against him in that his ideas ran far ahead of his public exposition of them. Marsh's methodical approach—slowly accumulating a sequence of horse-like bones and refining it into an easily comprehensible lineage—succeeded much better in getting attention. Cope never presented his own horse genealogy with anything remotely approaching Marsh's elegance. The paragraph and crude diagram that he devoted to it in his 1884 "Bible" fail to show the progressive changes of equid teeth and feet that make Marsh's diagrams so convincing. Cope didn't even classify the genera that preceded *Equus,* like *Anchitherium* and *Hipparion,* in the horse family.

Cope's challenge to natural selection was not easily deflected, however. His multisyllabic theories now seem quaint, and Gregor Mendel's experiments on garden peas were pointing the way toward modern genetics in the 1860s. But Mendel was an obscure clergyman, and neo-Lamarckism was one of the nineteenth century's more respectable explanations for biological variation. Albert Gaudry, who nominated Cope to the French Geological Society, called his classifications a "great work," and the German paleontologist Karl von Zittel wrote that the United States would "add a precious leaf to its wreath of fame" by publishing a second volume of his "Bible." In our own day, Stephen Jay Gould would call Cope "America's first great evolutionary theoretician."

Indeed, Darwin's uncertainties about the origin of variation had driven him toward a kind of neo-Lamarckism while Cope was still collecting salamanders on his father's farm. In the *Origin of Species'* first edition, he had speculated that an American black bear seen skimming insect swarms from a river's surface might be acquiring a habit that eventually could evolve into baleen whales' sieve-feeding method. He toned that down in later editions of the *Origin,* but in other writing, notably in *The Variation of Animals and Plants under Domestication* (1868), he developed a theory, "pangenesis," which posited vague "gemmules" as a mechanism for transferring acquired traits from parent to offspring. Nineteenth-century scientists didn't regard pangenesis much more highly than Cope's kinetogenesis, however, and many joined the younger naturalist in finding Darwinism unacceptably materialistic.

Darwin and Huxley had every reason to promote Marsh's career—he was one of their few really strong supporters. Henry Fairfield Osborn recalled that when, as a graduate student in London, he had met Darwin, the great man had "smiled broadly" and said that he hoped "Marsh with his students would not be hindered in his work." Marsh stiffened many younger evolutionists' resistance to Darwinism, however, by treating them as meanly as he had Cope. When Osborn and his Princeton classmate William Berryman Scott collected uintatheres at Fort Bridger in 1877, Marsh's henchmen went into action, and the Princetonians didn't forget it. They became Cope's henchmen, spying and tattling on Marsh, and they proved more effective at it than frontier thugs, since they were rising scientists. Scott would become Princeton's professor of paleontology; Osborn the American Museum of Natural History's paleontology curator.

Cope had increasing use for henchmen. His anger at Marsh's preemption of the geological survey turned to hatred after he lost his fortune in bad mining investments while trying to compete with Marsh's federally subsidized collecting. He underwent a personality change, from a bluff, bearded explorer in 1870s photos to a pointy-moustached plotter, a Dr. Moriarty to Marsh's Sherlock Holmes. Cope even managed to suborn some of the Peabody Museum staff, who had their own grievances against the Yale professor's autocratic ways. The plotting culminated when Cope accused Marsh in the *New York Herald,* on January 12, 1890, not only of lifting his horse genealogy from Vladimir Kowalevsky, who had "complained bitterly when here of Professor Marsh's theft of his important life work," but of plagiarizing his *Dinocerata* monograph—via a Peabody assistant—from Cope himself. "The generalizations were dictated by George Bauer," Cope alleged, "who repeated what he knew from my own work on the subject, Marsh changing the names of divisions of classification. This attempted theft of my work is making a laugh all over Europe."

The *Herald*'s position as the nation's leading daily made it an unexpected player in the exposure of paleontological scandals. Other papers had been uninterested when William Hosea Ballou, a freelance journalist allied with Cope, peddled his "Marshiana" around town. But Marsh's fame had liabilities. The *Herald*'s dictatorial publisher, James Gordon Bennett Jr., was a pioneer sensationalist who liked to feature explorers in desperate situations that he contrived, as when he sent Henry Morton Stanley to Africa to "find" Livingstone. He envied his subjects' celebrity, however, and subsequently turned against Stanley and tried to smear his private life. Bennett's envy was naturally attracted by someone as prominent as Marsh, and the latter's brusque public pronouncements on various subjects nettled the publisher. Marsh's 1869 debunking of the "Cardiff giant" had especially stung Bennett, who had just taken over the *Herald* and had run a long article calling the hoax an "unmistakable antiquity." Bennett thereafter sought to turn Marsh's fame against him whenever possible.

"The vertebra of the sea snakes in the Yale Museum show the kind of success with which their investigations were crowned," the *Herald* had cooed apropos of Marsh's first student expedition in 1870, "and we may expect in the scientific journals articles of learned length and thundering sound similar to those which Professor Marsh has already published. . . . In science as well as in fiction a well-connected tale can be built on very slight foundations." A few months later, a *Herald* editorial had accused the Yale faculty of having "no learning."

After cutting Marsh down to size in 1870, Bennett had resented having to hail him as an American Darwin in his 1876 coverage of Huxley's horse lecture. When the Peabody Museum opened in 1877, an event covered at length in other papers, the *Herald* had noted only that Marsh paid $1,000 for a pterodactyl. And Bennett clearly still resented Marsh in 1890, because his paper filled column after column with Cope's attack through the month of January. "SCIENTISTS WAGE BITTER WARFARE," headlines proclaimed, "LONG SMOLDERING EMBERS OF HATRED." Promising its readers "Sensational Disclosures," the *Herald* ran editorials hinting at an impending congressional investigation and sent reporters as far as Canada to seek scandalous copy. It fanned the flames by goading Marsh and his allies to retaliate, and Marsh obliged, reiterating his 1873 accusation that Cope had committed "a series of blunders, which are without parallel in the annals of science" in his uintathere papers. "I never saw Kowalevsky's work until my own was completed and partly published," Marsh insisted in response to Cope's accusations of genealogy-theft, and accused Cope of being Kowalevsky's twin "in work and methods" of stealing museum fossils. "Kowalevsky was at least stricken with remorse and ended his unfortunate career by blowing out his brains," Marsh thundered. "Cope still lives, unrepentant."

Prodded by a *Herald* reporter the next day, Cope shot back that Kowalevsky was "the ablest paleontologist" Europe had produced, and that "his 'depredations' consisted, like my own, in seeing things in specimens that the owners did not see." He insisted that Marsh's Dinocerata work was "absurd," consisting "mainly in the attempt to substitute his own names for those of Leidy and myself," and that Marsh had stolen his horse genealogy not only from Kowalevsky, but from Huxley and another Darwinian,

William Flower. Cope even accused Marsh of faking his own horse genera: *Orohippus, Miohippus, Pliohippus,* and *Eohippus.* "Although it is sixteen years since the first three names were proposed nobody knows what they are," Cope told the *Herald,* "and they have been finally, in the latest European works, relegated to the limbo of useless rubbish. . . . Thus it is clear that Professor Marsh's record on the horse question is in bad shape. What is true in it is not new, and what is new is not true."

Cope's charges backfired to some degree. Joseph Leidy told the *Philadelphia Inquirer* that his "consuming restlessness" had gotten him into "hot water with his associates" and "caused the deepest regret among the scientific men of the country." Cope tried to dismiss this, writing Osborn that "Poor old Leidy has come out against me, just as he has always done," but Leidy typified scientists' reactions to Cope's antics, and the *Inquirer* speculated that they might lose him a recently acquired University of Pennsylvania professorship. Still, the *Herald* scandal tarnished Marsh's luster. A June 1890 cartoon in the British humor magazine *Punch* implied a distinct decline of scientific grandeur. It showed Marsh as a portly ringmaster, bedizened in the stars and stripes, putting a top-hatted uintathere through its paces. The caricature seems good-humored enough, and Marsh professed himself "delighted" by it, but the giant skeleton has a baleful air as it rears over his pear-shaped figure.

The *Herald* was less benign when it later lampooned the paleontologists while promoting a prizefight between "Gentleman Jim" Corbett and "Lanky Bob" Fitzsimmons. "Think of Mr. Big Tailed Little Headed Megatherium Americanum, sixty feet in length and weighing a few odd tons, out on the warpath for Professor Heavy Voiced Curve Horned Mylodon, of equal proportions when stripped," it jeered, "and the tiny baby prize-fighter of the modern day sinks into utter, absolute insignificance." The article complained that Marsh had "ruined what might have been a good story" by showing that "giant primitive man" footprints reported near the prizefight's Carson Valley, Nevada, location actually were those of the ground sloth, *Mylodon.* Then it concluded: "Who cares about Carson Valley, buried in the misty haze and sediment of geologic time?"

Although the *Herald's* mudslinging did not involve theory, even fewer evolutionists believed after 1890 that natural selection was "the most potent cause for many changes of structure in Mammals during the Tertiary." Odoardo Beccari, an Italian who explored southeast Asia soon after Alfred Russel Wallace conceived his natural selection theory there, was typical. "I have not a blind faith in the slow and progressive evolution of organisms,

Figure 9. Cartoon of O. C. Marsh with a uintathere and a titanothere, *Punch,* September 13, 1890, p. 119.

and in the formation of species as a result of continuous but insensible variation from pre-existing forms," he wrote. "I am more inclined to admit the sudden appearing of some principal adaptation forms."

Even Marsh may have had his doubts about Darwinist "real progress" after 1890. Cope's *Herald* attack had bared political as well as scientific soft spots. Congressional enemies of the U.S. Geological Survey and its director, John Wesley Powell, considered government science an extravagance. They fastened on Marsh's lavish fossil monographs as a bizarre waste of taxpayer's money, and in 1892, they exploited the issue in a successful campaign to cut Powell's appropriation. The cuts included the paleontology section, and Marsh was suddenly no longer lord of the western fossil beds. He had to stop most collecting and make do with the fossils he had, a mere million and a quarter dollars worth at nineteenth-century prices. He also had to mortgage his New Haven mansion to avoid bankruptcy, joining his rival in genteel poverty.

Cope and Marsh died two years apart before the century's end, but their feud lingered, generating strange legends related to a society, founded by Joseph Leidy, for the study of brains other than the paupers' or felons' usually available. "There are three hundred men in various parts of the United States who have bequeathed their brains to science," reported the *Herald* in 1898. "These men are neither cranks nor fools. They represent the cream of American professional life. Science with them is paramount. Nevertheless, it is a grewsome organization, this American Anthropometric Society." The *Herald* printed a drawing of Leidy's brain, calling it "unusually heavy and richly convoluted," and added: "of the men who followed Dr. Leidy, there was one who was more interesting in many respects than he. The late Professor Edward Drinker Cope, the foremost paleontologist of the country, whose unequalled collection of fossils the Metropolitan Museum made strenuous attempts to acquire, bequeathed not only his brain, but his skeleton, as well, to the scientific world."

The *Herald* didn't ascribe ulterior motives to Cope's bequest, but others did. One rumor said that he had donated his brain to prove it was larger than Marsh's; another that he had left bones as well as brain because he aspired to be the human "lectotype," the specimen by which *Homo sapiens* is described and named. Neither rumor is likely. Marsh wasn't an Anthropometric Society member, so his brain was not among those measured by Edward Spitzka, an anatomist, for a 1903 study. (The rumor perhaps arose from a bet about brain size that one member, John Wesley Powell, made with another.) And although Linnaeus did not designate a type specimen for our species when he named it, there is no record of Cope wanting to be the human lectotype. He had lost most of his teeth, for one thing, which would have made him a poor one.

Despite their unlikelihood, the rumors involved Cope's skeleton in macabre legerdemains after the Anthropometric Society dissolved and the University of Pennsylvania's Museum of Anthropology acquired it. According to Cope's will, it should have been kept in a "closed cabinet," but the anthropologist Loren Eiseley took it into his office and made a kind of mad scientist's fetish of it, toasting it at faculty luncheons, decorating it with flowers and tinsel, and also embellishing an earlier story. In 1914, a Smith College professor, H. H. Wilder, had borrowed the skull for an experiment in facial reconstruction. Eiseley's biographer Gale Christiansen wrote in 1990 that Eiseley had lent Cope's skull to the Museum of Natural History for a facial reconstruction sometime in the mid 1970s, and that it had then disappeared.

After Eiseley's death, the skeleton, complete with a skull, went into the Anthropology Museum's collection, but it did not rest there. In the early 1990s, a photographer collecting Cope-Marsh memorabilia for a dinosaur book borrowed the skull and used it as a conversation piece when he interviewed paleontologists. His book repeated the legends, and when he visited Philadelphia on a promotional tour, the *Inquirer* ran a front-page feature that attacked the long-dead bone hunter as though he was still upsetting local applecarts. Sensationally calling Cope's Victorian prejudices "Hitleresque," it used a rumor that Eiseley had replaced his lost skull with another of "undetermined race, ethnicity and gender" as a pretext to deride his rumored wish to be the human lectotype. "And so a century-old act of unparalleled egoism receives its poetic comeuppance," the feature concluded.

But Cope's skull had not been lost. The cranium that the photographer eventually returned to the museum matched a drawing made for Spitzka's 1903 study. If Cope got a "poetic comeuppance," it was from the circumstance that the *Inquirer's* 1994 smear on him was in the same vein as the *Herald's* 1890 one on Marsh. Resting quietly in a New Haven cemetery since 1899, Marsh certainly would have agreed.

Marsh also would have agreed with taxonomy's final choice of *his* label over Cope's for the giant Eocene beasts that he had been the first to find— now officially the order Dinocerata. And he would have been equally pleased to know that *Pantolambda* is not considered ancestral to the Dinocerata, but is placed in another order, the Pantodonta. So much for Cope's moonshine, by gad. It is harder to say how he, or Cope, would have felt about the fact that uintathere evolution remains little understood. Taxonomists are pretty sure that they were not closely related to living hoofed animals. They may, if anything, be closer to the order Lagomorpha, making them, in a sense, "giant horned bunnies." But, as of this writing, nobody knows.

It is also hard to say whether Cope's main challenge to Darwinism— that it failed to address the relationship between the "development of the bodily structure in higher animals" and "the evolution of mind"—was the moonshine Marsh thought it to be. Notwithstanding its rejection of Cope's neo-Lamarckism, biology never has found another way to explain how consciousness interacts with organic evolution. Instead, it has left consciousness to philosophy and psychology, as though "mind" had no prehistory. Evidence of nonhuman animal consciousness keeps growing, however, so purely mechanistic models of "the development of the bodily structure" begin to seem incomplete. In this sense, that old troublemaker Edward Drinker Cope has never had his "comeuppance."

SEVEN

Fire Beasts of the Antipodes

FAITH IN PALEONTOLOGICAL PROGRESS was undergoing another test at the time of Marsh's death. If there was one question that the new science's first century seemed to have answered, it was that of when modern placental mammals had originated. Marsupial and monotreme beginnings might remain lost in the mists of time, but the work of Cuvier, Owen, Marsh, and Cope seemed to point to the same start for horses and other familiar creatures. Although the cause of their evolution might be obscure, they definitely seemed to have appeared after the Mesozoic's saurian world had vanished. It was a story that such ambitious, innovative men could understand. Once the fossilized old guard was gone, the new generation could thrive.

But what if modern mammals had lived with dinosaurs? The Peabody Museum would have very different wall decorations, for one thing. The latter part of the *Age of Reptiles* might look like something from Sir Arthur Conan Doyle's sensational 1912 novel about an expedition to South America, *The Lost World*. "His appearance made me gasp," the young journalist narrator says of the expedition's leader, Professor Challenger, who seems a composite caricature of Victorian evolutionists. "I was prepared for something strange, but not for so overwhelming a personality as this . . . a stunted Hercules whose tremendous vitality had all run to depth, breadth, and brain." The hairy, bellowing Challenger drags the journalist and sundry cardboard characters on a quest for a mysterious Amazonian pla-teau where an American artist named "Maple White" had made drawings of strange animals. When they get there, after hair-raising interludes with jungle drums and a previous explorer's skeleton, the narrator describes scenes that Zallinger might have painted in a parallel universe:

> Two creatures like large armadilloes had come down to the drinking place, and were squatting at the edge of the water, their long, flexible tongues like red ribbons shooting in and out as they lapped. A huge deer, with branching horns, a magnificent creature which carried itself like a king, came down with its doe and two fawns and drank beside the ar-madilloes. . . . Presently, it gave a warning snort, and was off with its fam-ily among the reeds, while the armadilloes also scuttled for shelter. A new-comer, a most monstrous animal, was coming down the path. For a moment, I wondered where I had seen that ungainly shape, that arched back with triangular fringes along it, that strange birdlike head held close to the ground. Then it came back to me. It was the stegosaurus. . . . The ground shook beneath his tremendous weight, and his gulpings of water resounded through the still night.

Conan Doyle's *Lost World,* which has dinosaurs living, not only with glyptodonts and deer, but with "ape-men" who bear a studied likeness to Professor Challenger, was the prototype of a thousand pulp versions of prehistory. Conan Doyle was breaking new ground, however, and his ro-mance reflects some acquaintance with early 1900s evolutionary science, which, as a spiritualist, he wished to lampoon as well as exploit. He had evidently studied Charles Knight's paintings, and his thriller *Lost World* re-flected a real "lost world" of a sort. Nobody ever let loose a pterodactyl at a London "Zoological Institute" meeting as Challenger does in the novel's climax, but disturbing fossils had been coming from South America.

That continent poses problems for thinking on mammal origins because its fauna is so unlike North America's or Eurasia's. I experienced this viv-idly one evening in the Inca ruins of Machu Pichu when, turning a corner, I encountered a beast about the size of a hare, but with softer, denser fur and a head more like a deer's. It eyed me calmly, crouched on a low wall. Another appeared, but it was shyer and soon dropped out of sight, flour-ishing a long, bushy tail. They seemed to incorporate elements of squirrel as well as of hare and deer, and when the first had tired of eyeing me, it hopped away like a kangaroo.

Creatures like the viscachas (they are related to chinchillas) I saw at

Machu Pichu have puzzled Northern Hemisphere naturalists since 1492. Buffon thought Old World elephants and pigs might have changed into New World tapirs and peccaries through a degenerative process after migrating to South America's unhealthier climate. Cuvier dismissed such ideas, observing that the huge *Megatherium* he described in 1796 certainly had not degenerated from a Eurasian giant, being unique to the Americas. Charles Darwin's Argentinean fossils provided further evidence that South America had supported a giant fauna of its own, not entirely unique, since he had found mastodon and horse bones with the strange *Toxodon, Macrauchenia,* and *Glyptodon,* but hardly a degenerate version of the Old World's. "If Buffon had known of the gigantic sloth and armadillo like animals, and of the lost pachydermata," Darwin wrote, "he might have said with greater semblance of truth that the creative force in America had lost its power, rather than it had never possessed great vigor."

Zallinger's art for Time/Life's book on the *Beagle* voyage includes a magnificent tableau of the Ice Age that Darwin's fossil quadrupeds inhabited, one that indeed possesses "great vigor." Above a high Andean valley, horses and mastodons stride along a grassy ridge, wild dogs harass a glyptodont, and a saber-toothed cat brings down a guanaco. On the valley floor, a puma grabs a capybara, while peccaries, deer, and tapirs gambol in the background and armadillos, opossums, mice, and skunks swarm in the foreground. Despite this animated throng, however, it is Darwin's great southern beasts that catch the eye. A black *Toxodon* wallowing in a marsh is the painting's largest mass, like a cave opening at its center. Beside it, *Macrauchenia* and *Megatherium* are the tallest, stretching trunk and claws to browse in the treetops. The two giants, towering above the mountain horizon, look more at home than the mastodons on the valley rim, one of which raises its trunk in a pose rather like the Peabody mammoth's, but more hesitant, as though daunted by the strange monsters below.

Northern Hemisphere paleontologists continued Buffon's tradition of regarding South America as a faunal backwater, however. They thought the mastodons, horses, and other familiar creatures that turned up in recent deposits had replaced the natives after immigrating from North America. Marsh, ever the Yankee booster, even thought that the edentates—sloths, anteaters, and armadillos—that thronged South America had evolved in North America. He found "many curious resemblances in the skull, teeth, skeleton, and feet" between edentates and tillodonts—the strange bear-rats he had dug from the Fort Bridger Eocene beds.

Marsh never traveled south of Panama, and few northern paleontologists visited South America after Darwin's 1830s visit. Thus the appearance, at the 1878 Paris Exposition, of a young Argentinean with a fossil collection from the pampas around Buenos Aires was a novelty. The Argentinean, with the euphonious name of Florentino Ameghino, sold fossils to paleontologists, including Edward Cope, who paid $5,000 for a collection of giants, although he was too busy to study them and they stayed in crates until the American Museum bought them in 1897. The proceeds of Ameghino's Paris sales allowed him to spend the next four years in France, where he acquired a wife and a network of correspondents that included Gaudry and von Zittel. He also published several books and articles about South American paleontology. When he returned home in the 1880s, he was enough of an authority to unnerve the local experts, with whom he intermittently quarreled for the rest of his life. They saw him as an upstart because his parents were poor Italian immigrants, although he and his younger brothers, Juan and Carlos, were Argentinean-born.

Florentino worked occasionally for universities and museums, but his professional conflicts forced him to spend much of his life running bookstores in partnership with his family. The first of these they named *El Glyptodon* after the giant armadillo relative that Darwin had discovered and Owen named. The situation was burdensome, but the Ameghinos made the best of it. Juan ran the store, which made enough profit to send Carlos, a sturdy, self-reliant individual a decade younger than his brothers, on expeditions to remote parts of the country. There he collected huge numbers of fossils, not only from the Ice Age beds whence Cuvier's, Darwin's, and Cope's specimens had come, but from older ones, allowing the first glimpses of how the continent's mammals had evolved.

Florentino studied the fossils, and named over a thousand species, so many that he ran out of greco-latinate words and started naming them after esteemed colleagues, creating genera like *Edvardcopeia* and *Othnielmarshia*. It was an unusual number of species, implying that a vast amount of evolution had occurred. Since Florentino was an enthusiastic Darwinian, it suggested to him that, given the gradual rate of species formation Darwin assumed, the South American mammals must be very old.

Florentino's Darwinism didn't convince him, however, that the mastodons, horses, and other "great quadrupeds" in Argentina's Ice Age deposits were superior immigrants from the north. Indeed, his own collection's abundance suggested to him that North America's Ice Age mammals might just as well have immigrated from the south. And in the older strata Carlos had found, Florentino saw many fossils that seemed to him ancestral, not just to South America's living edentates and marsupials, but to Ice Age mastodons, camels, horses, pumas, and wolves. Hadn't the great Richard Owen identified Darwin's Argentinean *Macrauchenia* as an extinct relative of llamas, and his *Toxodon* as a forerunner of the capybara, a living rodent? And some of Florentino's fossils suggested links to even earlier mammals. Ancient lupine creatures, which he named borhyaenids, had marsupial traits, and he theorized that they were in the process of evolving from the small marsupials common in South America's early Tertiary toward the wolves and foxes of its recent strata.

Florentino felt vindicated in 1886, when Carlos began collecting in Patagonia, Argentina's remote south. On his first trip, the youngest Ameghino brother went to the Bay of Santa Cruz, where Darwin had described strata full of giant oysters called the Patagonian formation. Above the oysters, Carlos found terrestrial deposits from which he collected over two thousand specimens of edentates, marsupials, ungulates, and other creatures. Florentino classified these into 144 species in 83 genera, almost all previously unknown, and apparently so primitive that he decided they must have lived just after the dinosaurs. They thus would have been as old as the New Mexico fossils Cope was describing at the time, yet they seemed more advanced than Cope's early Eocene marsupials, multituberculates, and pantodonts. Some of the Santa Cruz fossils belonged to the same group as the Ice Age *Toxodon,* and their high-crowned teeth were far from the primitive molars of Cope's *Pantolambda.*

Florentino sent Carlos back to Patagonia often, and the young man became an expert bone hunter, leading gaucho assistants and packtrains of thirty or forty horses on long treks through the chilly, arid country. In 1894, he made a discovery there that seemed not only to support Florentino's ideas of South American antiquity, but to stand paleontology on its head. Traveling north along the Deseado River into central Patagonia, Carlos saw Darwin's Patagonian formation, with its giant oyster shells, give way to even older strata. These consisted mainly of volcanic rock, but they also contained clay deposits, in which Carlos found "great quantities of remains of mammals . . . the remains of an extensive terrestrial formation, perhaps destroyed for the most part by the invasion of the Patagonian sea." The fossils had similarities to the postmarine Santacrucian ones that Florentino had identified as early Eocene, but there were significant differences. One of the more striking of these was the presence of large, previously unknown ungulates that appeared to have had trunks. Florentino named them pyrotheres, "fire beasts," for the volcanic matrix around them.

The implications that Florentino drew from the discovery were sensational. If the mammal beds above the Patagonian formation oyster shells were from the early Eocene, then the pyrothere beds below them might be from the late Cretaceous, the very first such beds discovered containing large, highly evolved mammals. They might be evidence that modern mammals had originated not after, but during, the dinosaur age.

Then, in 1898, penetrating to the remote Lake Colhue-Huapi in central Patagonia, Carlos found what the paleontologist George Gaylord Simpson later would call "the most imposing and important single known fossil mammal locality in South America, and one of the most important in the world." This was the Great Barranca, a cliff containing not only the Santacrucian, Patagonian, and Pyrothere beds, but two even older ones, which Carlos named the Astraponotus and Notostylops beds after Florentino's names for mammals common in them. (The names, meaning "southern starlike" and "southern pillarlike," refer to bone characteristics.) On February 15, 1899, Carlos wrote to Florentino that the two oldest formations were "definitely Cretaceous," conformable with other known Cretaceous strata. Florentino had written Carlos earlier that one of his rivals, F. P. Moreno, director of the La Plata Museum, had been collecting pyrothere fossils and had found some "near a deposit of dinosaurs," another suggestion that the beds were Cretaceous.

Florentino made his only trip to Patagonia in 1903 to examine the sequence of the Great Barranca fossil beds. Three years later, he published a profusely illustrated 568-page book maintaining that all the strata beneath the marine Patagonian formation were Cretaceous, that they contained dinosaur as well as mammal bones, and that beasts like the trunked pyrotheres and borhyaenids were evidently the distant ancestors of living elephants and dogs. Florentino even maintained that humanity's ancestors had probably evolved from Patagonian marsupials. South America's ungulates, carnivores, primates, and other placental mammals had spread out to occupy the rest of the world, he concluded, after the sea had covered Patagonia at the end of the Cretaceous, some 65 million years ago.

Needless to say, Florentino's ideas raised eyebrows in the Northern Hemisphere. While not inconceivable, such a vision of prehistory challenged all the previous century's notions of the past. The prospect was viewed with concern, and ambitious young paleontologists headed south, urged by bemused old ones.

When Albert Gaudry, recalling his Pikermi adventures, exhorted a young French emigrant to Argentina named André Tournouer to "undertake some excavations . . . for the honor of French science," Tournouer replied: "I am going to Patagonia, the Paris Museum will have some fossils." Gallic hugs perhaps ensued—Tournouer had no paleontological training, so the enterprise was something of a *beau geste*. Reaching his bleak destination in 1898, he luckily encountered Carlos Ameghino, who took pity on the helpless neophyte and told him how to look for fossils at the Great Barranca. Tournouer searched intermittently during the next five years, and found not only the Pyrotherium and Notostylops beds, but also some entirely new mammals, perhaps making Carlos think twice about helping pitiful foreigners. Yet Tournouer managed to charm even stately Florentino when the latter made his 1903 Patagonia visit.

Tournouer's Patagonian fossils dismayed the octogenarian Gaudry. Like other Old World paleontologists, he had assumed that the past he had found in places like Pikermi was largely the same everywhere. "Evolution, we thought, had run its course through the ages without anything stopping it," he wrote in his *Fossiles de Patagonie*. "Now behold that Patagonia teaches us that that what has seemed true within the limits of the Northern Hemisphere ceases to be true in the Southern Hemisphere, and perturbs our belief in the similarity of the course of evolution in the world as the whole." It was as though Cuvier's ghost were rising from the Paris gypsum to haunt transmutationists once again with drowned continents and mysteriously appearing faunas. Still, Gaudry remained skeptical of Florentino's dinosaur age mammals. He thought the Argentinean had named far too many species, and doubted that the Patagonian beds were as old as Ameghino believed. Gaudry's own detailed study of pyrothere skeletons suggested that they were not elephant ancestors but creatures with their own peculiar characteristics, unique to South America. He planned to look for such characteristics in other South American mammals, but he died in 1908 before he could do so.

Argentinean claims to mammalian priority disturbed North Americans even more than Europeans. Manifest Destiny was still in the air, and empire on the horizon. Princeton's William Berryman Scott and the American Museum's Henry Fairfield Osborn yearned to send expeditions south, but the fossiliferous United States remained absorbed in its own bones, and institutional pursestrings had tightened after the Cope-Marsh debacle. A paleontological paladin stepped forward, however, an aggressive Iowan named John Bell Hatcher, with bone-hunting credentials as solid as Tournouer's were slight. The scion of a respectable but poor family, Hatcher had made a fossil collection while working as a coal miner and had parlayed that into a position as O. C. Marsh's graduate assistant at Yale. He had risen to be Marsh's chief field-worker, and had discovered many of the most spectacular giant mammals and dinosaurs before Marsh had to let him go when his U.S. Geological Survey empire collapsed.

Hatcher went to work for Scott, but the lost world of Patagonia beckoned, and he proposed a Princeton expedition in 1894. Scott couldn't finance it, but he encouraged Hatcher, who finally went south on his own in 1896, taking only his brother-in-law, a zoologist named O. A. Peterson. Osborn, even more frustrated than Scott at not being able to field an expedition, suspected his Princeton friend of colluding in Hatcher's, and it remains unclear just how Hatcher, who had a family to support, paid for it. Legend says he funded himself at least partly with winnings from naïve Argentinean poker players, but money from the financier J. P. Morgan was probably more crucial.

Reaching the antipodes in April, Hatcher was "overcome and enraptured by the quite unexpected novelty of the situation. From childhood I had thought of the coast of Patagonia as visited by almost perpetual storms. . . . How different was our actual experience. . . . The complete quiet . . . was comparable only with that described by Arctic travelers." April is the start of the Patagonian winter, however, in a latitude comparable to southeast Alaska's, and the local custom of inhabiting unheated houses soon tempered the Yanquis' enthusiasm. Nevertheless, they stayed for over a year, collecting fossils around the port of Gallegos and in the interior near Lake Argentino. Sea cliffs at a place called Corriguen Aike offered the greatest "wealth of genera, species, and individuals" Hatcher had ever seen, although he had to dodge the icy South Atlantic to get them.

"On walking over the surface at low tide," he wrote, "there could be seen the skulls and skeletons of those prehistoric beasts protruding from the rocks in varying degrees of preservation . . . a skull and jaws of the little *Icochilus* grinned curiously, as though delighted with the prospect of being thus awakened from its long and uneventful sleep. On one hand, the muzzle of a skull of one of the larger carnivorous marsupials looked forth, with jaws fully extended and glistening teeth, the characteristic snarl of the living animal still clearly indicated, while at frequent intervals the carapace of a *Glyptodon* raised its highly sculptured shell, like a rounded dome set with miniature rosettes, just above the surface of the sandstones. . . . At times, when delayed longer than we had anticipated in the excavation of a particular skeleton . . . this work of transporting the fossils to shore became a work of rescue in every sense of the term."

Despite such gratifications, Hatcher failed to find the supposedly Cretaceous Pyrotherium or Notostylops beds. The closest he came were some bluffs along the Rio Chico in east central Patagonia. "Not a day passed that I did not find remains of dinosaurs," he wrote, "but never the smallest fragment of a mammal." The antipodal fossil formations seemed to mock the famous bone hunter, as though to refute all North American pretensions of superiority. "The weather conditions were always the same," he complained of the Rio Chico bluffs, "swept by a cold piercing blast that never ceased to come from the snow-clad peaks and ranges of the Andes." All the mammal fossils he found came from the much more recent Santacrucian beds above the Patagonian formation.

Hatcher's virtual ignorance of Spanish didn't help, and even his bone-hunting expertise may have worked against him to a degree. He was less willing to take advice than Tournouer, and when he encountered Carlos Ameghino, he failed to get the help that the young Frenchman would later receive. Whether this was because he couldn't understand what Carlos said or because Florentino had told his brother not to cooperate with the Yanqui is unclear. Hatcher had a prickly personality, and after his return in 1897, he wrote a paper seriously questioning the Ameghinos' Cretaceous designation of the Pyrotherium beds, noting that he himself had been "unable to identify the beds at all." In a later paper, he sarcastically doubted whether "even so capable a man as Dr. Ameghino" could determine "the exact sequence of strata in Patagonia from the window of his study, situated in La Plata or Buenos Aires."

Hatcher was persistent, however. In November 1897, he was back in Patagonia, where he stayed almost another year. This time he looked near the headwaters of the Deseado River, where Carlos first had discovered *Pyrotherium,* but he again found no trace of fire beasts. He had meant to spend the Patagonian winter in the field, but snowstorms changed his mind, and when he decided to return to the coast in April 1898, he was so ill with rheumatism that he could barely move, and he had to be nursed for six weeks by his assistant, a taxidermist named A. E. Colburn. Back at Santa Cruz, Colburn wisely decided to wait for a boat, but Hatcher insisted on riding 125 miles north along the coast alone in hope of finding more fossils, undergoing more blizzards and rheumatism attacks before finally starting home. Nevertheless, he returned to Patagonia in December of that year, accompanied by Peterson and a young paleontology student named Barnum Brown whom Osborn had sent along to collect for the American Museum. Resenting Osborn's interference, Hatcher consigned

Brown to driving the supply wagon with Peterson while he went off alone, again searching for the Pyrotherium and Notostylops beds, and again failing to find them.

Hatcher had so little luck, he might just as well have been looking for griffons and basilisks, but he did not give up. In 1903, he wrote Florentino asking him or Carlos to accompany him on yet another expedition "to go over together the complete Mesozoic and Tertiary collections as represented in the region." Florentino was noncommittal, replying that he was too busy, but that Carlos might go if he had the time, and that they would be glad to supply him with the information, anyway. This temporizing doubtless did not discourage Hatcher, and he probably would have embarked for the south again if his exertions hadn't ruined his health. He died of typhoid fever a few months later, at the age of forty-two.

Hatcher's Patagonia expeditions eventually did help to solve the *Pyrotherium* problem, however. He had left Princeton in 1899 because of unspecified "treatment" he had not "merited," but his fossils remained there, and Scott studied them over the next three decades. He, too, went to Argentina in 1900, not to dig more fossils, since his Marsh-influenced western experiences had left him with a distaste for bone-hunting's discomforts, but to examine the ones already found. There, unlike Hatcher, he became an Ameghino confidante. "I was received with the utmost cordiality and all the resources of the collections were placed unreservedly at my disposal," he recalled. "Ameghino said that he was glad I had come to study his fossils, 'because now the paleontologists of Europe and America will recognize that I have done my work loyally.'" At the time, the La Plata Museum had barred Florentino and his brothers from admission, so Scott spent his mornings studying its collection and his afternoons studying the Ameghinos'.

Scott's courtliness toward the contentious Ameghinos seems incongruous with the bare-knuckled rivalries of Gilded Age paleontology, and his character was unusually convivial for the time. The scion of an old but poor Yankee family, he never had any feuds during one of paleontology's longest careers, except for his secondhand one with Marsh. Scott was not afraid to criticize, however, once startling colleagues at an American Philosophical Society meeting by remarking, after a paper was read: "That was well done—exactly as I heard it ten years ago." Despite their friendship, he proceeded to demolish Florentino's Patagonian genesis.

Scott dealt the first blow when he gave seashells Hatcher had collected from the Patagonian formation to Princeton's invertebrate paleontologist,

Arnold E. Ortmann. By comparing them with shells from elsewhere, Ortmann determined that they were not from the Cretaceous, as Florentino thought, but from the early Miocene. This meant that, despite their strangeness, the Santacrucian mammals Hatcher had collected from above the Patagonian formation were not Eocene, but from the Miocene as well. And if the Santacrucian mammals were Miocene, Scott could be pretty sure—despite Hatcher's failure to find them—that the *Pyrotherium* and *Notostylops* faunas just below the Patagonian formation were not really Cretaceous, but Cenozoic—Oligocene or Eocene. So the big Patagonian mammals had not lived with dinosaurs after all, but, as everywhere else, had become large and dominant only after the dinosaurs had disappeared.

Like Gaudry, Scott thought that Ameghino had named far too many species, and that the haunting resemblance of pyrotheres and borhyaenids to elephants and dogs was an illusion. Borhyaenids seemed doglike because they were large, active predators like dogs, but they were not a link between early marsupials and dogs. They were marsupials that had filled the large predator niche in the absence of placental carnivores. Scott decided, indeed, that from at least the end of the Cretaceous until the Pliocene epoch, South America had been an isolated continent where mammals had evolved into a fauna that might look like the "Old Continent" one but was only distantly related. All big South American mammal predators had been marsupials, and although the ungulates were placentals, they also were peculiar to South America. Along with pyrotheres, they included Darwin's *Macrauchenia* and *Toxodon,* which were not related to llamas and capybaras as Owen had thought, but, respectively, to indigenous ungulates called *Theosodon,* "fortunate tooth," and *Thomashuxleya.* Only after a Central American land bridge formed in the Pliocene had llamas, dogs, horses, cats, and mastodons reached South America. The mastodons in Zallinger's Andean tableau might well have hesitated at the sight of *Macrauchenia* and *Toxodon.* By evolutionary standards, they were strangers in a strange land.

Florentino died in 1911, so he was spared Scott's final report, not published until 1932. He probably wouldn't have believed it, anyway. As Scott observed, Florentino's vision was "so passionately held that it amounted almost to an obsession." Carlos, practical field-worker that he was, seems to have been more open-minded. When George Gaylord Simpson visited Argentina in 1931 and discussed Florentino's ideas with him, he didn't seem to mind Simpson's objections. Simpson called Carlos "a great man, whose greatness was not recognized during his lifetime by anyone, not even himself."

Although the Ameghinos' dinosaur age pyrotheres and borhyaenids evaporated scientifically, they lingered culturally. Simpson found Florentino a national hero, an "Argentinean Darwin," of whom choirs of schoolchildren sang "Gloria, gloria a Ameghino" in ceremonies commemorating the anniversary of his death. And one schoolchild of two decades earlier may have given his saga to world literature. A quest for fossil beasts that become more elusive the more they are hunted might be a *ficcione* by Jorge Luis Borges, the chronicler of hopelessly paradoxical researches. Born in Buenos Aires in 1899, Borges surely knew the Ameghinos' story, although he never wrote about it. In his *Book of Imaginary Beings,* the only "Argentine fauna" are a "sow harnessed in chains" and werewolves that, "since no wolves inhabit these regions . . . are supposed to take the shape of swine or dogs." But the fire beast story is so "Borgesian" that its very absence from his work seems suspect, as though the writer, who insisted that "what is truly native can and often does dispense with local color," chose to mask this antipodean source for his tales of searchers lost in labyrinths that resemble rock strata as much as buildings.

"I have said that the City was founded on a stone plateau," relates one baffled seeker. "This plateau, comparable to a high cliff, was no less arduous than the walls. In vain I fatigued myself: the black base did not disclose the slightest irregularity, the invariable walls seemed not to admit a single door. The force of the sun obliged me to seek refuge in a cave; in the rear was a pit, in the pit a stairway which sank down abysmally into the darkness below." Borges identifies the seeker as "Cartaphilus . . . tribune of a legion quartered in Berenice, facing the Red Sea," but he sounds to me like John Bell Hatcher.

EIGHT

Titans on Parade

ALTHOUGH ZALLINGER'S ICE AGE MAMMOTH dwarfs most of the *Age of Mammals* beasts, another creature, midway in the mural, is almost as impressive, and serves to balance the composition. A huge, whitish ungulate, it looks rather like a giant rhino and also has a horn, although not like any rhino's, since it is bluntly forked at the end. Named *Brontops* ("thunderous looking"), it has the mammoth's muscular presence, and stands head and shoulders above a nearby early rhino, *Subhyracodon*. Its pallor glows like warm ashes beside the surrounding brown, tan, or rufous creatures—even the uintatheres seem almost unobtrusive by comparison.

Zallinger painted *Brontops* and *Mammuthus* in the same foursquare pose, facing futureward, but with an eye to the beholder. While the mammoth is calm, however, the ashen *Brontops* is distraught, bellowing with legs splayed and snout upraised, tiny eye unfocused. Zallinger emphasized its frenzy by painting the mural's only erupting volcano behind it. Eocene *Oxyaena* may be the painting's meanest-looking beast, but *Brontops* seems the most dangerous, ready to break into a thundering charge with a rhino's unpredictable speed. Its frenzy is puzzling, however, because it has no apparent reason. No animal threatens or challenges it. It is the only beast of its kind in the mural, and it seems anomalous, as much an interruption of the evolutionary frieze as a participant. If the mammoth embodies vertebrate paleontology's confident beginnings, then *Brontops* reflects a bafflement that engulfed it in mid-career, as mavericks like Cope and Ameghino showed that Cuvier's techniques for "bursting the limits of time" did not bring automatic agreement or understanding of prehistory.

The *Brontops* bafflement began in 1846, with the first mysterious "palaeothere" jawbone that Hiram Prout sent Joseph Leidy from the White River badlands. The jaw actually came from one of Zallinger's ashen monsters, and when Leidy eventually saw that it belonged to a new animal, he renamed it *Titanotherium* ("giant beast"). But then Marsh had returned from his 1872 expedition with the first near-complete skeleton and given it a name of his own, *Brontotherium* ("thunder beast"). A year later, Cope named his own specimens of the titans *Megaceratops* ("great horned looking"). Whatever their name, the beasts would prove even more exemplary of evolutionary confusion throughout the next half-century than uintatheres and pyrotheres.

In 1874, Marsh found a huge deposit of the giants during his strenuous November expedition to the Oglala reservation's White River badlands. "At one point, these bones were heaped together in such numbers," he wrote, "as to indicate that the animals lived in herds, and had been washed into this ancient gully by a freshet." One species, *Brontops robustus,* had been "the largest creature of its time." His rivalry with Cope was in full bloom, and Marsh saw the find as a triumph over the upstart, who had been crowing about his *Megaceratops* discoveries. "One morning Marsh and I went from camp on our usual exploring trip for fossils," recalled a member of his military escort, Major Andrew S. Burt.

> We hadn't gone far when, passing by a high-cut bank, I happened to glimpse a fossil bone sticking out of that bank. I pointed it out to Marsh. The instant his eye caught the object, he wheeled his horse, dismounted, and rushed to what afterward looked to me like the leg bone of an ox. Marsh dug around the specimen awhile, then suddenly seemed to have grown crazy. He danced, swung his hat in the air, and yelled, "I've got him, I've got him."

According to Marsh, the party escaped with wagonloads of brontothere fossils just ahead of a war party that suspected them of prospecting for gold. The episode became famous when, disturbed by reservation conditions, Marsh led a campaign to overthrow President Grant's corrupt interior secretary, so it is a tribute to his acquisitive secretiveness that he not only got all the "heaped" bones, but managed to conceal their original location. It

Figure 10. *Brontops* and *Palaeolagus,* with *Archaeotherium* (Oligocene) from Zallinger's *Age of Mammals* mural. Courtesy Peabody Museum of Natural History,
Yale University, New Haven, Conn.

probably was in the present Badlands National Park, but nobody knows
where. In later years, John Bell Hatcher collected hundreds of the giant
skeletons in Colorado White River beds, and Marsh published thirteen papers on them, tracing them from a small Eocene creature named *Limno-
hyus* ("pond pig") to the larger *Diplacodon,* then through a succession

of genera to a single giant species, whose extinction ended the line. He planned a lavish brontothere monograph during his U.S. Geological Survey overlordship, and spent over $12,000 on illustrations, but he postponed writing it because he had become involved with dinosaurs and then had to abandon it after Cope's attacks in the *New York Herald* cost him his government job.

If Marsh had written his monograph on brontotheres, he would have interpreted the group in Darwinian terms. His papers classed it with horses and rhinos as perissodactyls, "odd-toed" beasts, and he assumed that they had risen and fallen through competition for the giant mammal niche, replacing the dinosaur-brained uintatheres and in turn replaced by bigger-brained giants such as mastodons. Yet, as with uintatheres, Marsh was hard put to explain the monsters' truncated fossil record in terms of natural selection and adaptation. Lord Kelvin's restriction of the earth's age to 100 million years still prevailed in the 1890s. Although they had more of a fossil pedigree than uintatheres, brontotheres still had entered and exited the fossil record too abruptly, and their oddly horned gigantism made too little adaptive sense to fit easily into Marsh's gradualist progressivism. Like the Dinocerata, his greedily accumulated thunder beasts returned to haunt him in what biographers charitably called his "sunset years."

"It has been well said that what one truly and earnestly desires in youth, he will have to the full when no longer young," Marsh mused at age sixty, "and my own experience thus far has proved no exception to the rule." *Punch's* 1890 "Ringmaster Marsh" cartoon shows a brontothere skeleton rearing behind its top-hatted uintathere: both grin sardonically at their pudgy keeper.

Marsh does seem to have "gotten" his archrival in the case of the pallid titans. Cope never used the group effectively in attacking Darwinism, and even his names for it were invalid. Cope's young henchmen and spies had their own ideas about the strange monsters, however, and they recognized no "American Darwin." A Goreyesque 1880s photo shows wing-collared, frock-coated Peabody assistants grouped sullenly around one of Marsh's boulder-sized thunder beast skulls, as though preparing to march away with it. That, in effect, was what Cope's two chief henchmen did, although they marched in a different direction from their mentor.

W. B. Scott and Henry Fairfield Osborn had espoused Cope's theories in the 1880s, but they had dropped them after encountering the work of August Weismann, a German zoologist. In an 1887 experiment, Weismann had cut off the tails of multiple mouse generations to test a neo-Lamarckian prediction that mutilations could be inherited. "Thus 901 young were produced by five generations of artificially mutilated parents," he observed, "and yet there was not a single example of a rudimentary tail." Weismann concluded that the sexual cells that transmit inheritance are isolated from the rest of the body, and that only changes in the former could be inherited. Scott and Osborn accordingly rejected inheritance of acquired characteristics, although they otherwise disagreed with Weissman, a Darwinian who thought that mingling of sexual cells provided the variation necessary for natural selection. The younger men doubted that changes in sexual cells over a few million years could have engendered the skeletal kaleidoscope in the fossil beds of the American West. It was all very well for lab biologists to trumpet "the omnipotence of natural selection." They weren't up to their necks in bizarre bones.

Scott and Osborn thought they saw something else in the bone beds that natural selection couldn't explain, something that Albert Gaudry had perceived when he drew his first horse evolution chart in 1866. Gaudry had drawn three parallel lines of horse descent that independently led from *Hipparion* to *Equus*, implying that *Equus* had evolved at least three times from separate *Hipparion* populations. If so, *Equus* was not the result of one ancestral line of stallions and mares responding to variation and natural selection as Darwinism would have it. It was a "grade" that many different *Hipparion* lineages had attained on a parallel course, which raised the very un-Darwinian idea that there was some inherent tendency driving *Hipparion* to evolve into *Equus*.

Scott and Osborn believed so, and not without apparent cause. The fact that so many genera had occurred in both Old and New Worlds, for example, seemed evidence that evolution runs on parallel courses, with advanced traits such as single toes and high-crowned teeth appearing independently in separate populations. If horses did have such inherent tendencies, then germ plasm was not simply natural selection's pawn. It somehow contained plans, not just for the embryonic development of the next generation, but for the evolution of future ones, a concept comparable to the older idea that each sperm cell holds a preordained number of ever-tinier "homunculi," one within the other, waiting to enter generation after generation of ova. In the Scott-Osborn version, however, each tinier homunculus would be slightly different from its predecessor, ranging, for example, from *Eohippus* to *Equus*.

Of course, Scott's and Osborn's paradigm was not so quaintly literal. Scott expressed it in respectably empirical terms in an 1891 article. "[W]hen examining an extensive series of fossils reaching through many horizons," he wrote, "it is difficult to escape the suspicion that individual variations are not the material with which natural selection works, so steadily does the series advance toward what seems almost like a predetermined goal." As Gaudry's horse chart shows, the idea was widespread in the nineteenth century, probably more so than Darwinism. Called orthogenesis, "straight origin," in reference to the apparently parallel evolution of genera like *Hipparion* and *Equus,* it was a handy way of sorting out the bewildering diversity of evolutionary phenomena. Indeed, some hard-to-explain phenomena still suggest orthogentic tendencies, like the nearly simultaneous, apparently unrelated appearance of civilization in the Old and New Worlds.

Scott and Osborn elaborated on Gaudry's equine lineage in a number of horse studies. Osborn was particularly taken with orthogenesis, and developed his own version, which he called "aristogenesis." This referred to what he saw as an inherently progressive evolutionary trend, but also reflected his social background. Heir to a great railroad fortune, young Henry combined Cope's patrician hauteur with Marsh's parvenu pomposity, and he naturally saw evolution as an inevitable rise, with himself at the top.

Orthogenesis seemed particularly promising for explaining the evolution of Marsh's anomalous brontotheres, and thus one-upping the domineering Yale professor. Osborn's account of the beasts in his 1896 *American Century* article that featured *Mesonyx* was the antithesis of Marsh's gradual survival of the smartest:

> [H]ardly had the uintatheres gone to earth, when the Titanothere family, unmindful of the fate attending horns and bulk, began to develop horns which sprouted like lumps over the eyes. . . . Finally these horns attained a prodigious size in the bulls, branching off from the very end of the snout, unlike anything in existing nature. In the mean time, this "Titanbeast," as Leidy well named him, acquired a great hump upon his back, nearly ten feet above the ground, while he stretched out to a length of fourteen feet, and expanded to a weight of two tons. He increased in numbers also, as one sees in the scores of his petrified bones. This prosperity was, however, fatal, for in the stratum above not a trace of this family remains. It is difficult to assign the cause of this sudden exit: it was certainly not lack of brains.

After Cope's death, Osborn became Marsh's chief rival, and a more effective one, because his hatred of the Yale tyrant was almost as strong, and he wielded it more subtly. He had refused to attack Marsh in the 1890 *Her-*

ald scandal, but he later wrote to a friend, Iowa Senator William B. Allison, condemning Marsh's Geological Survey work. Named the American Museum's curator of vertebrate paleontology in 1891, Osborn used magazine and newspaper articles to tout Cope at Marsh's expense and condoned the journalistic intrigues of William Hosea Ballou, the hack who had colluded in the *Herald* affair. "Professor Marsh seems to spend a great deal of his time complaining of being misrepresented," he gloated in one letter to Ballou, "but he appears to enjoy very little sympathy, a fact I think is becoming pretty widely known."

Osborn took full advantage of Cope's fossil arsenal in his campaign against Marsh, not only persuading the American Museum's president to buy his impoverished friend's collection, but hiring Cope's expert collector, Jacob L. Wortman, to augment it. Wortman had worked for Cope in Oregon's John Day fossil beds as a student in the 1870s, and he had continued as his preparator and expedition leader after moving east and acquiring an M.D. "It contains a good many species, a good many new," Cope had written enthusiastically of a Wortman shipment in 1881, "but the most important things are nearly complete skeletons which permit of full and satisfactory classification." Cope had frustrated Wortman's own scientific ambitions by insisting on publishing on all the fossils himself, however, so he gladly moved to the American Museum, where Osborn let him publish, or at least co-publish, his finds.

Wortman led yearly expeditions to Tertiary fossil beds through the 1890s. Osborn recalled:

> It was indeed a rare bit of one's education as a paleontologist to see him return to camp on a cold night after a hard day's fossil hunt, huddle as close to the fire as possible, and tell glowingly of the day's discovery or lament bitterly the fatigue and exhaustion of a fruitless search. After a large draft of hot coffee and perhaps a good supper, the casualties of the day were forgotten and the Doctor would begin to philosophize or discuss some favorite hobby of his in comparative anatomy or to expound some theory of mammalian descent, stimulated perhaps by some outstanding "find" of the season.

The wily Osborn co-opted even Marsh's own fossil arsenal. As soon as Marsh died in 1899, Osborn maneuvered to take his place in government science, writing to John Wesley Powell's Geological Survey successor, Charles D. Walcott, to urge completion of the brontothere monograph. Sensitive to Osborn's growing influence, Walcott appointed him to head the survey's vertebrate paleontology section, the job Osborn had helped to push Marsh from a decade earlier. Osborn got access, not only to Marsh's carefully sequestered Yale collections, but also to John Bell Hatcher's expertise, which must have been particularly pleasing, since an attempt to hire Hatcher away from Marsh in 1891 had failed.

Osborn used Marsh's co-opted arsenal to do a deft diplomatic hatchet job on his brontotheres, which he called titanotheres, on the grounds that Leidy had named them first. Although allowing that Marsh had "made the largest and most valuable contributions to our knowledge of this family and its evolution," he maintained that "Marsh's detailed systematic work on the titanotheres was less than fortunate" and, indeed, that he had "failed to distinguish the sexes as well as the separate groups or phyla of titanotheres." Most important, Osborn denied that Marsh's Darwinian struggle for existence had played a major role in the titanothere's career. On the contrary, it was an inherent tendency to evolve in certain ways that had led to both success and failure. The beasts had succeeded because their tendency to gigantism allowed them to fill the niche vacated by the overspecialized uintatheres and had failed because specialization prevented them from continuing to fill the giant mammal niche. That all this had occurred in less than a million years, by Osborn's chronology, seemed not only to allow for inherent tendencies, but to require them.

Osborn's 1896 *Century* article featured a Knight painting of a "Titanothere Family—Bull, Cow, and Calf" so vivid that the horns seem to leap stereoscopically from the page. In the next three decades, he and his anatomical assistant, William King Gregory, used the beasts in some of the American Museum's most popular exhibits. These included not only skeletons, but life-sized clay reconstructions by Erwin S. Christman, a sculptor, including a sequence of trophylike heads showing the growth from a horse-sized species, *Eotitanops gregorii,* to the elephantine *Brontotherium.* "There, modeled in clay, was Osborn's view that evolution resulted in ever larger, more complex, and more baroque skull and horn structure," the historian Ronald Rainger observes. Another Christman reconstruction, a full-scale statue of *Brontotherium platyceras* in a frenzied pose, was the source for Zallinger's *Brontops.*

The museum's titanothere work lasted twenty years, "occupied dozens of workers, and cost thousands of dollars," with the result that, before 1960, titanic mammals evoked prehistory in the popular mind almost as much as terrible lizards. Many museumgoers probably perceived little distinction between them. I don't remember doing so when I visited the American

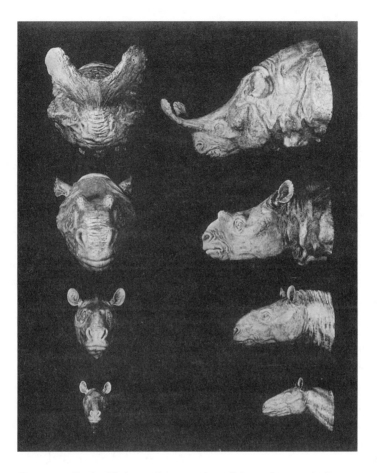

Figure 11. Erwin Christman's restoration of titanothere evolution.
Courtesy American Museum of Natural History Library.

Museum on an elementary school field trip in the early 1950s, except that
the titanothere statues perhaps impressed me more than the dinosaur
skeletons, because they seemed more real.

As Osborn's influence grew, his orthogenetic outlook widened. One aspect of seeing inherent tendencies in evolutionary phenomena is that their
origins assume a grail-like quality. They seem to hold a key to the present
and future as well as the past. If the inherent nature of extinct groups like
titanotheres could be sought in their ancestors, then that of living groups
might too. "One of the most fascinating of the many problems of paleontology," Osborn wrote in an article that the *Century* published in 1907, "is

to ascertain the birthplace of each of the great animal groups—the fertile or arid nursery of their infant history wherein they first took on their peculiar and characteristic form." The article concerned a highly gratifying solution to one such problem:

In the year 1899, paleontology had advanced to such a point that the origin of many families was known, but there remained in doubt the group of elephants, certainly the most paradoxical in structure of all quadrupeds, and appearing in the Lower Miocene period, both in Europe and North America, fully formed, as if from the sky or by a fiat of the Creator. No one knew whence they came nor how their remarkable characters had evolved. . . . In the year 1900, I ventured a prophecy which placed the original home of the elephants and of several other groups in Africa. At the time it seemed a curious fact that this possible theater of evolution of mammals had not been sufficiently considered. . . . In discussing the matter, I admitted that Africa was the "dark continent" of paleontology, that it had no early fossil mammal history . . . but I held that this absence of evidence was not proof of the absence of life in Eocene Africa, because geological records are proverbially as incomplete as the torn chapters of a book, and the chances are always against the burial of land animals in such a manner as to be preserved for the future record.

Osborn predicted, in effect, that the "African" aspect of Pikermi's famous Miocene fauna in Greece would prove to be more a product of immigration from the south than of evolution in the north. "I described this African 'Garden of Eden' as the probable nursery not only of the elephants, but of the Hyracoidea, or rock conies; of the Sirenia, or sea-cows; and, among larger quadrupeds, of the antelopes, giraffes, and hippopotami, as well as numerous smaller animals."

It was a bold prediction, as Osborn acknowledged, although modestly noting that two European paleontologists had also made it the same year. It proved to be a lucky one. "So far is the scientific imagination tempted to roam when once released," he wrote. "It may be correct, yet forever lack direct confirmation; but in the present instance the confirmation came almost immediately."

Again, the British empire came to paleontology's aid. It had established an Egyptian geological survey after taking control of Suez, and in 1898, H. L. Beadnell, a survey geologist, had found land animal fossils in an Eocene river delta of the Fayum Basin about fifty miles southwest of Cairo. As Osborn was making his prediction, Beadnell and C. W. Andrews of the

British Museum were digging out a diverse fauna of African ancestors, and they announced the discovery in 1901. "It has proved to be epoch-making," Osborn wrote:

> marking a turning-point of our knowledge of the history of the earth, and arousing such a wide spread interest that for the time north Africa becomes the storm center of paleontology. Between 1901 and 1905 the exploration and collection of these fossils continued under Messrs. Beadnell and Andrews. One unexpected discovery succeeded another. Africa, far from being a continent parasitic upon Europe, was proved to be a partly dependent, but chiefly independent center of highly varied life.

The problem with this exhilarating discovery, of course, was that the British Museum had made it. Already frustrated by Patagonia, Osborn seethed with envy from 1901 to 1905, although his *Century* article was typically suave: "The reader can imagine . . . with what very personal keeness this brilliantly successful exploration was followed by myself. The temptation to go to the desert at once was very great, but the well-earned prior claims of the English precluded any thought of our visiting this region as long as the English exploration continued." Beadnell and Andrews finally published their discoveries in 1906, however, "and almost at once came to mind the possibility of an expedition of our own to the Fayum desert. Now that many of the great extinct animals of the American Museum paleontological collections had proved to be of remote African origin, what a temptation to secure some of the diminutive ancestors, to place beside their American descendants, the majestic mammoths and mastodons!"

With help from the museum's millionaire president, Morris K. Jessup, Osborn "almost at once" succumbed to the temptation. He reached Egypt in January 1907, although the expedition's arrival may have seemed anticlimactic. Weather and labor problems delayed their camel caravan's journey to the fossil escarpment, and when they arrived, "heaps of bones which had been rejected in previous excavations" seemed to mock their aspirations. "One quarry was about seventy-five feet in diameter, and great heaps of sand bore witness to the extent of previous excavation," Osborn wrote. Examining the quarries to "ascertain whether they had been exhausted," he and his assistants, Walter Granger and George Olsen, found that "the bone layers were from four to six feet below the surface and that large quantities of overlying sand would have to be removed before we could reach them. . . . Good and important specimens were rare, and it soon became evident that work must be done rapidly, on a very large scale, and also with the utmost care."

Their Egyptian workforce, which hitherto had shown alarming enthusiasm for hacking at the fossil beds, then decided on "a general strike for double the ordinary rate of wages." Osborn put them back to work by raising their pay to forty cents a day, consoling himself that Egyptian rustics at least showed more respect for bone hunters than American ones. "One must find coal, or oil, or gold," he mused, " to command the admiration of our good-hearted, but too practical-minded, countrymen." The workers learned to go easy on the fragile bones, but, for awhile "the results were decidedly disappointing and the success of the expedition hung in the balance."

After two weeks, however, "one of the prospecting party" found a "splendid skull of the *Palaeomastodon,* belonging to the second stage in the remarkable history of the evolution of the elephant." A week later, "the same prospector discovered a skull of a *Moeritherium* . . . which represents the oldest known or first distinctive stage of the evolution of the elephant." Osborn decided that the two fossils—along with those of early whales, hyraxes, sirenians and other creatures—amply justified the considerable expense underwritten by Jessup. "The finding of the two heads was the chief object of the expedition from the purely museum standpoint," he exulted, "and the discovery of both within thirty days of our arrival was an extraordinary piece of good fortune."

A tapir-sized creature, *Moeritherium* ("Lake Moerus beast") had a basic elephantine feature, tusks formed from incisor teeth. (Tusks of ungulates, such as pigs, are canine teeth.) *Moeritherium*'s tusks were small, however, and it lacked a trunk. "[W]e cannot regard the little *Moeritherium* as the direct ancestor of the elephants," Osborn wrote; "it is rather a type which constitutes a missing link between the elephants and other quadrupeds, and stands near the true and direct ancestral line."

Osborn had no doubt, on the other hand, that the rhino-sized *Palaeomastodon* ("old teat tooth," referring to its udder-shaped molars) was "the direct forerunner" of the elephants. "Both the teeth and skull are prophetic, as well as the limbs," he wrote. *Palaeomastodon*'s incisors had decreased to four sizable tusks, two in each jaw, while its upper lip had elongated into a short trunk. "Since the fore and hind limbs were elongate, with proportions very much like those of a young modern elephant, it was probably difficult for this animal to reach the ground with its head. We are thus forced to the conclusion that *Palaeomastodon* possessed a proboscis which extended at least some little space beyond the forwardly protruding tusks, because without this most unique and marvelous prehensile modification of the upper lip and nostrils combined, the animal could not have seized its food." *Palaeomastodon,* then, was "the first definitely known member of this great order," which, "superbly equipped for foreign travel and conquest," had later, "as secure as the Romans with their phalanxes and their legions . . . marched over one or other of the land bridges as soon as they were formed" to occupy Eurasia and the Americas.

"In the whole history of creation no other animal, with the single exception of the horse, accomplished such feats of travel," Osborn concluded, implying that no other had accomplished such feats of orthogenesis. It was as though the speed and thoroughness of elephant conquest had obviated any real need for Darwinian adaptation to terrain or climate.

On the Peabody's wall, the mammoth's majestic calm seems to reflect such accomplishments. Unlike the frenzied titanothere, it is no isolated anomaly. In the Miocene, four-tusked *Gomphotherium* ("bolt beast"), the first proboscidean to invade America, rests with its calf under a ginkgo tree. (Zallinger adapted it from a Knight illustration of *Palaeomastodon* in Osborn's 1907 *Century* article.) In the Pliocene, the shovel-tusked *Amebelodon* browses at a red maple, unfazed by proximity to a grumpy-looking *Teleoceras* rhino. The mammoth itself shares the Ice Age with *Mammut,* the mastodon, which brandishes tusks and trunk toward the past as though to proclaim the penultimate hegemony *Mammuthus* embodies. And the two genera do seem to have ruled harmoniously over the Ice Age. With high-crowned molars, mammoths could graze tough grasses, while mastodons browsed on softer fodder with more primitive, cusped molars. The mural never shows them or their relatives in confrontation with each other, or with anything else, a patrician serenity that Osborn would have understood.

NINE

Five-Toed Horses and Missing Links

OSBORN'S FAYUM EXPEDITION WAS AN ACCOMPLISHMENT, but it was not quite the adventure that his *Century* article made it seem, not for him, at least. He didn't mention that he had traveled with his family, stayed at a separate, deluxe camp, and gone home after three weeks, leaving his assistants, Granger and Olsen, to keep digging in the Egyptian spring's sandstorms and flies. As they began a planned return in late April, moreover, they got a cable from Osborn commanding them to stay in the desert and search for more specimens. They obeyed, but didn't find many, and Granger had to spend two weeks in a Cairo hospital recovering from septic flea bites before they finally sailed home in June. A lean Vermonter who'd started as an apprentice taxidermist in 1890, Walter Granger had become an expert paleontologist under Jacob Wortman's tutelage, although he had never graduated from high school, and Osborn increasingly depended on his technical expertise. Indeed, he used Granger's Fayum notes to write his *Century* article.

The middle-aged Osborn had less enthusiasm for the strenuous life than his mentor, Cope, who in his early fifties had spent two summers driving a bone wagon across the plains. Osborn's 1896 *Century* article on Rocky Mountain mammals was eloquent on the subject. "You are off in the morning, stiffened by a frosty night," he wrote. "You know by sad experience that the ice in the basins does not promise a cool day. Your backbone is still freezing while the sun begins to broil and blister your skin, and you are the living embodiment of the famous dessert served by the Japanese—a hot crust without, an ice within . . . your soul abhors a fossil, and longs for the green shade of the east, and watermelon."

Osborn remained firmly committed to fossil discoveries, as long as he didn't have to look for them. "The fossil hunter is predestined to his work," he proclaimed. He also had firm ideas about who this fossil hunter should be—one of his own staff—although his 1907 *Century* article glossed over something else about his Fayum triumph. In fact, the "member of the prospecting party" who found the two skulls that justified Jessup's investment was an Austrian collector named Richard Markgraf, not a museum staffer. In Egypt for his health, Markgraf had been collecting Fayum fossils since 1903, when a German scientist, Eberhard Fraas, had interested him in bone-hunting. The Americans didn't meet him until February 16, when seeing his camp "worried" Osborn's wife during a quarry visit. Osborn offered to hire him temporarily when Markgraf paid a call the next day, but they didn't agree on financial terms until March. According to Granger's notes, he produced a "fine skull of *Paleomastodon*" for the museum on February 21, and a "very good skull of *Moerotherium*" on March 15, then went back to work for Fraas and other Germans.

Osborn brushed aside such logistical details in pursuit of an enterprise that grew increasingly grandiose. Named to succeed Jessup as the American Museum's president in 1908, he began to see himself in regal terms, and tales of his vainglory are legion, as when he published a book devoted entirely to his own publications and honors, or serenely bid his chauffeur to drive against traffic on one-way streets. He ruled the museum magnanimously on the whole, however, and his institutional megalomania had its creative side. He behaved like a scientific King Arthur, sending paleontological paladins after orthogenetic grails.

The true grail for Osborn *rex,* as it had been for Owen and Huxley, was early mammals, not just the ancestors of elephants, but of all the modern beasts. His old foe Marsh had considered Mesozoic mammals a *mysterium tremendum* and had written Owen in 1877 that he was "bound to have" their fossils from his spectacular Jurassic dinosaur site at Como Bluff in Wyoming. When his collectors began to find their jaws and teeth a few months later, Marsh had gloated characteristically that they were worth more than their weight in gold, and he had named some of them "pantotheres," basic beasts, thinking them ancestral to living marsupials and placentals. Jacob Wortman's discovery in 1882 of the first known Cretaceous mammal fossils, two teeth and part of a leg from South Dakota's Laramie Formation, had been a setback for Marsh, since they had gone to Cope. But seven years later, his own man, John Bell Hatcher, had begun picking Cretaceous mammal pieces from anthills near Buck Creek, Wyoming. "[T]eeth, fragmentary jaws, and even skeletal parts began to pour into Marsh's eager hands," Charles Schuchert notes. "In July 1889, he had a paper in print describing 12 genera and 18 species, and 2 succeeding parts of this paper brought the total to 19 genera and 39 species, classified in 6 new families." Let Cope and his boys top that, by gad!

Driven to top the professor, Osborn had sent a party to Como Bluff in 1897 to find and reopen the mammal quarry, but with an unexpected result. The party had found sauropod dinosaur bones, and in August, Walter Granger had stumbled on a new deposit of the huge skeletons in the nearby Bone Cabin Quarry. This had furthered an already unfolding development. Osborn's popular 1896 *Century* article on Rocky Mountain mammals had encouraged him to cooperate with Cope, William Hosea Ballou, and Charles Knight to publish a lavishly illustrated dinosaur piece, which piqued public interest as never before. In 1898, the yellow press snatched up the potential sensation, and invented dinomania. "MOST COLOSSAL ANIMAL EVER ON EARTH JUST FOUND OUT WEST," William Randolph Hearst's *New York Journal* blared, and Joseph Pulitzer's *World* blared back with a lavish spread on Osborn's Wyoming fossils headlined "NEW YORK'S NEWEST, OLDEST, BIGGEST CITIZEN, MR. C. DINOSAUR."

This had been heady stuff, especially to one member of the 1897 Como Bluff party, young Barnum Brown, whose sojourn seeking Patagonian mammals as John Bell Hatcher's unwelcome assistant during the next year probably had come as a letdown. Anyway, he had thrown himself wholeheartedly back into dinosaur-hunting after his return from Argentina. Brown had a degree from the University of Kansas, but he was unscholarly, and became a kind of paleontological bloodhound who, as Osborn said, seemed to find fossils by smell. He sniffed after the most spectacular ones, and was highly competitive, later telling an assistant, Roland T. Bird, that Wortman and Granger had been "just getting down to good bones" after three seasons at Bone Cabin, and that Granger had spent his time duck-hunting. Brown's single-mindedness had paid off in 1902, when he had discovered the first *Tyrannosaurus rex* skeleton in Montana Cretaceous beds. Osborn gave him free rein thereafter, which brought the museum many famous dinosaurs and some significant, if vaguely classified, mammals. Bird recalled a trip to Big Bend, Texas, during which Brown identified "camel" and "ground sloth" tracks in Paleocene strata—finds that would have pleased Florentino Ameghino.

Brown's meteoric rise had contributed to yet another unexpected development on the early mammal front. Wortman had wanted Osborn to send *him* to Patagonia, having cultivated a specialty in South American edentates, and he increasingly had resented the monarch's favoring of the young Kansas upstart. He also had thought his mammal expertise qualified him to succeed Marsh in Yale's paleontology chair, an odd idea given his long association with Marsh's archenemies, but one he had taken seriously enough to leave the museum in 1899, when he had felt that Osborn wasn't supporting him for the job.

Dinomania and Wortman's departure had removed some scientific focus from the museum's early mammal quests, particularly in the American West. The grail of placental ancestry continued to obsess Osborn, however, and he had sent Granger back to Paleocene and Eocene deposits after two more seasons at Bone Cabin. As it grew during the next two decades, the obsession drove Osborn to dispatch his paladins to remoter places than the West, or even than Patagonia and Egypt, as though far travels in space would somehow help to bring the distant past closer.

The most tempting destination was Asia, which, as the largest continent, was expected to have produced the most evolutionary activity. The fact that many key genera—*Equus, Hipparion, Coryphodon*—had inhabited both Europe and North America suggested that they had come from somewhere in between. Osborn wrote hopefully:

> In the dispersal center, during the close of the Age of Reptiles and the beginning of the Age of Mammals, there evolved the most remote ancestors of all the higher kinds of mammalian life which exist today, including, for example, the five-toed horses, which have not as yet been discovered in either Europe or America. That the very earliest horses known in either Europe or America were four-toed indicates that their ancestors may have lost their fifth toe while still resident in the Asiatic homeland.

Osborn's hopes for Asia extended to early humans. He believed that man had evolved, not from arboreal African apes, as both Darwin and Marsh had thought, but from bipedal, tool-making creatures that had inhabited "the relatively high, invigorating uplands of a country such as Asia was in the Miocene and Oligocene time—a country totally unfitted for any form of anthropoid ape, a country of meandering streams, sparse forests, intervening plains and meadow lands." He envisioned an orthogenetic human fossil succession like that of his horses or elephants, with parallel lineages of inherently manlike species growing steadily larger and smarter. Osborn's devout Presbyterianism contributed to his objection to an arboreal African ape for an ancestor, but his ideas were scientifically respectable in the 1920s. Asian fossil evidence existed. The Dutch anthropologist Eugene DuBois had found primitive humanoid bones in Java in 1891, and Max Schlosser, a German who had become interested in primates while briefly working as one of Marsh's assistants, had obtained a human tooth from ancient Chinese deposits.

H. G. Wells, who had studied with Huxley in the 1880s, summed up the case for a terrestrial "dawn man" with a pedigree reaching back to the early age of mammals. Conceivably, wrote the author of *The Invisible Man,* "the human ancestor at the opening of the Cainozoic period was a running ape living chiefly on the ground, hiding among rocks rather than trees. It could still climb trees well . . . but it was already coming down to the ground again from a still remoter, a Mesozoic arboreal ancestry."

A problem with the Asian origins hypothesis was that, although fossils of prehistoric mammals were common from China and India by the early 1900s, none, except some rhino teeth, were known from the continent's center. So Osborn was intrigued when a handsome young museum mammalogist, Roy Chapman Andrews, proposed a series of expeditions there. Osborn may also have been surprised. Although Andrews had a popular reputation based on East Asian exploration and whale research, he was not a brilliant investigator in the Cope and Marsh mold, or even the Barnum Brown one. He might have been Walter Granger's less-bright younger brother. He, too, had begun his museum career as an apprentice taxidermist, but although he had a master's degree from Columbia, he lacked Granger's scientific flair. His first two Asian expeditions had been scientifically mediocre hunts for living mammals, and he knew little about paleontology. Yet prehistory fascinated him. "Ever since 1912, when I began land exploration in Asia," he wrote. "Professor Osborn's prophecy as to the Asiatic origins of mammalian life had been on my mind, and the determination to test that theory became stronger as my travels and experience increased."

Andrews shone as a promoter, projecting a rugged but romantic image that inspired men and—despite a celebrated 1914 marriage to a society beauty—attracted women ("My Gawd," wrote one admirer, "you can park your shoes next to mine anytime you feel the urge . . . "). With Osborn's support and contacts, he eventually raised $600,000 to fund a 1920s series of expeditions to Mongolia. Osborn saw to it that the media touted the project, although that had drawbacks for Andrews. "When the plans for the expedition were made public," he complained, "the world press seized upon the possibility of our finding primitive human remains as a feature of rare news value. We were somewhat appalled to find that we immediately became known as the 'Missing Link' expedition, and that the broad scientific aspect of our intended work was entirely lost. . . . At first I was indignant, but my protests were futile. Moreover it did have the advantage of creating an enormous public interest which otherwise would have been lacking."

Andrews's ambivalence about publicity reflected a larger anxiety. He had to handle the discomforts from which Osborn shrank. "Ninety-nine out of a hundred persons think that hardships are an essential part of an explorer's existence," he wrote, "but I don't believe in hardships. They are a great nuisance." He tried to minimize them, and the Central Asia expedition was the most lavish bone-hunting foray of all time, with camel caravans, retinues of servants, a fleet of custom-built cars and trucks, an arsenal of expensive guns, and enough tents, food, and other supplies to create small towns where they camped. Yet Andrews's heroic reputation put him in a double bind. "The newspapers . . . hoped for some thrilling stories of the dangers and hardships that we would encounter in the Gobi Desert," he grumbled, "and when I said that we did not expect to have either they seemed to think it would not be a real exploring expedition." He vacillated between insistence that it was all just well-planned research and sensational yarns, once scoffing that he could remember "just ten times when I had really narrow escapes from death" in "fifteen years of wandering into the four corners of the earth," then describing each narrow escape in thrilling detail.

Mongolia was a place for hardships. The Gobi Desert at its center has winter temperatures dropping to minus 50° Fahrenheit, summer ones rising to 110°, and sandstorms that can take the paint off a car. Half as large as the continental United States, the country was isolated from modern civilization in the 1920s, except that Russian and Chinese incursions were destroying the traditional Buddhist lamasery system. The expedition found ancient communities abandoned or disintegrating: "At the base of a hill upon which lama city is built at Urga there are hundreds of human skulls and bones, gruesome reminders to the living priests of what their own fate will be," Andrews wrote. "Great black dogs skulk about this 'burial ground' and fight over the bodies that are dragged from the city. They live almost entirely on human flesh and are terribly savage. It is certain death for a man to pass near the spot at night unless he is armed." One of his "ten narrow escapes from death" had come when a dog pack "gathered for a feast" on him and his wife, Yvette, while they slept.

Mongolia's scenery and wildlife enthralled the explorers, however. Crossing the Altai Mountains:

> There was not a sign of human life, but a dry lake bed ran the entire length of the valley, which swarmed with antelope and wild ass. . . . I have never seen such a concentration of game in a small area. Antelope were running beside the car and crossing our course every moment. . . . Herd after herd of wild ass pounded along beside us unable to tear themselves away from the fascination of the car.

Much of the Gobi Desert is badlands, and they quickly found evidence of ancient wildlife. Walter Granger and his assistants located fifty pounds of rhino teeth and bone fragments at a place called Iren Dabasu on the fourth day of the expedition's first collecting foray in 1922. Andrews recalled his own bone-hunting initiation breathlessly. "In a spot only a few yards from the tents," he wrote, "my eyes marked a peculiar discoloration in the gray upper stratum, and bits of white, which looked like crumbled enamel. Scratching away at the soft claylike earth, I exposed the grinding surface of three large teeth and felt sure that it was an important specimen. . . . My initial experience as a paleontological collector stimulated me to spend every leisure moment in wandering over the badlands hunting for new treasures. The veriest fragment of exposed bone might lead the way to a skull or skeleton: a single specimen might turn one more page in the pre-history of Central Asia."

This enthusiasm became a trial for Granger. "Our efforts to discover fossils met with an approval that did not in the least apply to our efforts to remove them," Andrews confessed. "I was inclined to employ a pickax where Granger would have used a camel's hair brush and pointed instruments not much larger than needles. When a valuable specimen had been discovered he usually suggested that we go on a wild ass hunt or do anything that would take us as far as possible from the scene of his operations." At the museum, the staff began calling damaged fossils "RCAed" ones.

Despite its amateurish aspects, the expedition eventually found large mammals from every Cenozoic epoch, including horses, mastodons, coryphodons, and titanotheres that ranged from small, early species to huge, late ones. Some of these remain unmatched in size, including the largest land beast known, a hornless Oligocene rhino now called *Paraceratherium* (which means "related to *Aceratherium,*" another hornless rhino). Paleontologists had found fragments in India and Turkestan, but a skull and other bones Andrews and Granger collected in 1922 allowed the first clear idea of its dimensions. "It required three months of the most skillful work at the museum laboratories to restore the skull and jaws," wrote Osborn. "From the first, the animal seemed incredibly large; it was hard to believe that it was actually a reality." A specimen was seventeen feet high at the shoulder and twenty-four feet long, with a skull five feet long. Twice as big as an elephant, it probably had browsed in the treetops for a living. It may have been as large as a land mammal can grow, since a larger one might not be able to eat fast enough to maintain its high metabolism. "*Paraceratherium* was one of the few mammals that made a living like the browsing dinosaurs," a paleontologist wrote in 1994. "Not surprisingly, very few mammals have tried it before or since because it is a very different lifestyle in terms of bioenergetics."

In 1925, the expedition found the legs of a *Paraceratherium* that apparently had smothered in quicksand still standing. "To one who could read the language, the story was plainly told by the great stumps," Andrews recalled. "Probably the beast had come to drink from a pool of water covering the treacherous quicksand. The position of the leg bones showed that it had settled slightly back on its haunches, struggling desperately to free itself from the gripping sands. It must have sunk rapidly, struggling to the end, dying only when the choking sediment filled its nose and throat." When he teased Granger for not getting the whole skeleton, the paleontologist replied, perhaps gently mocking the explorer's self-regard, that it was Andrews's fault for not bringing them there 35,000 years earlier, before the hill with the rest had eroded away.

During the second field season, in June 1923, they also found the biggest known meat-eating land mammal. In a place known as Irdin Manha, "Jewel Valley," because of its multicolored quartz pebbles, Granger's Chinese assistant discovered a massively fanged skull, which William Diller Matthew, the museum's curator of vertebrate paleontology, identified as a mesonychid, like Marsh's Peabody skeleton. It was a spectacularly grander one, however. In comparison with the yard-long cranium, a wolf's skull looks the size of a rat's. Although the rest of the skeleton remained (and remains) unknown, a museum artist, E. R. Fulda, reconstructed it as a low-slung, almost crocodilian beast, grimacing as it stalks a herd of the seventeen-foot *Paraceratherium*. Osborn named it *Andrewsarchus* ("Andrews's ruler").

Such treasures enticed Osborn from his throne in September 1923, and his visit must have seemed a regal progress indeed. Andrews gushed:

> It was one of the greatest days of my life and of the expedition when the man, whose brilliant prediction had sent us into the field, stepped from the car at our camp in the desert. The next days were like the fulfillment of a dream [to] both of us and to the professor. Granger had discovered a splendid Titanothere skull and left it in the ground partially excavated so that Prof. Osborn might actually see in position one of the animals that he had prophecied would be found in Central Asia.

"This is the high point of my scientific life!" Osborn exclaimed. So much for Marsh's White River brontotheres.

The progress touched magically on an even older discovery. "A specimen in which he was greatly interested was a single tooth representing an archaic group of animals known as the Amblypoda," Andrews continued.

> None of these great ungulates had hitherto been found in Eurasia, excepting *Coryphodon* of the lower Eocene of France and England. The single pre-molar tooth was the only specimen of the group we had discovered in two month's search. Prof. Osborn considered it so important that he asked. . . . to have me photographed on the spot where I had picked up the tooth. Later we drove ten miles down the valley and stopped for tiffin. Prof. Osborn pointed to a low, sandy exposure a half mile away and said: "Have you prospected that knoll?"
>
> "No," I said, "it is the only one in the basin that we have not examined. It seemed too small to bother about."
>
> "I don't know why," said the Professor, "but I would like to have a look at it. Do you mind running over?" When we stopped at the base of the hillock, I did not leave the car, but Prof. Osborn and Granger walked out to examine the exposure. As he left, the Professor turned to me with a smile and said: "I am going to find another *Coryphodon* tooth."
>
> Two minutes later, he waved his arms and shouted: "I have it—another tooth!"

Osborn must have felt like a gasoline-powered Cuvier at that moment. A year later, he wrote, with Cuvierian confidence: "Discovery of the an-

Figure 12. E. R. Fulda's restoration of the giant Mongolian mesonychid *Andrewsarchus*. Courtesy American Museum of Natural History Library.

cestral five-toed horses, tapirs, rhinoceroses, and titanotheres, which we are confident will be found on the border line between the Age of Reptiles and the Age of Mammals, is another of the main reasons for the continuation of the Third Asiatic Expedition. . . . If we find hoofed animals at this period, we may surely anticipate that they will have five digits on both the front and hind foot and we will be in the presence of the greatly desired life zone of the five-toed ungulates."

And the mammal quest took an even more promising turn the year after that. The most publicized fossils in 1923 were dinosaur eggs found at a place called Shabarakh Usu, but Granger also found a tiny skull, which he assumed was reptilian, since it was from the Cretaceous. When they were in the same area during the 1925 season, however, he got a letter from W. D. Matthew telling him that the skull was mammalian, and imploring him to find more. Whole skulls of Mesozoic mammals were very rare— Cope and Marsh had found none. Granger walked to the base of a formation called the Flaming Cliffs, according to Andrews, "and an hour later was back with another mammal skull." In the next week, the expedition found six more in sandstone concretions that had eroded out of the cliffs.

Matthew decided that the skulls belonged to several species. They were "among the earliest placental mammals which are related to existing groups," Andrews exulted. "[I]t was one of the most valuable seven days of work in the whole history of paleontology." Central Asia seemed ready to yield up not just the five-toed ancestors of horses and rhinos, but the ancestors of all placental mammals.

TEN

The Invisible Dawn Man

ANDREWS'S ELATION ABOUT THE MONGOLIAN fossils was premature, however. Osborn's quests began to reveal a certain vacuity after 1925. In fact, he had reached his scientific, if not his administrative, peak with his 1907 probiscidean parade. King Henry thereafter found it increasingly hard to march titans orthogenetically about the globe (despite the Peabody mammoth's placidity, its eye has a wayward gleam), and his regal pretensions assured that there would be Mordreds as well as Lancelots at his Round Table. In the end, the anomalous, whitish titan that he had set against Marsh would return to haunt him in his turn.

Spectacular as they were, the Mongolian fossils were not that much more scientifically informative than less exotic ones. The titanothere fossil that marked the "high point" of Osborn's career in 1922 did not hold the key to aristogenesis, nor did Mongolia's Cretaceous fossils reveal the secret of modern mammal origins. Even more baffling, the confidently awaited five-toed horses remained utterly elusive in Central Asia.

"The earliest Tertiary fauna is the Gashato, of Paleocene age, and of curious and rather unexpected character," Matthew and Granger confessed in a 1926 report. "We had expected to find in this horizon the ancestors of the four-toed horses of North America and Europe. Instead we found the ancestors of certain South American ungulates whose origin had been equally obscure, but which we had expected to find in North America rather than in Central Asia." Horses didn't even appear in later deposits. "[I]t is remarkable that we find no traces in the Eocene of Mongolia of the four-toed horses of the American, or the Paleotheres of the European Eocene," mused the puzzled curators. "In their places occur various small, slender-limbed animals related partly to tapirs, partly to rhinoceroses, but not closely to either group."

Worse from a publicity viewpoint, the expedition failed to find an early primate "missing link," or any early primates at all. Despite Andrews's demurrals about "broad scientific aims," this was a blow to Osborn's fondest hopes. It was also ironic, because Osborn's first foreign expedition site—in Africa—had long since yielded early primates. Before his death in 1916, the Austrian collector Richard Markgraf had found a number of anthropoid fossils at Fayum, which he had sent to Max Schlosser and others in Germany for study. Schlosser thought at least one of the fossil genera, which he named *Parapithecus,* "half ape," was an early human ancestor. Indeed, Markgraf had given Granger part of a jawbone in 1907 that Granger thought might be a primate. Osborn had thought it more likely to be a small artiodactyl, however, and "in reference to the sacred bulls of Apis," he named it *Apidium.* The "reference to the primates" was "uncertain," he had concluded, because the teeth didn't "agree with any Eocene or Oligocene primate known." He made no mention of African primates in his 1907 *Century* article on Fayum, and he continued to downplay them.

Then an even more dismal failure compounded the situation. Excavations in western Nebraskan Miocene deposits in 1922 had turned up a worn molar that Osborn had identified as an ape's. He had named a new genus after it, *Hesperopithecus* ("western ape") Since ape fossils were unknown in the Western Hemisphere, the tooth had raised huge implications. It had seemed to substantiate Osborn's idea that human ancestors had inhabited the northern temperate plains at a very early date. "The animal is certainly a new genus of anthropoid ape, probably an animal which wandered over here from Asia with the large south Asiatic element which has recently been discovered in our fauna," Osborn had written to the tooth's discoverer, a rancher and geologist named Harold Cook. "It is one of the greatest surprises in the history of American paleontology."

A similar tooth already in the museum's collection had seemed to corroborate the discovery, and newspapers had carried illustrations of "dawn men" striding about the western plains with rhinos and camels. When Osborn sent expeditions to the Nebraska site in 1925 and 1926, however, they found no sign of apes, and the teeth proved to be from a fossil peccary. It was an honest mistake; worn pig and hominid molars are notably hard to distinguish. But it was still a great gaffe, greeted derisively by a growing biblical creationist movement.

Osborn's regal self-confidence insulated him from embarrassment, or even much discouragement, at such debacles. His photos always look complacent, even in the dusty Gobi, and a 1926 article entitled "Why Central Asia?" radiates his faith in his ultimate vindication:

> It was at the dramatic moment of the close of the season of 1923 when the discovery of the dinosaur eggs gave our expedition a world wide fame that the writer joined Andrews' party in the east central Gobi and began to visualize the life environment of Tertiary time as ideal for the early development of the dawn men or the direct ancestors of the human race.
>
> This high plateau country of central Asia was partly open, partly forested, partly well watered, partly arid and semi-desert. Game was plentiful and plant food scarce. The struggle for existence was severe and evoked all the inventive and resourceful faculties of man and encouraged him to the fashioning and use first of wooden and then of stone weapons for the chase. . . .
>
> *We prophecy that the Dawn Man will be found in the high Asiatic plateau region and not in the forested lowlands of Asia, but many decades may ensue before this prophecy is either verified or disproved* [emphasis in original].

The invisible dawn man also would have drawn a complacent response from some of the museum staff, but for a different reason. Like Marsh, Osborn had a flair for surrounding himself with resentful subordinates, and his two chief assistants particularly came to dislike his theories and personality. Their photos tend to look peevish rather than self-satisfied, but they must have found Osborn's Mongolian deflation gratifying.

William K. Gregory, the anatomist who helped develop the life-size titanothere exhibit, would have taken particular pleasure in the dawn man's invisibility, even though Osborn saddled him with announcing the *Hesperopithecus* debacle to the media. Gregory's pet project was demonstrating that humans *had* evolved from forest apes, and that our nearest living relatives are gorillas and chimpanzees. He considered the dawn man an unscientific concession to creationists, and he discreetly said so in a "dissenting opinion" appended to Osborn's 1926 credo. "If the family of mankind has *always* been superior to the family of the anthropoids," Gregory inquired, "why do modern anthropoids and men, in spite of their widely diverse specializations, still resemble each other so profoundly and in so many directions that even the 'man in the street' recognizes the almost human qualities of his despised relatives?"

Osborn's curator of vertebrate paleontology, William Diller Matthew, grew to dislike him so intensely that it was as though O. C. Marsh's spirit possessed the man. A Canadian with a Ph.D. in geology, Matthew joined the museum in the 1890s, and his expertise, combined with Wortman's and Granger's collecting prowess, led to important work on the succession of western fossil strata, work that Osborn used extensively for his 1910 tome *The Age of Mammals.* Matthew agreed with Osborn that Asia was a center of mammal, even of human, evolution, but he came to differ fundamentally with him on evolutionary theory. Noting an enormous variation in fossil mammal skeletons, with no two individuals of the same species quite alike, he decided that natural selection was, after all, the main evolutionary force. A 1902 speech by Matthew to the Linnean Society in London reads uncannily like Marsh addressing the American Association in Nashville twenty-five years earlier:

> The main trend of mammal evolution was not a predeterminate one, carried out on certain lines inherent in the organism, but was an adaptation to changing external conditions. As the climate changed, and with it the conditions of their habitat, the animals changed correspondingly, and those races which were best fitted for the new conditions, either because of natural adaptability or because in their original habitat the new conditions were earlier reached than elsewhere and they thus had a handicap over their competitors, these races spread widely and became dominant.

Convinced that survival depended on adaptation, Matthew denied that traits could evolve in the straight lines of orthogenesis. Although he had done the stratigraphic work for Osborn's takeover of Marsh's planned titanothere monograph, Matthew essentially reached the same conclusion about the titans' bizarre evolution that Marsh would have. Titanothere horns had not grown longer because of any inherent tendency, but because they were useful, even though that use might not be obvious. Indeed, the challenge in the Peabody *Brontops*'s frenzied pose implies a competitive function for the horn, although no rival is in sight. Matthew probably contributed ideas to the Erwin Christman statue of *Brontotherium platyceras* on which *Brontops* is based.

Matthew thought he did see an obvious use for another of the orthogeneticist's favorite traits, the canine teeth of saber-toothed cats. He thought the canines had grown down below their jaws as an adaptation to piercing the ever-thicker hides of ungulate prey. Saber-tooths had not become extinct because lengthening fangs got in the way of attacking their prey, an absurdity that even orthogeneticists like William Berryman Scott rejected, but perhaps because of prey scarcity. Matthew considered it "utterly impossible" that a species could "continue specializing in some particular direction beyond the point where specialization is of use."

Inspired by *Smilodon* skeletons from the La Brea tar pits in southern California, Matthew wrote a vivid, if anthropomorphized, re-creation of saber-tooth behavior that mocked Osborn's personality as well as his theories. His saber-toothed protagonist is not only a tyrant but a snob who thinks himself "a really high class animal" and scorns camels because they drink at muddy pools instead of his own clear mountain brooks. "It all belonged to him . . . to him and his race. Individually or jointly none of the inhabitants of the plain dare dispute their sovereignty; bloody and merciless tyrants though they were, none could successfully resist them. Well might he stand, fearless and majestic, viewing the scattered, timid groups of great pachyderms from which he intended to select his next victim."

Victim selection is not without complications, however. The snobbish *Smilodon* first considers a herd of bison, "immigrants from some distant region who had crossed the mountain passes to the north, and were becoming more and more numerous in the valley, ousting many of its former inhabitants." He finds their communal defense methods "annoying and quite incomprehensible," however, and attacks a familiar native instead. "This was his favorite prey—the big, clumsy, slow moving ground sloth that waddled around in such stupid confidence that its heavy hair and thick bone-studded skin made it invulnerable. So it was to ordinary animals, but not to him. He could pierce that tough skin with tremendous hammer blows of his great dagger teeth, and tear wide gashes in neck and flanks until the beast bled to death."

But the tyrant's attack brings down a fate that would have gratified "timid groups" of museum employees as well as pachyderms. Like another animal bully, Br'er Fox, he traps himself in tar:

Strive as he might he could not release more than one foot at a time, and that but for a moment. He forgot all thoughts of prey and turned with a choking snarl to drag himself out. But it was too late. The fierce sabre-tooth, the tyrant of hill and valley, the dreadful scourge of the prehistoric world . . . was hopelessly doomed to follow his intended victim to an awful and lingering death in the black and sticky depths of the asphalt pool, from which rose now, faster and faster, bubbles of oil and malodorous gas as the struggling animals sank lower and lower beneath the surface.

Osborn's reaction to his right-hand man's thinly veiled lampoon is hard to imagine, but he doubtless read it, since he published it in the museum's magazine. He would have gotten the scientific message, if perhaps not the social one. *Smilodon* had evolved saber-like canines, not in the process of becoming a "really high class animal," but as an adaptation to preying on ground sloths and similar victims. It might have become extinct as better-defended immigrants like bison replaced its adaptive prey. Inherent "aristogenetic" tendencies dreamed up by wealthy egotists had little, if anything, to do with it.

More than fossils justified Matthew's reversion to Darwinism. Pierre Curie had announced in 1903 that the radioactive decay of certain elements generated heat, and three years later another physicist, Lord Rayleigh, had shown that radioactive heat in the earth's core could counteract Lord Kelvin's cooling effect, thus confuting the latter's 100-million-year estimate of the earth's age. It took a while for these discoveries to reach, much less convince, orthogeneticist paleontologists. Osborn still referred to a two-million-year Cenozoic in his 1907 *Century* article on the Fayum expedition. By the 1920s, however, geologists were developing radioactive dating techniques that suggested that the Paleozoic Era had begun half a billion years ago, increasing natural selection's time frame to something like Darwin's first estimates.

Osborn wouldn't let a mere hundredfold increase in life's age trouble his complacency, and he tolerated Matthew's apostasy, praising his fieldwork and publications. He was more exacting about museum displays, however, insisting that they reflect his ideas. In turn, Matthew became increasingly vocal in opposition to Osborn's theories, and in 1918, he withdrew from a joint research project on horses on the grounds that their views on variation were too divergent. Osborn took this calmly, but he may have had a subtle revenge several years later when he made Matthew curator not only of vertebrate paleontology but of geology, mineralogy, geography, and invertebrate paleontology. He told Matthew that it was "one of the great opportunities" of his life, but the younger man found it a crushing burden. His main interest was research, and he had health problems, another source of friction with Osborn, who liked lecturing employees about their

Figure 13. *Bison* and *Smilodon* (Pleistocene) from Zallinger's *Age of Mammals* mural. Courtesy Peabody Museum of Natural History, Yale University, New Haven, Conn.

personal habits. In effect, the old tyrant had told Matthew (whose photos do suggest a camellike obstinacy) to drink from clear mountain brooks instead of muddy pools, that is, to give up caffeine and tobacco.

The paleontological Camelot was a tense place as Granger and Andrews pursued their Mongolian errantry. Gregory called Osborn "a terrific problem for all of us," and bemoaned his own "smiling hypocrisy" in staying loyal to him. The expedition was probably part of the problem as far as he and Matthew were concerned. Both co-wrote papers about its finds with Granger, but neither got to go along, and they must have resented watching Andrews's expensive antics from the sidelines. Matthew finally did travel to China to join the 1926 field season, but political unrest forced its cancellation. A year later, he quit the museum to head the University of California at Berkeley's paleontology department, and Osborn appointed Barnum Brown to replace him as curator of vertebrate paleontology, perhaps reflecting the fact that dinosaurs were less elusive and laden with evolutionary perversities than mammals.

Matthew might have been well advised to take Osborn's advice about tobacco and caffeine, since he died in 1930, aged only fifty-nine. That year, Chinese and Mongolian political turmoil finally terminated the Central Asian Expedition. Andrews's exploring career ended with it, along with his glamorous marriage, his socialite wife having found solace with their financial adviser during his long absences. Asian exploration was doomed anyway. The Wall Street crash cut the museum's endowments, as well as the contributions from Osborn's upper-class friends that largely had funded fieldwork. In 1932, the museum prohibited use of its funds for expeditions.

Osborn retired unwillingly the following year, and he died two years later. His downfall was not as humiliating as Marsh's had been. No newspaper scandal was involved, although he had been using endowment funds questionably for pet projects. But Osborn and Marsh underwent similar "sunset years." Both had created a milieu for younger paleontologists by promoting collections and institutions of major importance. The younger paleontologists scorned both, believing that they had failed in the central work of showing how mammals had evolved.

ELEVEN

A Bonaparte of Beasts

NOT ALL OF THE ARRESTING figures in the Peabody's *Age of Mammals* are huge or fierce. One of the more conspicuous is an amber-colored herbivore named *Aepycamelus* standing near the early mastodon, *Gomphotherium.* It emphasizes the "African" aspect of North America's Miocene plains, because its neck is so long that it might be a giraffe. But *Aepycamelus* was not a giraffe. As its name implies, it was a "highest camel," a long-necked member of a family that, like horses, evolved in North America and spread almost worldwide, then became extinct in its home continent, and survives largely in domesticated form, as dromedaries and llamas.

Zallinger's mural features a succession of camel genera, beginning in the Oligocene with one, *Poebrotherium,* that looks only a little bigger than that epoch's *Mesohippus.* A somewhat larger genus called *Procamelus* shares the Miocene savanna with *Aepycamelus,* and, although it looks less remarkable (not unlike the living South American guanaco), it is unusual too. It is the only genus in the mural that appears in two epochs. Two *Procamelus* also stand in the Pliocene's autumn woodland, although their position on the wall shows they lived perhaps ten million years after the Miocene *Procamelus.* Ten million years is a long time for a mammal genus to last. Then, right at the end of the Pleistocene, another camel pair rests under a red maple tree. They are named *Camelops,* "camel-appearing," and so they are, humps and all. They would not look out of place in the Sahara.

Figure 14. *Aepycamelus* and *Gomphotherium* (Miocene) from Zallinger's *Age of Mammals* mural. Courtesy Peabody Museum of Natural History, Yale University, New Haven, Conn.

The camels' succession is a striking contrast to that of the mural's horses. While the equine contingent parades along getting steadily larger and more horselike, a sequence explicable either through Marsh's progressivist Darwinism or Osborn's orthogenesis, the camels seem noticeably less goal-oriented. The earliest ones shown, *Poebrotherium,* are not tiny and primitive like dawn horses but "already very progressive in their loss of digits, elongation of the legs, and in the closure of their bony eye socket." *Procamelus* and *Aepycamelus* seem not so much advances on them as variations, as though camel evolution was more of a meander than a march. Indeed, while Zallinger's horses tend to be galloping or trotting, his camels are just standing or lying around, as camels will do if given a chance.

I can attest to the aptness of Zallinger's characterizations. I once went on a pack trip with a llama, and I have seldom met a less forward-looking mammal. It took life entirely as it came, pausing to nip daintily at appealing plants along the trail, with no apparent sense of destination. It was very agreeable about carrying our equipment, indeed surprisingly likeable. It never spat at us. If not gently encouraged to proceed, however, it simply would not. On the other hand, a horse I rode on another backcountry trip couldn't wait to advance along the steep, muddy, rocky trail, although preferably when I was off its back. It clearly had a destination in mind, to which my presence was incidental. Once I'd tired of its impatience on a particularly scary stretch near the biological station we were headed for and dismounted for good, it took off for the station's corral without a backward glance.

The camels' meandering evolution—pausing to grow a long neck here, a hump there, sometimes neglecting to change at all for millions of years—suggests a transmutational mode different from Marsh's and Osborn's. Indeed, it seems not unlike the "law of higgledy-piggledy" that Darwin himself proposed in his original "descent with modification." In the 1920s, as Osborn's orthogenetic balloon slowly deflated, this mode was about to get a more energetic, if less genteel, champion than the squire of Downe House, and it eventually would become the dominant way of looking at mammal evolution. Zallinger's mural seems to recognize this in an understated way, because its first and last camels, *Poebrotherium* and *Camelops,* might be ironic footnotes to the mural's dominant titanothere and mammoth. They stand and lie in the background, looking retrospectively at the giants, as though thinking: "Well, you were imposing, but *we* are here." *Camelops* is the final figure on the wall, an evolutionary last word.

The deflation of orthogenesis reflected vertebrate paleontology's overall status in the 1920s. It was no longer at evolutionism's cutting edge. After biologists began paying attention to Mendel's work, the laboratory had become the main arena for ideas about variation. Thomas H. Morgan, a Columbia University professor, studied the genetics of fruit flies and regarded the American Museum's horse and titanothere exhibits as little more than mass entertainment. "My good friend the paleontologist," he wrote in 1916, "is in greater danger than he realizes when he leaves description and attempts explanation. He has no way to check up on his speculations and it is notorious that the human mind without control has a bad habit of wandering." Morgan saw random genetic mutation as the force driving evolution, and he found Osborn's orthogenetic theories ludicrous. "I am sorry to hear that the mammals have not evolved by mutation," he taunted his rival in 1918. "It would be too bad to leave them out of the general scheme . . . and I cannot but hope that you will relent someday and let us have the mammals back."

William Diller Matthew agreed with Morgan about the cause of evolutionary change. He thought the enormous range of variation he found in horse skeletons arose at least partly from genetic mutation, like the varied eye colors of fruit flies. "The nature of these variations is much better understood than it was in Darwin's time, thanks especially to the researches of T. H. Morgan and his school," he wrote. "Some of them are inherited according to certain definite laws. They are 'mutations' of the same nature as the larger, more conspicuous, and more occasional mutations which geneticists have studied in detail." Yet, according to Ronald Rainger, Matthew "never systematically applied genetics to the fossil record, and he did not develop a comprehensive theory of evolution." Like most field naturalists of the time, he was untrained in the quantitative techniques that genetics-based theory required.

Matthew remained unschooled in other new technologies. Despite his dedication to fieldwork, he shrank from a machine that made it more efficient. "One of Matthew's oddities," recalled his assistant on a 1924 collecting expedition to the Texas panhandle, "was that he had not learned to drive a car." The assistant, 22-year-old George Gaylord Simpson, was in awe of his boss, then "at the height of his powers and his fame." He didn't let that stop him from rushing in where Matthew feared to tread. Since Matthew needed an assistant who could drive, Simpson got the job by pretending that he could. "Our car was delivered to us in a crowded garage in Amarillo," he recalled. "Someone else cranked it. I had seen someone drive a Model T once, so my lie was not absolute: I had a vague notion of what to push, pull, and turn. Off we went somewhat jerkily into the traffic, and I believe that Matthew never realized how close he came to ending his career at that high point."

Far from ending anybody's career, the car facilitated a successful trip to Pliocene badlands that Cope had discovered at Mount Blanco in 1892. There, after minor setbacks, which included Simpson stepping on one of Matthew's prize fossils, they found two horse species that proved to be "missing links" between *Pliohippus* and *Equus.* The ambitious assistant applied the same seat-of-the-pants approach to another transport problem that summer, when Matthew sent him to explore badlands in New Mexico. The Model T didn't work off the road, so they hired horses. "My horse somehow guessed that I was neither an Indian nor a horseman, and that there was a simple way to rid himself of the burden," Simpson recalled, but he also turned that to professional advantage. "Out in the badlands the beast decided that if I wanted to go down a steep slope I could do it without him, which I did headfirst and landed beside the best fossil we found all summer."

Young Simpson's headlong approach also landed him in some hot water. The fossil was that of a large Miocene carnivore, and he and an assistant did a good job of excavating it. "Months later this block was received and opened up at the museum," wrote Childs Frick, a research associate and museum trustee. "Before us lay a nearly complete skeleton of *Hemicyon,* revealing a beast with somewhat tiger-like proportions and dog-like teeth." Frick, an heir to steel-making millions with such a passion for Tertiary mammals that he personally funded a laboratory for the museum, was delighted, because the genus was known only from a few European jawbones. He called it "this first great trophy of our most complete specimen of any of those ancient animals." Frick was less pleased when the go-getting Simpson submitted a paper on it for publication. The millionaire's touchiness about fossil ownership rivaled Marsh's, and he was paying for Matthew's expedition. Matthew shelved the paper, to the chagrin of Simpson, who professed ignorance that Frick had been furnishing his paycheck.

It was an audacious performance for a neophyte collector, but typical of Simpson. Although he came from a modest background, the son of a Colorado businessman and land speculator, he somehow had as strong a sense of his own gifts as the lordly Osborn. "I was an unexpected last child," he recalled, "and being different from the other members of the family, and indeed from any known relatives both in temperament and in appearance, I was something like a cuckoo in the nest. It seems symbolic that unlike any known ancestors (or, now, descendants) I had bright red hair." Simpson did work a lot harder than Osborn at demonstrating his abilities, perhaps spurred by an unprepossessing physique as well as humble origins. During the Cope-Marsh scandal, a *Herald* reporter had likened the six-foot Osborn to a college athlete; Simpson was a scrawny five foot seven, with protruding ears.

The youthful Simpson had tried business, once setting out to walk from Chicago to New Orleans to promote a shoe-liner brand. Disliking commerce, however, he had gone instead to the University of Colorado, where he first studied to be a writer, but then, deciding he needed something to write about, cultivated an interest in geology and paleontology. After transferring to Yale when his family's fortunes improved, he had discovered Marsh's Peabody fossils, which were still largely unstudied. "I went collecting in the basement," he recalled, "and decided that the most important fossils there were the exceedingly rare remains of the earliest mammals." His advisor, Richard S. Lull, trembled at putting the precious jaws and teeth into a brash neophyte's hands, but Simpson had his way after his productive 1924 summer in Texas and New Mexico. He wrote his Ph.D. thesis on them, and it was the best study of American Mesozoic mammals to date, taking up the subject where Marsh and Hatcher had left it. Simpson became an authority so fast that in 1925, the *New York Times* gave him a page of its Sunday edition to publicize his tiny fossils.

The *Times* found the subject so marginal that it illustrated Simpson's article only with a picture of "a typical dinosaur," but the 23-year-old authority was not daunted. First he used "ponderous dinosaurs fourscore feet long" to get readers' attention. "Immensity is never uninteresting, whatever the specific emotion it arouses," he wrote. "Perhaps that is why the scientist, when he wishes to place before the public the great past which his labors have revivified, does so with the statement that there were indeed giants in those days." Then he abruptly changed focus. "Yet there were also dwarfs, and if they were less awesome and pretentious, still were they more significant than the reptilian lords. Theirs was the promise of the meek: they were to inherit the earth." The article jauntily reviewed the short history of Mesozoic mammal discoveries, from Owen's Stonesfield jaws to Marsh's Como Bluff ones, then gave an approving nod to its author's own efforts. "Professor Marsh, unfortunately, was destined never to complete the studies which he projected, and for twenty-five years after his death his unrivaled collection lay in drawers in New Haven and Washington. It is only this year that his plans are beginning to be fulfilled, and that the majority of these priceless relics are being made known to the scientific world."

Aside from the accident of finding Marsh's basement treasures, Simpson's reasons for seizing on mammal evolution are less clear than those of his paleontological predecessors. The heroic age of Cenozoic discovery was past, and dinomania was in the ascendant. Yet Simpson was peculiarly brilliant as well as headstrong, and he would have perceived that mammals remained key to the intractable evolutionary problems that had baffled his mentors. He also would have seen that he had little chance of competing for sensational bones with the likes of Barnum Brown. His own egotism, although as energetic as Brown's, was more complicated, including neurotic perversity as well as powerful intellectual insight. The combination set the tone for an ambivalent career.

"The very thought of Mesozoic Mammals . . . makes me feel a trifle pale around the gills," he wrote his sister Martha not too long after his jaunty *Times* debut. " I've belabored the filthy beasts for about four years now and I hope to heaven no one finds any more during my lifetime—I've studied all there are in the world now and I couldn't bear it if I had to study any more. I'm going to dinosaurs or something like that now, or so I hope." But Simpson's complaints were seldom a measure of his actions. After dealing with the New World's Mesozoic mammals, he proceeded to deal with the Old World's, getting a grant to study the British Museum's collection, then traveling around the continent to see the other specimens. And if any more of the filthy beasts were to be found, G. G. Simpson would be the one to find them, godamnit.

His goal at the time was to join the Yale faculty, but one of the situations arose that occasionally make his life seem like something in an Ayn Rand novel. He had married and fathered two daughters at Yale, but he and his wife Lydia were such an epic mismatch that, although they did have in common the awkwardness of being westerners in New England, their pairing seems fundamentally mysterious. "We quarrel rather continually and she regularly threatens to leave me," he wrote his sister in 1926, "sadly enough, I'm afraid I wouldn't care much if she did. . . . In spite of which we are going to have another child next July. . . . Odd. I can't begin to express how the whole thing hurts and disgusts me." Lydia refused to accompany him during his research in England, and settled with the children in France, while he lived in London in a rented room. Applying for the coveted Yale position in 1927, Simpson was perplexed when his application went unacknowledged. Then his advisor, Lull, arrived in London and explained that the university was investigating Lydia's complaint "that I was not supporting her and that she was in desperate need while I lived in luxury apart from her." Simpson convinced Lull of his version of the situation, and Yale offered him the job, but its former willingness to believe him a "scoundrel" so affronted him that he took Matthew's offer of an assistant curatorship at the American Museum instead, arriving on the job as Matthew left for the University of California.

Osborn had hesitated to hire Simpson—he wanted dapper employees like Brown and Andrews for public relations. Simpson's ongoing enmity with Osborn's fellow patrician Childs Frick also worked against him. As it happened, however, the freckled upstart seems to have gotten along better with King Henry than Matthew or Gregory, recalling sympathetically that Osborn had demanded civility from his employees, not subservience, although "some were not civil behind his back." Like the others, Simpson had to bear with presidential conceit. When Osborn presented him with a freshly signed copy of his titanothere monograph, Simpson reached for a blotter, but Osborn stopped him, saying: "Never blot the signature of a great man." Such airs evidently amused Simpson more than they irritated him, however. He probably rather identified with them, as one Atlas to another.

Simpson stayed with his wife until 1929, and fathered two more daughters, although squabbles continued and even drew in the museum staff, including Margaret Mead, "one of a small coterie of women anthropologists who took what I considered a mistaken and invasive interest in my marital problems." The turmoil didn't stop him from expanding his Mesozoic mammal empire. He seized on the fossils that Walter Granger brought back from Mongolia and wrote several studies of them with Gregory. "[T]hey are generally considered the greatest paleontological discovery of the present century, so far," he wrote his mother, "and having my share in them is almost unbelievable good fortune as my reputation will be established at once." The Mesozoic mammal bubble burst, however, when the belt-tightening museum refused to send Simpson to look for more. "Thus I was out of a specialty and had to extend my scope, which I was glad to do," he wrote. . . . "A logical extension, having studied the oldest known mammals, was to go on with the more abundant mammals of the Paleocene."

It was, after all, just the other end of the basic problem of where the ancestors of living mammals had come from. Simpson studied fossils that Barnum Brown had brought back from Montana Paleocene beds, and in the summer of 1929, he led an expedition to New Mexico's San Juan Basin, the site of Cope's epoch-making Puerco formation. A whirlwind flattened his camp on arrival, and a flash flood demolished his truck on departure, but he came away with Paleocene mammals.

That was just an appetizer to a pair of expeditions to Patagonia in quest of the South American Paleocene, the source, presumably, of the Ameghinos' antipodal bestiary. "The study of origins is always peculiarly difficult," Simpson mused, "and I suppose that is a large part of its fascination for me and for many others." Although impressed with Carlos Ameghino's early mammal finds, he thought them scrappy and poorly preserved. "So it fell to our expedition to take the next step in this important and interesting work." The possibility that he might not find what Carlos had spent thirty years finding, and John Bell Hatcher had spent over two years not finding, evidently didn't trouble him. "I am quite sure he felt he could do anything," his daughter once remarked. The Depression-strapped American Museum would provide only moral support, but a wealthy New Yorker, H. S. Scarritt, put up the money after Simpson had gained his goodwill through some lengthy cocktail sessions. Simpson said he only regretted that he had but one liver to give for his museum.

He and an assistant, neither speaking Spanish, arrived in Argentina as a revolution was under way. Strolling across Buenos Aires, Simpson recalled, he encountered a cavalry charge. "[I]t was apparent that I had not yet found the proper atmosphere for study and meditation," he wrote. "I departed, rapidly, running down the Avenida in search of shelter." He hid in a clock shop, a shuttered storefront, and a bar as the government *Escuadron de Seguridad* sabered and machine-gunned civilians on the streets outside. A few days later, a pair of revolutionaries stuck guns in his stomach, thinking him a government spy. "I wish I could say that I looked the lads coldly in the eye and told them in flawless, fluent Spanish to go about their business," he wrote. "I did nothing of the sort. I fully expected to die immediately and messily, and the thought almost paralyzed me." His Spanish was already improving, however, and he convinced them he was harmless.

Their arrival in Patagonia was scarcely more encouraging. "[T]here were no trees or anything green," Simpson recalled. "It was bleak . . . and I may as well start repeating it now: Patagonia is bleak." The revolution pursued them. When Simpson's assistant, a relative of Scarritt's named Coley Williams, whimsically discharged his shotgun in the air after a late night party, he ran afoul of martial law, and Simpson had to petition the authorities to get a firing squad reduced to a fine. Later the expedition's Argentinean cook took such a dislike to Williams and his *yanqui* disdain for garlic that Simpson had to beg *him* not to execute his assistant. And that wouldn't have worked, Simpson claimed, if he hadn't tricked the cook by making him promise to wait two weeks, then firing him before the sentence could be carried out.

Simpson brushed aside such obstacles, learning Spanish from a Patagonian assistant, adapting to a diet of cold mutton and stale bread, and driving the ramshackle expedition truck through huge mudholes and across trackless plateaus. He soon found the Notostylops beds that had baffled Hatcher, reconstructing the genus as probably having looked and acted like a jackrabbit, although it had hoofs. "The animals buried here would look absurdly strange to our modern eyes," he wrote of the Great Barranca of Colhue Hualpi, where they dug fossils from ledges hundreds of feet up cliff faces while nesting eagles dove at their heads. "Here in the oldest deposits, the Casamayor Formation, which particularly interested our party, is indeed a lost world. There are just two sorts of animals that would not look wholly unfamiliar today: opossums and armadillos." Some of the finds recall Zallinger's Paleocene figures. They dug up the bones of "one of the largest snakes, living or fossil, ever recorded, if not the largest," and the commonest mammals they found sounded like the Peabody's *Barylambda*. They were homalodontotheres, "smooth-toothed beasts," which had "five toes, each ending in a hoof, stocky legs, long heavy tails, and disproportionately large heads."

Simpson later modified homalodontotheres a bit, cutting the seven syllables in their "monstrosity" of a name to the five of "homalodotheres," and reducing their five hoofed toes to four clawed ones. Despite such minor difficulties, his work was so successful that in 1982, Larry Marshall, a paleontologist, still called it "the authoritative study of Patagonian land mammal faunas." During two six-month sojourns in the bleak region, he combed the landscape until he found an even older formation than the Notostylops beds. Petroleum geologists had encountered mammal fossils that they considered Cretaceous in age, but Simpson established that the bones, mainly of small, primitive browsers, weren't part of the Mesozoic's "orderly evolutionary sequence of relatively primitive forms." He considered them late Paleocene, and named them after the Rio Chico, an outlet of Lake Colhue-Hualpi, near where he found them. They still weren't old enough for Simpson. "Only a few tantalizing scraps and not even a proper sample of a fauna have so far turned up from rocks older than our Riochican," he complained, "and considerable evolution had already occurred within the fauna of the island continent by Riochican time."

Mongolia, where the Central Asian expedition had found Paleocene *and* Cretaceous beasts, seemed the next place to look for modern mammal origins. It apparently didn't occur to Simpson that he might not be able to rush in where Andrews's heavily armed columns now feared to tread. Andrews had made a last attempt to return to Mongolia via the Soviet Union instead of war-torn China in 1933, traveling to Moscow and offering to fund a joint expedition, but he soon abandoned the project after the friend who had suggested it advised him that the Russians were "without exception the greatest, most enthusiastic, long-distance liars the world has ever known." But that still didn't discourage Simpson. "I had long looked forward to this and had made some preparation for it," he wrote, as though that should have decided the matter. The preparation included teaching himself Mongolian, since no instruction was available in the United States, and devising a Monopoly-like parlor game he called "Going to Mongolia" in which "the winner arrived at Shabarakh Usu and found numerous Mesozoic mammals there."

After his second Patagonian expedition in 1934, Simpson headed straight for Moscow. The United States had just reopened diplomatic relations, and support from the new ambassador and consul general buoyed his hopes of using Russia as a springboard to the Gobi. Even his wife Lydia's vagaries had not prepared him for Stalinist bureaucracy, however. His Soviet visa was for Moscow only, and Immigration told him he needed to get a Mongolian one before he could travel to the border. The Mongolian embassy then told him that he certainly could have a visa, but that he first needed Russian permission to travel to the border. As he wriggled in this double bind, agents bugged his hotel phone so clumsily that he overheard them, an "ex-American young communist" shadowed him, and "a handsome female army parachutist" appeared in his room "to see if she could help me in any way."

"Thus I was batted back and forth for weeks without result," Simpson complained, "while I dangerously overstayed my visa to the U.S.S.R. itself." This eventually unnerved the ambassador and consul general, who told him to leave. Simpson seems to have regarded this as a temporary setback at the time. "My retreat from Moscow was late enough in the summer to make field work impractical that year," he remarked. He continued hunting Paleocene fossils in the West, and, eventually would visit every continent on paleontological forays, although never Mongolia.

The Moscow retreat must have shaken him, however, because he afterward diverted much of his previous drive for digging up the fossil keys to modern mammal origins into another, albeit equally ambitious, campaign. "I had long been intensely interested in and personally concerned with the theoretical aspects of evolution," he wrote. "Now . . . I dared to take modest personal steps in the direction of principles and theory. I was planning and working toward larger steps." In Napoleonic style, the new campaign of conquest would involve even *l'amour.*

TWELVE

Love and Theory

SIMPSON REMARRIED IN 1938, in his usual epic mode. A childhood friend named Ann Roe had charmed him when they had met again in New York in 1926. "[S]he interests me tremendously," he had written his sister. "Her judgement and taste, once so far behind, are catching up with her remarkable intellect. She has abandoned her bigoted pseudo religion and her ridiculous priggishness and is really human and sensitive—all of this to my surprise for I had not seen nor heard from her for several years. Physically she has ripened and is really striking in appearance." Ann had married someone else, and Simpson had gone to Europe and Patagonia, but they had kept in touch, and Ann's marriage had collapsed at the same time as his. They had started living together after Simpson and Lydia separated in 1932 but were forced to await his long-contested divorce before they could legalize the relationship.

Simpson's autobiography gives no details of this long clandestine romance beyond a 1935 snapshot of the slouch-hatted paleontologist swigging from a bottle while his paramour kneels demurely on a picnic blanket. In any case, the second marriage proved as happy as the first had been miserable. "My new wife is a lanky blonde hoodlum name of Ann," he wrote his sister, "and I find being married to her curiously exciting and soothing at the same time." It was an intellectual as well as romantic affair. Simpson had mathematical talent but had never studied statistics, whereas

Ann had, since she was a clinical research psychologist. "This complementary relationship," he wrote, "suggested that development of the application of statistical and related quantitative methods to zoological studies could be forwarded by pooling our knowledge, a sort of figurative marriage of minds."

Evolutionary theory had not always charmed Simpson. "The highest possible scientific motive is simple curiosity and from there they run down to ones as sordid as you like," he had written his sister in 1926. "And all our scientific interpretations and theories are simply meaningless. There are facts, of course. . . . but with us as with artists and other impractical people here facts are considered as only such mud and straw unless they can be piled up into a hypothesis, gaily stuccoed and concealed with theory." His public utterances about evolution remained a conventional melange of Darwinism and orthogeneticism into the 1940s. "The most basic mammalian characteristic is intelligence," he wrote in a 1942 magazine article. "Small in size, without armor, without large fangs, the earliest mammals survived largely because they used their heads. Unable to outfight dinosaurs, they outsmarted them. The essential upward trend in mammalian history is an increase in mental power, in grade of intelligence, culminating (up to now!) in man." Marsh and Osborn both would have found something to like in that.

But Simpson had a better theoretical mind than either Marsh or Osborn. "The level at which Dr. Simpson operated was, scientifically speaking, up in the stratosphere," his American Museum colleague Edwin Colbert recalled. "His mind worked in a way that can only be a mystery to most of us: he saw problems and he promptly saw their solutions in a manner that could only inspire awe." And theory must have seemed a ray of light in mid-Depression, when the museum had discontinued fieldwork, fired much of its staff, and cut salaries. Roy Chapman Andrews had assumed the presidency, but proved no better as an administrator than as a paleontologist. Walter Granger had to shore him up, but Granger also hated office work, and he had developed heart disease from a combination of stress and weight gained in the Asian expedition's luxurious Peking winter quarters. "Work at the Museum proceeds, but . . . it worries me to see our department, once incomparably the best in the world, going steadily downhill," Simpson grumbled in 1937. . . . "Granger is a broken man, Brown is a jealous egomaniac, Frick rides roughshod over us all, and no one even tries to guide the rudderless ship." When Albert Parr, the Peabody's erstwhile director, replaced Andrews as president, he planned to abolish the fractious Paleontology Department and transfer Simpson to Mammalogy.

Outside the museum, on the other hand, the deadlock between lab and field biology that had prevailed since 1900 was loosening. Geneticists were increasingly willing to address evolution in terms of Darwinian adaptation as well as mutation. This led "to what is now commonly called the synthetic theory, because it became a synthesis from all the many branches of biology." But paleontology still lagged. Although W. D. Matthew had seen that Mendelian genetics might elucidate the variation on which natural selection operated, neither he nor his pre–World War I generation colleagues had the mathematical background to interpret fossils in a genetic light. Simpson later recalled:

> [A]s I saw a possible synthesis beginning to take form, I also saw a serious gap in it. Paleontology, as the study of the history of life, should provide if not *the* at least *a* touchstone for the nature and validity of evolutionary theory. . . . My zeal was awakened . . . by . . . Dobzhansky's book *Genetics and the Origin of Species.* That great seminal work showed me that genetics is indeed consistent with and partially explanatory of nonrandom explanations of evolution. In 1938, I began to work on a book that would relate paleontology to this point of view.

Simpson's newfound statistical powers allowed him to draw some novel perceptions from the study of fossil populations. It had seemed self-evident to Osborn, even to Marsh, that size had increased more or less steadily throughout horse evolution. Such an increase was in keeping both with orthogenesis and progress-minded Darwinism. Statistics gave Simpson a different picture. "During the Eocene, the record, contrary to a rather general impression, does not show any net or average increase in size," he wrote. "In fact, the known late Eocene horses average rather smaller than eohippus in the early Eocene." Size had continued to fluctuate throughout the Tertiary, and statistics showed that horse evolution's other supposed "inherent tendencies"—increasingly high-crowned grinding teeth and decreasing toe numbers— had also fluctuated. High-crowned teeth had evolved mainly in the Eocene, but toe numbers had changed hardly at all in that epoch. Three-toed feet had evolved rapidly in the subsequent Oligocene, then had remained largely unchanged until the late Miocene, when one-toed horses had begun to appear.

The horse "lineage" that emerged from Simpson's statistical approach was very different from previous, vertically ascending ones. It sprawled like a lopsided cactus. Genera such as *Anchitherium* and *Hipparion,* directly ancestral to *Equus* in older genealogies, were side branches in Simpson's. Indeed, Simpson's genealogy showed *Equus* itself as a side branch, which happened to widen in the late Pliocene to fill the worldwide equine niche, then narrowed to the Old World in the late Pleistocene.

Side-branching was the main pattern of Simpson's evolutionary model. According to it, *Eohippus,* the first known equine genus, was so successful that it had spread through the Old and New worlds. Over time, this widespread *Eohippus* population "fragmented into numerous small isolated lines of descent," and these lines then diverged through what Simpson called "inadaptive differentiation and random fixation of mutations." A few of the divergent populations happened to have traits adapting them to environmental changes, and they survived and grew into large worldwide populations as their less adaptable relatives died out. *Eohippus* diverged into *Palaeotherium* and related forms in the Old World, and into *Orohippus* and its relatives in the New. *Palaeotherium* failed to evolve the high-crowned cheek teeth that adapted horses to a cooling, drying world, and its descendants disappeared at the end of the Oligocene epoch. *Orohippus* evolved high-crowned grinders by the mid-Eocene, and its side-branch *Anchitherium* eventually spread to the Old World.

According to Simpson's model, it was impossible for *non*adaptive traits to evolve, since the "strong selection pressure" always working on small, diverging populations would eliminate them quickly. Simpson applied his rule particularly to Osborn's favorite supposedly nonadaptive trend, the titanotheres' progressively enlarging nose horns. "Argument that this trend was not oriented by adaptation depends on the assertion that the incipient stages of bone thickening were of no use or advantage to the animals possessing them," he wrote. "There are nevertheless at least two alternative explanations, both of which could make the trend adaptive throughout." First, even a slight thickening of the skull in the butting region could be adaptive in defense or sexual rivalry, "and many studies have now shown that in populations of medium to great abundance any appreciable advantage, even though exceedingly slight, may be surprisingly effective in producing further changes of the same sort in future generations." Second, horn size might have been linked to an increase in overall body size, and "development of both horns and body size would be accelerated because both are advantageous, adaptive trends and the two go together."

"Thus the whole trend is adaptive from beginning to end," Simpson concluded. He dealt just as summarily with that orthogenetic chestnut, saber-tooth fangs:

> Even today, writers not familiar with the actual fossils continue to say that the canines became steadily, "orthogenetically" larger through the history of the group and finally became so overgrown that the sabertooths became extinct. As a matter of plain fact that is completely untrue. Students of the sabertooths have been pointing this out in vain for at least forty years: another example of the durability of error. The earliest sabertooths had canines relatively about as large as the last survivors of the group. For some forty million years of great success the canines simply varied in size, partly at random and partly in accordance with individual advantage to species of various sizes and detailed habits.

Simpson ultimately saw no orthogenesis in evolution, physical or mental, dismissing even the popular assumption that mammals are inherently smarter than other classes:

> The little eohippus lived about 125,000,000 years after the mammals arose, and it belonged to one of the specialized 'modern' orders of mammals and to the same family as its descendant, the horse, which is a very brainy animal as non-primates go. Yet the brain of eohippus was on about the same structural grade as that of a modern reptile, and eohippus cannot have been much, if any, brighter than its reptilian forebears in the Triassic. . . . A typically mammalian level of intelligence, then, is not an original or early characteristic of the Mammalia, but has developed independently and rather differently in each of the different groups of mammals that show it.

Simpson never seems to have used camels as a case in point against orthogenesis, but his whimsical description of the guanacos he encountered in Patagonia demonstrates his thinking. "The guanaco is a most improbable creature, apparently put together in haphazard fashion and in questionable taste," he wrote. "If actions are a fair guide, his mind is often vacant, sometimes hysterical, and always stupid." Yet stupidity didn't preclude Darwinian survival. "Patagonia is the guanaco's country. He is better adapted to it than we are and he belongs there. . . . A patagonia without its guanacos would be dismal, indeed, unthinkable. Ungainly, ugly, and irrational as he is, the guanaco lends to the region much of its life and interest." Mammals not only aren't inherently smarter than other animals, Simpson implied, they don't necessarily need to be smarter in order to survive. They just have to adapt, which may entail becoming stupider, if that is adaptive.

Simpson concluded that "the history of life, as indicated by the available fossil record, is consistent with the evolutionary processes of genetic mutation and variation, guided toward adaptation of populations by natural selection." It was a big conclusion, as Simpson himself acknowledged. "I am currently (when I can bear to work) cleaning up all the mysteries of evolution, you'll be glad to hear," he wrote his sister in 1940. His major book, *Tempo and Mode in Evolution,* came out during World War II, and his ideas took a while to permeate the scientific establishment, but they eventually helped free vertebrate paleontology from the scientific isolation of Osborn's reign.

"Simpson's task," the historian Peter Bowler wrote in a 1983 study of evolutionary thought,

> was to demonstrate the plausibility of the claim that macroevolution as revealed by the fossil record took place through the accumulated effect of microevolutionary processes that were now being studied in modern populations [i.e., fruit flies]. No proof could be offered; but it was possible to show that the available evidence from paleontology was at least consistent with the new theory, despite the anti-Darwinian claims of an earlier generation of biologists. Simpson used a quantitative analysis to show that major evolutionary developments took place in the irregular and undirected manner predicted by Darwin.

A younger colleague of Simpson's, Stephen Jay Gould, was more effusive in the 1990s, calling him "unquestionably the greatest vertebrate paleontologist of the twentieth century, perhaps the greatest of all time . . . a brilliant theorist who brought a conceptually backward field of traditional paleontology into synthesis with the neo-Darwinian consensus that solidified in the biological sciences during the 1930s."

Simpson probably would have agreed with Gould's estimation, except for the part about a "neo-Darwinian consensus." He disliked being called a neo-Darwinist, pointing out, with historical exactitude, that neo-Darwinism properly referred to an early 1900s faction that had tried to reassert natural selection against the anti-Darwinism then prevalent. The neo-Darwinist label stuck, however, and with reason. The original "neo-

Darwinism" was simply paleo-Darwinism revived, since it didn't incorporate Mendelian genetics. The 1930s synthesis really did create a new kind of Darwinism, and simply calling it "the new synthesis" was confusingly vague. Simpson might secretly have been pleased if people had started calling it "Simpsonism." He would have recognized that this was unfair, however, since he was only one of several scientists—including not only his mentor, Theodosius Dohzhansky, but the biologist J. B. S. Haldane, the zoologist Ernst Mayr, and the botanist G. Ledyard Stebbins—who had contributed to what now is called the neo-Darwinian synthesis.

Simpson's theoretical ascendancy was not an unmixed blessing. It partly displaced his original drive to seek new fossil evidence, and, although active, his fieldwork was less focused after the 1930s. He spent most of the early 1940s in the Army, behaving with typical perverse brilliance. His age and parental status would have exempted him, but he enlisted and passed a six-week military intelligence course in a week, entering active duty with the rank of captain. He spent two years in North Africa and Italy, and although he expressed disdain for "warfare," the Army promoted him to major and awarded him two Bronze Stars. His favorite martial feat seems to have been defying General George Patton's order that he shave off his beard by going over the famous martinet's head to the U.S. supreme commander in Europe, Eisenhower.

When Simpson was invalided out of the army with hepatitis, the American Museum's President Parr changed his mind about abolishing the Paleontology Department and made him its chairman. Simpson had the usual field paleontologist's aversion to administration, however, and soon came into conflict with his staff, who found him so remote as to be virtually invisible and resented his delegating routine tasks to them while he went traveling. He also resumed his feud with the millionaire Childs Frick, whom he accused of excluding him from his enormous mammal collection. Frick did rebuff Simpson, suspicious that he was trying to spy on his research, but Simpson provoked Frick by using funds earmarked for mammal collecting to finance reptile and amphibian fossil hunts. Simpson claimed that Frick had threatened to withdraw his funds from the museum unless Simpson was fired, but that "the director" had refused.

Despite his unfailing energy and self-confidence, Simpson's later career seems somehow touched with desperation. His autobiography's post-1950 chapters read like Christmas letters, melanges of exotic destinations and hobbyhorse enthusiasms—anthropology, Antarctica, penguins, South Sea islands. The Ayn Rand aspect cropped up. During an unsuccessful 1956 expedition to collect Cenozoic mammals in Amazonian forests (Barnum Brown, by then an octogenarian, was collecting them successfully in Guatemalan forests), a falling tree injured him gravely, breaking multiple bones. In the hospital, Simpson proudly wrote his sister that a Peruvian geologist had visited him just to see "someone who broke a leg there and is still alive. He says they've lost more damned geologists that way!" It took him two years to recover, however, and when he finally returned to work, he claimed, President Parr told him that, given his decrepitude, he need only show up daily until due for his pension. Simpson apparently exaggerated the slight. Colleagues recalled that Parr had simply asked him to resign the chairmanship, since he clearly preferred research to administration. Whatever the reason, the episode so offended Simpson that he dumped twenty-two years of seniority and moved to Harvard's Museum of Comparative Zoology. There other conflicts emerged, and he then went on to the University of Arizona. Difficulty in walking and episodes of heart failure compounded his troubles, although Simpson kept working until his death in 1984.

Simpson was notorious for reticence about his feelings, but he left a kind of personal testament, although a typically ambiguous one. Sometime during his last years, probably in the early 1970s, he wrote a novel about a scientist accidentally caught in a time-travel experiment who finds himself in New Mexico's Late Cretaceous. This would have been a good way to speculate on the questions about Mesozoic mammals that he had posed in the *New York Times* a half century before. "The known span of these beasts was at least twice as long as the whole 'Age of Mammals,' yet they remained small and relatively simple and undiversified," he had written in his 1925 article. "How many problems are evoked by this strange fact! What kept them small, insignificant in prowess and number? Could true members of the highest group of animals subsist for such ages without progress or assertion?"

The novel (first published in 1996) does not address such questions, however. Instead, its seems to explore Simpson's feelings about his turbulent career. The narrator, Sam Magruder, is trapped alone in prehistory, and Simpson evidently saw parallels with his own fate. "My life seems to be turning in on itself almost viciously," he had written his sister in 1926. "I literally haven't a single friend in the world." If his theoretical work had helped end paleontology's isolation, it hadn't ended his own. His loneliness as an early mammal specialist was unrelenting, and his growing eminence did not dispel it. In 1962, lecturing at a New Mexico university, he wrote his sister: "The faculty and student attitude is one of utter indifference. . . . Incidentally, there has been no heckling about evolution because my audience—now down to 7 or 8, including only one junior biologist—simply doesn't react at all. They sit absolutely deadpan, then scatter as soon as I stop talking." Stephen Jay Gould, one of Simpson's students in the 1960s, recalled that he never seemed really happy or comfortable unless he was drinking and smoking.

Isolation, not curiosity, is the novel's theme. Early mammals play a part in it, but a largely unscientific one. Simpson doesn't describe the little Mesozoic beasts he knew so well in any detail, whereas he does describe dinosaurs, perhaps thinking that readers would be more interested in them (although there's no evidence he tried to publish the novel, which his daughter found among his papers after his death). But the novel even avoids speculating about dinosaurs, and after the excitement of seeing them wears off, the narrator loses much of his scientific enthusiasm when he finds that he has no way of communicating with other humans except to carve a stone tablet that someone may find eighty million years later.

Magruder's one faint link with humanity is the tiny early mammals, and they only visit him to raid his pantry at night. He first retaliates by making robes out of their pelts, then reflects that these are, after all, his ancestors. (Evidently, the pointy-nosed, bushy-tailed creatures are early placentals, although the novel doesn't say so.) When Magruder playfully tries to shake hands with one, however, it bites him, and he concludes that it is "one of the dumbest brutes I ever met. I could no more teach him tricks or even common docile civility than I could tame a dinosaur." Still, he thinks, their descendants "would inherit some spark that would keep them fighting in the long struggle for survival. . . . Is it not, indeed, that drive that animates me now?" He briefly toys with the idea of breeding the creatures selectively to "speed up" their evolution, but realizes the futility of this and releases them. Then he is completely alone, and a tyrannosaur kills him by tearing off his leg.

Simpson's refusal to speculate about mammal evolution's great questions seems as enigmatic as Cuvier's. But perhaps it reflected a sense that his theoretical successes could not compensate for his failure to find the famous fossils—the grail of five-toed horses, perhaps—that might have been his if the Kremlin hadn't defeated him. "Surely no one ever had such a trip to look forward to as I have for the next year—to both of the two most remote spots on earth, Patagonia and Mongolia, the West Pole and East Pole," he had written his parents in 1933. But the "East Pole" had eluded him, and what followed must have seemed anticlimactic. One of his novel's introductory passages describes New Mexico's San Juan Basin, the scene of many successful but unspectacular post-1934 bone hunts, with surprising gloom.

"Almost completely devoid of vegetation and carved into myriad fantastic erosion forms, this landscape is a barren and unearthly as the face of the moon. . . . The scene is more grim than colorful. Many badlands are characterized by varied and vivid colors, but these badlands are painted in tones of gray or dull yellow. When this picture was composed, the pigments in the Earth's palette were muddy and depressing."

THIRTEEN

Simpson's Cynodont-to-Smilodon Synthesis

GEORGE GAYLORD SIMPSON'S NOVEL IS BETTER than most of the paleontological pulp fiction that came after Conan Doyle's *The Lost World.* Some of the pulp evokes a livelier feeling for prehistory, however, like L. Sprague de Camp's wry story "A Gun for Dinosaur," whose "pukka sahib" narrator uses a time machine to guide rich sportsmen. Brandishing his doubled-barreled "sauropod gun," which fires .600 "Nitro Express" cartridges the size of bananas, he tells one would-be client: "I'm sorry, Mr. Seligman, but I can't take you hunting late Mesozoic dinosaur. I could take you to other periods, you know. I'll take you to any epoch in the Cenozoic. You'll get a shot at an entelodont or a uintathere. They've got fine heads. . . . But I will jolly well not take you to the Jurassic or Cretaceous. You're just too small."

The guide's admonition would have nettled Simpson, but it might have benefited his fiction. Uintatheres did have "fine heads," and an informed trip to the early Cenozoic could have been more interesting than yet another dinosaur story. Nobody knew more about the period than Simpson, and he might have evoked it better than the grim Mesozoic of *The Dechronization of Sam Magruder.* But his writing has curious lacunae. He wrote entertainingly about bone-hunting travels, and eruditely about ideas, but he seldom described excavating and reconstructing fossils. That subject occupies a few offhand pages among the adventures in his great Patagonia memoir, and when he did write popular articles about bone-hunting, they tended to be brief and dry. "There are as many different ways of collecting fossils as there are fossils," was about as juicy as he got in a piece about one of the most complete South American ungulate skeletons ever found.

Although Simpson did write a few articles about uintatheres and the like, he didn't try very hard to resurrect them for the general reader. In one article about Paleocene mammals, for example, he dismissed the carnivorous creodonts and largely herbivorous early ungulates called condylarths as looking "very much alike, differing only in size and proportions and in minor anatomical details clear only on careful study." This seems evasive; at least, the three condylarth genera in Zallinger's mural don't look "very much alike" to me. *Loxolophus* seems badgerlike, while *Tetraclaenodon* might be a small tapir, and a *Phenacodus* pair behind *Coryphodon* vaguely resemble long-legged rodents. None of them looks much like the wolverine-sized *Oxyaena*. Less gifted men like Osborn and Andrews evoked prehistoric life more eloquently than Simpson. It was as though his genius for the work somehow blocked his ability to communicate it.

Yet, despite his resistance to reviving lost worlds in print, Simpson's synthesis formed the armature upon which post–World War II paleontologists restored prehistoric creatures, particularly mammals. When Zallinger painted three saber-toothed genera in his mural, he showed a lopsided cactus of big cat evolution, not an orthogenetic line. Tawny Oligocene *Hoplophoneus* ("armed killer") is rangy and long-tailed, springing on an early artiodactyl under *Brontops*'s nose. (*Hoplophoneus* wasn't even a cat, but a member of an extinct catlike family called nimravids.) Stocky, rufous *Machairodus* ("sword tooth"), stalking horses in Miocene prairie, seems bearlike. Leonine *Smilodon* ("carving knife tooth"), is the classic saber-tooth, based on a Charles Knight drawing and possibly inspired by W. D. Matthew's tar pit fable, since, like that "really high class animal," it is pondering whether to attack a bison or a sloth. They all have saberlike canines of varying sizes, but there is no hint of hypertrophy. Although they catch their prey in different ways, the haste with which it tries to avoid them leaves no doubt that the fangs were "adaptive from beginning to end."

There is circumstantial evidence that Simpson influenced the Peabody mural, because, before he heaped scorn on saber-tooth hypertrophy, Zallinger rendered the beasts differently. His cartoon shows the same three genera, but only *Smilodon* flashes saberlike canines, and it seems less comfortable with them than its mural counterpart. *The World We Live In*'s text describes *Hoplophoneus* as "saber-toothed . . . a powerful predator with two-inch fangs," but the cartoon doesn't show them, and *Machairodus* is described only as a large cat. Not only does the cartoon not convey the point that saber teeth were "adaptive from beginning to end," it hints that gaping *Smilodon* might have had a problem.

Lopsided cactus evolution predominates in the mural, and most genera show no trend toward anything except adaptation. The camel family first appears with a moderately long neck in *Poebrotherium*, then again with longish neck in Miocene *Procamelus*, then with a very long neck in *Aepycamelus*, then with longish one again in Pliocene *Procamelus*, and finally with a slightly shorter neck in *Camelops*. Rhinos remain gray and stumpy from Oligocene *Subhyracodon* to Miocene *Diceratherium* to Pliocene *Teleoceras*, then vanish, perhaps unable to adapt to climate change. Oligocene *Archaeotherium*, one of the enteledonts offered as a dinosaur substitute by L. Sprague de Camp's prehistoric safari guide, looks like its Miocene descendant, *Daeodon*, and that group also vanishes. Smaller ungulates—oreodonts, antilocaprids, protoceratids—flit through the landscape without pretense of doing more than fit in.

The mural does have trends toward larger horses and proboscideans, but to show actual size fluctuations, Zallinger would have had to exceed even the paleontological illustration's conventional overcrowding. As many as ten horse genera, ranging from fox to moose size, simultaneously inhabited Miocene savannas. One of the mural's strengths is a sense of spaciousness that Zallinger somehow gave its busy landscape. Too many horses would have tipped the scale toward a stockyard effect.

In any case, the mural's mild orthogeneticisms show neo-Darwinism's historical context. Osborn's and Scott's ideas might have been a too-straight version of actual evolutionary relationships, but they were the ladder from which Simpson constructed his branching one. His synthesis built on other important aspects of paleontological history.

Since Richard Owen's time, the jaw had been the main fossil indicator of a reptile-to-mammal transition. Reptiles have multiple bones in each half of their lower jaw, some of which form a complex joint with the cranium. Mammals have a single bone in each half, attached to the cranium by a simple joint. As early as the 1830s, anatomists had noticed that mammal embryos pass through an early stage in which they have multiple lower jaw bones. And one reason Owen's "mammal-like reptiles" seemed to link the two classes was that, in them, multiple jaw bones shrank through time, eventually resembling mammals' jaws. (Owen's tiger-crocodiles and weasel-lizards are no longer considered reptiles, in fact, but "synapsids," a group that diverged from reptiles in the Permian Period and that encompasses mammals.)

Naturalists had long disagreed on the meaning of the jaw changes. Did a transcendent reduction and displacement of jaw bones cause a changing way of life? Or were the reduction and displacement a transmutational adaptation to a changing way of life? Simpson saw mutation and natural selection behind the change. "The complicated reptilian jaws are . . . so put together as to permit only a limited repertory of motions," he wrote. "In mammals the lower jaw is a single bone, strong and simple, with very wide possibilities of movement." Combined with permanent, specialized teeth, wide possibilities of jaw movement allowed more efficient shearing and crushing of food. And the adaptations went beyond chewing. Reptiles have a single ear bone, transferring vibration from the eardrum to the inner ear. In early synapsids, two reduced jaw bones played a part in hearing while still attached to the jaw joint. In mammals, the two bones detached from the joint and became the middle ear's "hammer and anvil," another shift still repeated in mammal embryos. Environmental pressure for more efficient hearing evidently was selecting for reduced and relocated jaw bones.

Other changes in early mammals' skulls suggested that mutations were helping them adapt to a life increasingly attuned to smell and touch as well as sound. The nasal area became much larger and more complex, and a secondary palate grew to divide nose and mouth, assisting mammals to breath and chew food at the same time. This would have increased sensitivity to taste. Small openings in fossil snouts hint that whiskers may have helped guide early mammal movements. Like titanothere horns and saber-tooth fangs, Simpson might have said, the first mammal skulls were "adaptive from beginning to end."

The adaptations implied at least one environmental cause. Dinosaurs had begun to dominate the late Triassic's daytime, megafaunal niche, driving cynodonts and other early synapsids, some dog-sized or larger, toward extinction. The earliest reasonably complete mammal skeletons, of late Triassic creatures called morganucodontids, are a few inches long, with very large eye sockets, which would have combined with improved hearing, smell, and touch to facilitate a nocturnal life. No known Triassic mammals grew bigger than a rat, and their teeth and jaws allowed them to exploit a diversity of foods for which the larger dinosaurs did not compete.

Restriction to a nocturnal microfauna did not stop mammals from evolving, however. The beasts that thronged the Mesozoic may superficially have resembled shrews or mice, but they were as genetically diverse as dinosaurs, probably more so, since small size would have made them more widespread and numerous. Although Simpson complained in 1927 that known Mesozoic mammal fossils would fit in his hat, he and his colleagues picked apart their minuscule remains into an impressive range of taxa.

He regarded Reverend Buckland's Stonesfield slate "marsupials" as long antedating real marsupials, and he classed them in two groups. Some were triconodonts ("three cone teeth"), which Simpson considered too different from marsupials or placentals to be ancestral to either. He thought another Stonesfield species was at least remotely ancestral to marsupials and placentals, so he classed it as a pantothere, one of the "basic beasts" Marsh had named from the Como Bluff beds in the 1870s.

Como Bluff's mammals themselves encompassed more taxa than its famous dinosaurs, although none grew bigger than a squirrel. There were triconodonts and pantotheres. There were docodonts ("spear teeth"), mouse-sized, long-snouted creatures with molars specialized for grinding, as with rodents, but primitive incisors and canines. There were symmetrodonts ("equal teeth"), another snouty group that might have been ancestral to pantotheres, but whose fossil record was so sparse that paleontologists could only guess that they ate insects. The oddest Como Bluff mammals were the multituberculates, the order Cope had named from David Baldwin's San Juan Basin fossils. Multituberculates had large incisors like rodents and bizarre bladelike premolars as well as the many-cusped, or "tuberculed," molars for which Cope named them, so they seemed to have been plant-eaters, but nobody knew how they had evolved. Although Cope had first classed them with marsupials, then with monotremes, their relationship to other groups remained puzzling.

The Central Asiatic Expedition found over half a dozen species and four families of late Cretaceous multituberculates in the "flaming cliffs," showing that the anomalous creatures had thrived during the many millions of years separating them from Como Bluff's Jurassic ones. Other Mongolian fossils seemed so like living placentals that Simpson and Gregory called them eutherians ("true beasts"), a name paleontologists prefer to use starting with our Mesozoic relatives, since nobody knows when the placenta first evolved. One of them, euphoniously named *Zalambdalestes* ("storm thief") was rabbit-sized and long-snouted. Another, *Kennalestes,* was shrew-sized. Judging from their teeth, both probably mostly ate insects. The expedition's fossils also included a previously unknown mammal group including weasel-sized fanged creatures, which Simpson and Gregory named deltatheres ("triangle beasts") because of the shape of their upper molars. They had similarities to marsupials, although not enough for Simpson to be sure how to class them.

Marsupials themselves, also called metatherians ("beast relatives"), disappointed expectations that they would prove to be ancestral to eutherians. They appeared in the Mesozoic fossil record mainly as teeth, prompting one paleontologist ruefully to call them "an odontologist's delight, " and they appeared no earlier than eutherians. "It has been common practice for zoologists to regard the marsupials as an . . . intermediate step in the evolution of mammals between the ancestral mammals of the Jurassic period and the Cenozoic," mourned Simpson's American Museum colleague Edwin H. Colbert. "However, the evidence indicates that . . . the two groups probably arose independently from a common pantotherian ancestry, to evolve side by side."

Yet despite their bewildering fossil record, Simpson made neo-Darwinian sense of Mesozoic mammals. If they had evolved as small, nocturnal creatures to adapt to a dinosaur-dominated world, their smallness for the Mesozoic's remaining 150 million years was predictable, as was evidence that much of their evolution consisted of adaptations to smallness. "[T]hey were undergoing constant and fundamental evolutionary changes," Simpson wrote, "oppressed by myriad foes, learning perforce to survive by some means other than reptilian brute strength." Tiny teeth continued to display new refinements. In the late Jurassic, a new kind of molar appeared that distinguished deltatheres, marsupials, and eutherians. Simpson thought that it had allowed better shearing and grinding than before and described it as "tribosphenic [shear-grinding] . . . suggestive of a mortar and pestle." And as bone-hunting techniques improved, growing collections of leg and spine bones showed that locomotion had evolved from a sprawling, flat-footed gait toward a variety of nimbler running, jumping, and climbing ones. Even improved bone-hunting did not reveal how another important trait—internal gestation—had evolved, but its advantages seemed evident.

A replacement of early mammal groups like triconodonts by later ones like eutherians toward the Mesozoic's end also was predictable. Adaptations such as the tribosphenic molar, a more erect gait, and internal gestation would favor reproductive success. Even the survival of one egg-laying group, the monotremes, made sense, because, as Darwin had pointed out, Australian platypuses and echidnas are specialized and geographically isolated.

Indeed, mammal survival made better neo-Darwinian sense than their dinosaur oppressors' Late Cretaceous extinction. Marsh had speculated that a combination of climate change and competition from increasingly intelligent mammals might explain it, but there was little evidence, and orthogenetic explanations had prevailed afterward. Even while fashioning his synthesis, Simpson had blamed racial senescence, an orthogenetic idea, for the dinosaurs' demise. "[T]he main reason for that extinction," he had written in 1942, "was probably that the dinosaurs became too sluggish and too inadaptive to meet the conditions of rapid changes in their environment." He had soon abandoned that explanation, but, as his novel shows, he never really developed another. "I have studied the dinosaurs for years and brooded about their approaching end," his narrator, Magruder, muses, "but I cannot see any presage of it and cannot imagine a reason. . . . In short, I know no more about the causes for their disappearance than if I had never seen one."

Other neo-Darwinians papered over this theoretical hole with the idea that environmental change had caused a gradual dinosaur disappearance. "All that can be said is that conditions changed, and for some reason the ruling reptiles were unable to adapt themselves to the changing world," Edwin Colbert wrote in 1955.

> In middle and late Mesozoic times, much of the world was tropical and subtropical. Uniform temperatures, with but slight seasonal changes, ranged from the equator into very high latitudes, so that tropical plants and dinosaurs lived from northern Eurasia and Canada to the tips of the southern continents. . . . As the continents were uplifted and the new mountain systems began to grow, there were gradual alterations of the world environments in the direction of increased variety and differences. Climatic zones became established, and as time went on they became ever more sharply defined from each other.

Two decades later, Bjorn Kurten, a Finnish paleontologist, added evidence of increased volcanic activity as an extinction factor, but otherwise echoed Colbert: "That the Cretaceous fauna would be less resistant to climatic change than, for instance, that of the present day, is in itself quite plausible."

On the other hand, small, furry beasts with burrows and nests seemed predictable survivors of the changes Colbert outlined. Simpson wrote that the mammals prevailed "primarily because they were more adaptable. Their adaptability enabled them to survive the crisis of environmental change that slew the last of the giant reptiles. Once these reptiles were gone, this same adaptability enabled the mammals to multiply relatively rapidly and to adopt new modes of life, formerly closed to them by reptilian competition." Fossils proved that various mammalian taxa had lived through the dinosaurs' demise. The little *Cimolestes* in Zallinger's *Age of Reptiles* also left bones in the early Paleocene, and the multituberculate *Ptilodus* in his *Age of Mammals* had a close Cretaceous relative. Crouched on its "nest," *Ptilodus* is typical of the mural's early Paleocene fauna, inconspicuous compared to surviving reptiles, the giant serpent and the *Champsosaurus* "crocodile lizard," but thriving. Other small survivors further demonstrated adaptability by growing into the vacated megafaunal niche, from medium-sized mid-Paleocene *Loxolophus* and *Tetraclaenodon*, to the later epoch's big *Barylambda*.

Another Simpson colleague, Harvard's Alfred S. Romer, deepened the neo-Darwinian ambience by proposing that, far from being orthogenetically destined for dominance, early Tertiary mammals had to compete for it. In 1917, Granger and Matthew had named a "giant Eocene bird" from a skeleton found in Wyoming. "Our *Diatryma*," they had written, "must have been a truly magnificent bird—much bigger than an ostrich though not so tall, and more impressive because of its huge head and thick neck." They had not speculated about its way of life, but Romer decided that it must have been a predator of small Eocene mammals. "The surface of the earth was open to conquest," he wrote, "as possible successors there were two groups, the mammals, our own relatives, and the birds. The former group succeeded, but the presence of such forms as *Diatryma* shows that the birds were, at the beginning, our rivals."

Zallinger painted a baleful, black-and-red *Diatryma* strutting just behind his *Mesonyx-Uintatherium* confrontation, and though the eight-foot bird isn't attacking mammals, its beak and claws certainly raise the issue. Romer's scenario convinced most paleontologists that birds had challenged beasts. "Towering over the mammals of its day," wrote Bjorn Kurten in 1971, "this giant bird was fast enough to outrun them all and powerful enough to kill any one of them . . . in the late Paleocene we find them in both Europe and North America. They . . . were to remain a menace to the mammals well into Eocene times." In 1975, Adrian Desmond expanded on the idea. "Flightless birds continued to increase in size throughout the Tertiary," he wrote, "developing stout running legs, long necks, and good eyesight, and coming to resemble the bipedal dinosaurs. But it was not to last. Mammalian predators took their toll of the large flightless birds, and gradually restricted their range."

Yet if Simpson's synthesis explained why mammal evolution was a Cinderella story, it still had trouble explaining exactly how. The beasts that became large or otherwise impressive after the dinosaurs stopped dominating them were not simply overgrown versions of Cretaceous midgets. They shared primitive traits with known Mesozoic mammals, but they had others of mysterious origin. Many of the Paleocene beasts Zallinger painted had hoofed toes, for example, including Cope's tayralike *Pantolambda.* The hoofs clearly were an adaptation to bearing their weight. Flat-footed, cow-sized *Barylambda* needed all the support it could get. No known Cretaceous mammals had hoofs, however, and most other modern mammalian traits appeared mysteriously in the early Cenozoic—ungulate grinding teeth, carnivore shearing teeth, rodent gnawing teeth, even the reduced, peglike teeth of armadillos and sloths. The most spectacular trait, the bat's wing, appeared fully formed in the Eocene, as though its fossil possessor had landed from Mars. Bats evidently evolved to fill a night-flying niche, perhaps one vacated by pterosaurs, but known fossils provide no clue as to how it had happened.

Neo-Darwinian paleontologists assumed that the fossil antecedents of these traits were yet to be found, a reasonable assumption, but still one that put them back with paleo-Darwinists. Simpson's tentative solution to the apparently sudden appearance of four-toed *Eohippus* in North America was that its five-toed ancestors—Marsh's and Osborn's five-toed horses—had come from an early Paleocene Eden, probably in Asia. Indeed, according to Simpson, most of the sudden appearances in Zallinger's *Age of Mammals,* from *Eohippus* to titanotheres to gomphotheres, would have been immigrants from a neo-Darwinian Eden. "The next step—from theory to attested fact, would be the finding of the ancestry and the tracing of the actual steps of the invasion radiating from a center," he wrote. "Such a discovery has not been made. Perhaps it never will be; it is possible that the center of origin is now sunk beneath the sea or that no fossils survive in it. But in any case the theory is established and seems almost certain to be true." Osborn had said much the same, albeit more complacently.

Once the Cinderellas had escaped their long oppression, at least, they settled down to predictably neo-Darwinian evolution. Whether or not

Figure 15. *Diatryma* and *Mesonyx* (Eocene) from Zallinger's *Age of Mammals* mural. Courtesy Peabody Museum of Natural History, Yale University, New Haven, Conn.

environmental change had extinguished the dinosaurs, it evidently drove evolution in the Cenozoic, providing, among other things, the conceptual armature for the Peabody mural. Zallinger's mentor, Carl O. Dunbar, the Peabody's director, wrote:

> Following the major uplift of the Rocky Mountains near the end of the Mesozoic Era, much of the region had high relief and was generally timbered and moderately humid. Its early mammals, evolving in this environment were browsers, many of them were relatively small, and their teeth and feet were unspecialized. But by Miocene times the mountains had been worn low, the climate had become semiarid, and grasses had evolved to cover the wide prairies. In this environment the herbivorous forms lived in herds like those of modern Africa where fleetness was their best protection and long legs and specialized feet evolved rapidly. . . . Finally, in the Pliocene and Pleistocene epoch, the mountains rose again to their present height.

Neo-Darwinism also had an environmental explanation for the weird faunal similarities that fooled Darwin and Owen, misled the Ameghinos, and puzzled Gaudry and Scott. I remember my own puzzlement, after traveling a thousand miles to Guatemala's tropical forest, when the first mammal I saw resembled the groundhogs in U.S. temperate forest. It was an agouti, however, a South American rodent whose ancestors have evolved apart from North American ones for at least thirty million years. Yet, according to Simpson's synthesis, although groundhogs are related to squirrels, and agoutis to viscachas, there is no puzzle in the resemblance. They look and act alike simply because both inhabit burrows on the forest floor, and similar adaptations have caused their anatomy to evolve toward similar forms, to converge. Far from being ancestors of North American horses, camels, and dogs, as the Ameghinos thought, then, South American ungulates and marsupial carnivores simply evolved to look and act like them as forests shifted to savanna on both continents. One South American ungulate group called liptoterns even went horses one better, evolving a single-toed grazing genus, *Thoatherium* ("nimble beast"), millions of years before *Equus* appeared.

If environmental explanations of Cenozoic changes failed, Simpson could fall back on the other Darwinian mainstay, competition. South American ungulates and marsupial carnivores had vanished in the Pleistocene, after sixty million years of adapting to local environments, because North American ungulates and placental carnivores immigrating over the

Central American land bridge had shown greater fitness. "Those extant in the Plio-Pleistocene were the ones that had been successful in a long series of competitive episodes," he wrote of the North Americans. "They were specialists in invasion and in meeting competitive invaders. South American mammals had . . . met no impact from outside their own closed economy, and when it came, they had not evolved the required defenses."

Even Cenozoic evolution continued to have its mystery stories, however. Scott had been skeptical about the "invasion specialist" explanation for the replacement of South American beasts by North American ones, and Simpson admitted that the "ultimate factors" behind it were uncertain. There simply weren't enough fossils known to be sure why the marsupial carnivores and giant ungulates had disappeared. Like the bats' mysterious appearance, this uncertainty demonstrated a basic problem with his synthesis. It was a good tool for interpreting the mid twentieth century's known fossil evidence, but no better than previous theories at conjuring up the further evidence needed to verify many aspects of it. That would depend on the chancy, expensive legwork Simpson had partly relinquished while launching into theory, and, in important ways, the evidence was as scanty in 1965 as when Darwin had waffled a century earlier.

Although the cynodont-to-triconodont continuum provided a scenario of how mammals first evolved, the origins and development of most Mesozoic mammal groups still had the intellectual consistency of Swiss cheese. There was no evidence, for example, of why, when, or how some mammals had stopped laying eggs and started bearing live young. The origins of most major Cenozoic groups remained mysterious, and even later Cenozoic evolution contained gaping holes. If American, Asian, and African badlands provided some vivid glimpses of continua such as horse evolution, forested regions were persistently stingy with bones, and even some arid places like Australia remained largely enigmatic when Zallinger was painting *Peradectes,* a distant relative of the kangaroo.

To a degree, the neo-Darwinian consensus had simply come full circle from the paleo-Darwinian one in explaining things for which the fossil evidence remained sparse. The fossil evidence would improve in the next fifty years, but that would not save Simpson's synthesis from bruising encounters with literally earth-shaking scientific developments. Just as Zallinger's distraught, ashen *Brontops* had reflected the turbulence of twentieth-century mammal evolution studies, it would continue to do so after the millennium.

FOURTEEN

Shifting Ground

LIKE THE 100 MILLION-YEAR-OLD EARTH that Lord Kelvin had thrown in Darwin's way, the first challenge to Simpson's synthesis came from underfoot. Simpson followed Darwin as well as mentors like Osborn and Matthew in assuming that mammal evolution has occurred on a planet with relatively stable geography. All of them believed that the continents had occupied more or less their present positions at least since the Mesozoic Era, albeit with fluctuating coastlines as land rose and fell, sometimes allowing inland seas to cover large areas. It was a convenient way to regard biogeography, providing a kind of proscenium stage for dramas of migration and speciation.

It was not a universally accepted one. A welter of theories about mobile geography had grown up since Cuvier hypothesized the sea drowning lands and their fauna as other lands rose elsewhere. Some theorists had continents sprouting and crumbling like mushrooms; others had them drifting across the planet iceberg-fashion. Although they were sometimes harebrained, such ideas did provide explanations for puzzling aspects of the fossil record, like the discovery in Arizona of a Triassic fauna of early synapsids very like Richard Owen's "mammal-like" South African reptiles. This suggested either that a lost continent had once linked Africa to the Americas or that the two landmasses had been united, then had drifted apart.

As of 1950, however, few geologists supported mobile or sunken continent theories. The existing evidence was weak. The Royal Navy's *Challenger* expedition had dredged and trawled some 69,000 miles of ocean while circumnavigating the globe from 1872 to 1876 without bringing up bits of drowned lands. And not even Alfred Wegener, the German meteorologist whose 1915 book *Die Entstehung der Kontinente und Ozeane* (translated as *The Origin of the Continents and Oceans*) assembled the best case for "continental drift," could suggest a convincing mechanism whereby continents might move across the earth's crust. Minus such a mechanism, mobile geography theories resembled parlor games, with players making their own rules of shifting lands and seas to move organisms anywhere they pleased.

Such games threatened to play fast and loose with Simpson's ideas of species formation and dispersal. If a lost Atlantis had connected the Old and New Worlds during the Tertiary, and *Palaeotherium* and *Orohippus* had moved freely between America and Europe, what would happen to his carefully plotted episodes of horse evolution? With stable continents, on the other hand, he could reduce life's wanderings to a few basic categories consistent with mutation and natural selection. "The probability of spread of a group of animals from one region to another may have any level from nearly impossible to nearly certain," he wrote. "Although any degree of probability may occur and no sharp distinctions are possible, it is convenient to consider three main paths of faunal interchange."

First, Simpson envisioned "corridors," wide stretches of land or sea across which animals could migrate freely. Western North America had been a corridor throughout the Cenozoic, allowing most horse genera to gallop across. Then there were "filters," which allowed some organisms to pass at some times, but not at others. The Isthmus of Panama had been a filter for the past three million years, allowing a few equids to cross from North to South America. Finally, Simpson conceived "sweepstakes distribution," which allowed only occasional, chance movements, as during South America's isolation before the Panama isthmus formed. Horses never had reached South America by sweepstakes distribution, but primates and rodents had, possibly on rafts of vegetation drifting from Africa or North America.

The Peabody mural reflects Simpson's migration ideas. The continuous presence of horses, camels, rhinos, and various tapirlike beasts shows that western North America was a corridor for most of the Cenozoic. The mural's narrow, elongated shape suggests the prevalence of corridor migration.

The fewer beasts that arrived by filter bridges stand out, like the Miocene's *Gomphotherium* and the Ice Age's bison, both Bering Strait immigrants, while the *Megatherium* and *Glyptodon* that later came via Panama are even more anomalous. None of the mural's beasts are known to have arrived by sweepstakes distribution, but some relative of the little Eocene primate *Pelycodus* may have rafted to South America in the Oligocene epoch and become the ancestor of New World monkeys.

"Armed with quantitative methods for estimating from biological data whether past dispersals had been by corridor, filter bridge or sweepstakes," Malcolm McKenna, another American Museum colleague, writes, "Simpson proceeded to show convincingly that the data from mammalian distribution were not in harmony with the proposed Tertiary transoceanic bridges nor with transoceanic continents. Developing the ideas of Matthew, he argued persuasively for northern dispersal routes as sufficient to explain all mammalian distribution." Like Matthew, Simpson thought a "World Continent" of Africa, Eurasia, and North America, intermittently connected by filter bridges, had been the main mammal evolution corridor during at least the past sixty-five million years. Mammals had dispersed to other continents via more evanescent filter bridges or sweepstakes distributions, and in the case of Australia and South America, the links had been so sparse throughout the Cenozoic that a few early mammal groups had evolved unique faunas.

Simpson was less sure about distribution during the Mesozic, the first two-thirds of mammalian history. But then, its fossil record was much sparser, and the continents were thought to have been low-lying and partially covered by inland seas, which would have complicated mammal movements. "I did not deny the possibility of earlier effects of drift," he recalled, "but at the time I considered evidence for the drift theory so scanty and equivocal as to make it an unconfirmed hypothesis."

According to the geologist Leo F. Laporte, Simpson's authority, particularly on mammal evolution, allowed him to suppress the drift theory's paleontological side almost single-handed.

"Simpson was preeminent among American geologists who opposed drift and, in a series of papers in the 1940s, dismantled the paleontological arguments for continental drift as well as for transoceanic land bridges," Laporte writes. " From 1912 when Wegener first announced his theory, until the 1940s, when Simpson definitively rebutted the paleontological arguments, fossil data were central to the theory of continental drift. After World War II, fossils were either de-emphasized or not mentioned at all in support of drift. On the contrary, when fossils were considered they were used to support a stabilist position."

Simpson gave himself less credit for sitting on Wegener, recalling that his paleontological colleagues had been "almost unanimous in opposing, or at most not supporting, these revolutionary ideas" even before Wegener's death in 1930. "They found that the history of life, so far as it was known to them, could be explained just as well (and some points of it could be understood better) by stable continents," he wrote. "They also found that the supposed paleontological support for continental drift specified by Wegener and some other nonpaleontologists was misinterpreted or downright wrong."

Whoever did the sitting, Wegener stayed sat on for decades. A 1955 book on mammal evolution with six successive world maps on which continental outlines remain the same, while oceans and seas change, shows the dominance of the stabilist position. "Students of mammal origins tell us that many of the familiar animals we know originated on the Asiatic mainland and then migrated throughout the world," the text says. "At one time or another nearly all the major continents were connected for rather long periods. These dry-land connections between continents were broken by structural changes in the earth's surface. Water flowed in to seal off the breaks and isolate the animals involved by the change."

A minority of geologists clung to drift, however, and in the 1960s, new data began to support them. The stable continent faction always had assumed that the seabeds were ancient crustal rock overlain with deep sediments, but when post–World War II oceanographers on the *Glomar Challenger* expedition began drilling holes in the ocean floor, they found the opposite. Sediments were surprisingly shallow, and the rock underneath was at most a few hundred million years old. Indeed, the rock became progressively younger toward ocean centers, where long basaltic ridges interrupted the seafloor. The ridges were volcanically active, belching lava, and geologists realized that the earth was extruding new ocean floor on both sides of them. The reason for this extrusion was unclear and remains so, although a likely one is that convection currents in the earth's hot, plastic mantle cause lava to rise along cracks in the cooler, harder crust. The results were evident, anyway. Sea floor spreading from the ridges formed huge basaltic plates, which eventually would collide with plates spreading from other directions. The upshot was movement, both vertical and horizontal, as the plates scraped against or rode up on each other.

Science had discovered a mechanism for continental drift, which was more precisely, if less vividly, renamed "plate tectonics" ("tectonics" simply means the theory of building, or structure). The new paradigm saw continents, not as permanent bumps on the earth's crust, but as patches of granitic rock perched uneasily on the basaltic plates continually growing from spreading centers. Indeed, continents had no basic identity, because spreading centers could form under them just as well as under oceans, in which case long rifts would tear them apart as the plates they perched on diverged, and new seas would spill into the rifts. One of the first things deep-ocean drilling suggested was that a north-to-south spreading zone in what is now the mid-Atlantic had torn apart a continent that had existed two hundred million years ago.

Plate tectonics made sense of the similarities between South African and North American early synapsids, for one thing. Evidently the continents had formed a single giant landmass, called Pangaea, in the early Triassic, when early synapsids thrived, as discovery of still more of their fossils in South America and Asia helped to confirm. Riven by new spreading centers, Pangaea had later broken into two continents, Laurasia in the northern hemisphere and Gondwana in the south, then into a welter of smaller landmasses, which have been wandering like fractious families ever since. Gondwana's breakup was messy, with Antarctica abandoning Africa and South America for frigid isolation at the South Pole, while Australia and India fled toward Asia, and waifs like New Zealand, New Guinea, and Madagascar jostled about. Laurasia's was tidier, with Eurasia and North America parting more or less amicably along the mid-Atlantic rift, creating relatively few waifs.

The continental breakups raised potential headaches for Simpson's mammalian evolution synthesis. New species were supposed to evolve within their parent species' territory, then disperse as improved adaptations allowed them to supplant older species. If a new spreading center could split a species' range and drag half of it out to sea, dispersal might not be as adaptation-driven as neo-Darwinism proposed. Plate tectonics made the whole question of an organism's origins tenuous, since saying that mammals had originated in Asia could be meaningless if mammals had originated *before* Asia. This led one biogeographical faction to discard the model of species dispersing from areas of origin. "Vicariance" theorists argued that wandering continents made the origins question moot, and that organisms were better studied as though they had always lived everywhere they live now, even if their ranges are fragmented or spotty. Such reasoning annoyed Simpson. "It is absurd that these two principles or ways of looking at biogeography should be regarded as mutually exclusive alternatives," he wrote. "Yet some few enthusiasts have maintained that absurdity so emotionally as even to descend to personal vituperation, outside any acceptable discussion of scientific principles."

Overall, however, considering his previous role, Simpson took plate tectonics' triumph with good grace. "Although I am still sometimes cited as an opponent of continental drift, I fully accept this great addition to our knowledge of earth history," he wrote in 1978. But he added that it was still "highly probable that continental drift has had little effect on the geography of land faunas during most of the Cenozoic," and he had a point. Although relatively speedy in relation to the earth's five billion year history, crustal plates moved with Lyellian gradualism over the Cenozoic's sixty-five million years. A continental plate might move sixty miles during the average mammal species' lifetime, estimated to last two or three million years.

If tectonic theory seemed an alarm for Simpson's synthesis, it proved a false one. In 1971, when many geologists still doubted it, a paleontologist, Bjorn Kurten, hailed it, writing that the evidence had "established beyond doubt that great continental movements have indeed taken place in geological time." Kurten envisioned a bit more continental wandering during the Cenozoic than Simpson might have. He speculated that the earth's land masses had dispersed during the late Cretaceous into "eight separate areas, each . . . with its own flora and fauna," then had coalesced again, with the northern masses forming "what has been known as the World Continent, a gigantic evolutionary playground over which the terrestrial animals could move more or less freely." But that was not so different from what Simpson said.

Once they had learned the game of landmass musical chairs, his colleagues adapted it with surprising ease to Simpson's model of filters and sweepstakes linking northern continental corridors to southern ones. Malcolm McKenna simply added two more ways for organisms to wander—or to seem to wander. A "Noah's Ark" was a piece of landmass that broke from a parent continent and put out to sea with a load of organisms, which then would evolve in isolation unless collision with another landmass unloaded them. A "beached Viking funeral ship" was a load of *fossils* that had traveled from one continent to another on a Noah's Ark, giving the misleading impression that the new continent's evolutionary past had been like the original one's.

Plate tectonic interpretations of Mesozoic mammal evolution did grow more complicated, however, as new evidence came to light. Simpson's assumption that dinosaur age mammals had dispersed from the Northern Hemisphere to the Southern had prevailed when none were known from the south. But paleontologists have found a number of Mesozoic and early Cenozoic fossils there in recent decades, and they are not like northern ones. Early Cretaceous monotreme teeth turned up in Australia in the 1980s, and Paleocene ones emerged in Patagonia in 1991, all suggesting that the platypus's ancestors had a southern origin. Other new fossils seem even stranger than monotremes. At the U.C. Berkeley Museum of Paleontology in 2002, I encountered two casts of jaws from early Paleocene Patagonia that were unlike anything I'd seen before. One was tiny, but its molars looked strangely like a horse's, while I found the teeth in the other, larger jaw literally indescribable. Such mammals, called gondwanatheres, perhaps had some kind of common ancestry with triconodonts and other early mammals like those that inhabited Como Bluff in the Jurassic, but nobody is sure.

In the mid 1990s, paleontologists discovered something even more complicated, an Early Cretaceous Australian mammal, given the jaw-breaking name *Ausktribosphenos,* with the "tribosphenic" molars by which Simpson distinguished marsupials and placentals from older groups. Finding *Ausktribosphenos* was startling, since when it lived, an ocean had separated Gondwana from Laurasia, presumably cutting off migration between them. In 1998, another mammal fossil with tribosphenic molars—but a more pronounceable name, *Ambondro*—turned up in even earlier, mid-Jurassic sediments from Madagascar, also part of ancient Gondwana. In 2000, yet another tiny Jurassic skull with tribosphenic molar characteristics, named *Asfaltomylos,* turned up in Patagonia.

Since the earliest known Laurasian mammals with tribosphenic molars appeared in the Late Jurassic, the discoveries raised two unexpected possibilities. Either the first "tribosphenic" beast, the ancestor of marsupials and placentals, had evolved in Gondwana and somehow crossed the ocean to Laurasia, or the tribosphenic molar had evolved twice, in two different groups. Zhe-xi Luo and two co-authors favored the latter view in a 2001 paper, which located marsupial and placental origins in Laurasia, and placed *Ausktribosphenos* and *Ambondro* with the monotremes in a separate Gondwanan group. Living monotremes, the platypus and echidna, don't have tribosphenic molars because they don't have teeth, except as babies, but fossil ones had molars, which some paleontologists consider tribosphenic, although others do not. Other paleontologists believe all tribosphenic mammals had a Gondwanan origin, however, and the matter is far from settled.

Even those surprises weren't necessarily damaging to Simpson's synthesis. Evolution of two separate versions of a "higher mammal" definitive trait—the tribosphenic molar—may have an orthogenetic ring, but it equally can be seen as an evolutionary convergence. And if the alternate possibility is true, and the tribosphenic molar first evolved in Gondwana and then emigrated to Laurasia, plate tectonics might become an aid to Simpsonian distribution, with one of Malcolm McKenna's "Noah's Arks" perhaps rafting a tribosphenic-freighted bit of Gondwana north to Laurasia in default of a "sweepstakes" migration there. Gondwanan creatures probably wouldn't have survived a sweepstakes log raft trip to Laurasia, but they and their descendants might have thrived on a slow ark to China.

Indeed, far from interfering with Simpson's environment-driven model of evolutionary change, plate tectonics abetted it. Paleontologists interpreted Pangaea's coalescence as a cause of the planet's greatest known mass extinction, when, just before the Triassic, over 90 percent of living species disappeared. They thought the establishment of cool, dry continental climate over most land, combined with the recession of inland seas, which had covered much of the continents, when they were dispersed, had drastically reduced biotic diversity. Warm climate and inland seas had returned as the continents again dispersed in the Jurassic and Cretaceous, creating the dinosaur age's global hothouse. Then, as continents began to converge again in the Late Cretaceous, and as Antarctica's polar isolation changed ocean currents, climate had begun another inexorable cooling and drying phase, which climaxed in the Pleistocene.

The Tertiary cooling came a little late to account entirely for dinosaur extinction, but paleontologists cited various other tectonic events as possible factors. Enormous volcanic eruptions in India during the late Cretaceous might have caused a brief but drastic global cooling by blocking out sunlight with clouds of toxic gases. As well as affecting climate, the simultaneous recession of inland seas had destroyed much habitat of marine reptiles and other organisms. One dinosaur expert, Robert Bakker, thought migration over newly formed land bridges might have contributed to their extinction by spreading diseases among nonresistant populations.

A measure of plate tectonics' limited effect on Simpson's synthesis is that Zallinger's *Age of Mammals* needs no corrections to bring it into line with the theory. What is now the east slope of the Rockies probably looked very much as the mural shows it through the past sixty-five million years, with volcanism in the early epochs, the formation of high plains in the Miocene, and the sudden uplift of the present fault block range in the Pleistocene. The mural would have been more anachronistic if Zallinger had set it on the West Coast, since much of that didn't rise from the Pacific until the Eocene. But his boss, Carl O. Dunbar, wouldn't have wanted him to set it on the West Coast, which has a comparatively poor fossil record because of damage by the tectonic forces on the continent's edge.

FIFTEEN

————————

Dissolving Ancestries

ANOTHER CHALLENGE TO SIMPSON'S SYNTHESIS *might* have had a noticeable effect on the *Age of Mammals* mural if an idea related to it were shown. Zallinger's giant sloth, *Megatherium,* would be even more striking if, instead of gazing into the past, it were upending its neighbor *Glyptodon* as a prelude to ripping open its belly and devouring it. It is hard to imagine Zallinger's teddy bear monster preying on the Volkswagen-sized *Glyptodon,* and it would contravene what is known about the diet of living sloths. Still, carnivorous ground sloth theorists have a point when they say that the herbivory of living tree sloths doesn't prove that prehistoric ground ones ate only plants.

This other challenge to "Simpsonism" involved a dispute over fossil classification, as with the Cope-Marsh feud, but a more basic one. Cope and Marsh wanted to be the first to classify prehistoric beasts: to the extent that they agreed on anything, they agreed on the way to classify them. About the time that plate tectonics was coming to acceptance, however, some paleontologists began to think that the way of classifying fossils their precursors long had agreed upon was based on a false analogy between past and present. They thought traditional classification depended too much on history in trying to describe prehistory, as with assuming that extinct sloths ate plants because living ones do.

The aspect of traditional classification to which they most objected was the most pervasive one—the practice of drawing lineages of supposedly ancestral organisms, which had dominated schoolbooks and museums for a century. The early transmutationist Robert Chambers had begun it in his 1844 best-seller, *Vestiges of the Natural History of Creation.* "Previous taxonomists had taken parallelisms of structure, or organic affinities, as the basis of natural order," the historian James Secord writes. "But *Vestiges* now argued that this view was incorrect. The grouping together of similar species on the basis of affinities needed to be replaced by a system of genetic lines." Chambers was not a good naturalist, and some better ones, including Huxley and Lyell, at first disdained his idea as too speculative. Genealogical classification caught on, however, as the charts of horse ancestry that began to appear in the 1860s demonstrate. It seemed the natural way to show evolutionary relationships, and although there were confusions, as between branching Darwinian lineages and parallel orthogenetic ones, everyone understood the basic idea of the biological "family tree."

In the mid twentieth century, however, biologists began to criticize genealogical classification on much the same grounds as their mid-nineteenth counterparts—they thought it was too speculative. They argued that diagramming genealogies of fossil horses, for example, was treating them as though they were historical personages whose ancestry could be traced back through documents such as birth certificates. Of course, rock strata contain only bones, and although the fact that *Pliohippus* bones lay in strata below *Equus* bones had led Marsh to assume that *Pliohippus* was *Equus's* ancestor, there was no conclusive evidence for the assumption. There was evidence for a close relationship between the two genera, but nobody actually had witnessed *Equus's* "birth" from *Pliohippus.* It was possible, given the fossil record's imperfections, that *Equus* had evolved from some undiscovered genus with more similarities to it than *Pliohippus.* And, in fact, horse fossils closer to *Equus* than *Pliohippus* turned up.

Paleontologists opposed to such genealogical classifications proposed to discard assumptions of prehistoric ancestry and classify fossil creatures strictly according to the patterns of similarities ("parallelisms of structure") between their bones or other parts. An East German entomologist, Willi Hennig, had invented a technique for doing so in the 1950s. As with Darwin's "descent with modification," Hennig had given his exciting brainchild a dull name—"phylogenetic systematics," but Ernst Mayr, one of neo-Darwinism's architects, later coined a catchier term, "cladistics," from the Greek for "branch." In the 1960s, paleontologists who adopted Hennig's technique began diagramming evolution in "cladograms" that linked fossil species according to the number of "derived characters" they shared, instead of by assumed ancestry.

Rather than ascending straight up like Gaudry's horse diagrams, or branching upward lopsidedly like Simpson's, cladograms shoot off horizontally, with branches or "clades" sticking out to show derived lineages. A horse cladogram might show that *Pliohippus* and a genus called *Dinohippus,* its "sister taxon," share the most characters derived from *Merychippus. Equus,* in turn, might share the most characters derived from *Dinohippus* with its sister taxon, a South American genus called *Hippidion.* A hereditary relationship between *Pliohippus* and *Equus* might be inferred, but not necessarily a direct or close one, and it is not even assumed that *Pliohippus* preceded *Equus.* Unlike genealogies, cladograms don't show organisms' place in time, since long intervals can separate fossils with similar derived characteristics, as with the two *Procamelus* ten million years apart in Zallinger's mural.

Cladistics was a continuation of the trend toward quantified fossil analysis that had led to Simpson's synthesis. Counting derived characters and building cladograms from them lent itself well to computer modeling, and in the 1980s, the technique replaced the genealogical approach to diagramming evolution of both fossil and living organisms. Cladistic taxonomy helped to clarify many relationships, notably that between birds and theropod dinosaurs. Less famous cladistic analyses of unprecedentedly complete Mesozoic mammal fossils from China—a triconodont and a symmetrodont—suggested a tentative solution to an old problem of mammal origins. Monotremes are so different from marsupials and placentals that many paleontologists thought they evolved from creatures that had branched off from early synapsids separately—in other words, that monotremes are not really "mammals" in a strict phylogenetic sense. Cladograms based on the Chinese fossils, however, showed a series of sister taxa beginning with triconodonts and branching out to include monotremes, multituberculates, placentals, and marsupials. All apparently had a common origin in the Late Triassic—and all were thus phylogenetic mammals. Creatures that branched off earlier, although they may have looked and behaved similarly, are called "mammaliaformes."

The mammal cladograms would not have said much about origins, however, had they not been analyzed in relation to the geological data showing that mammal fossils first appear in Late Triassic rocks. After throwing out the old "family tree" diagrams, most paleontologists used cladograms as a way of testing the basic ideas of evolutionary descent that the old diagrams had illustrated. "In order to . . . construct a phylogeny, or family tree of ancestors and descendants through time," Bruce MacFadden, a modern expert on horse evolution, writes, "most cladists studying fossil groups then analyze the cladogram with regard to the stratigraphic distribution of taxa. Rather than considering the different geometries produced by the cladograms versus stratigraphically calibrated phylogenies as totally independent data sets, they find that these two methods often provide corroboration of each other."

A remark made by a colleague of MacFadden's, David Webb, seemed to typify how cladistics has affected paleontology. As we were discussing prehistoric mammal migrations between North and South America, he told me about a particularly rich fossil site in Florida and said that it "taught us more about fossil raccoons than anything else in the world and gives us, we think, a pretty good view of the ancestry—we don't use that word anymore, but you know what I mean, the primitive sister group—of the raccoons that first got into South America."

Some cladists wanted to throw out more than genealogical diagrams and words like "ancestry," however. If assumptions of ancestry were based on a false analogy between history and prehistory, then other paleontological assumptions might be as well. Indeed, radical cladists could argue that since no documentation exists for anything in prehistory, most of what evolutionists from Gaudry to Simpson have said about prehistoric events is more fictional than scientific. Such fictions might include the fundamental Darwinian notion, first conceived by Kowalevsky and reiterated in every evolutionary theory thereafter, that horses had changed with their environment, beginning in Eocene forests as small, many-toed browsers and eventually evolving into large, hoofed, Pleistocene prairie grazers. Nobody had been around to see *Eohippus* eating leaves, and no scientist had been around to see Ice Age *Equus* eating grass. One zoologist, Henry Gee, called Simpson's synthesis "useless" and claimed that cladistics had transformed paleontology "from a book of children's stories into a true science."

As though anticipating such attitudes, Simpson had dismissed cladistics as "not true" and "inane" in 1978. He was wrong about its potential as a classifying technique, and paleontologists have become increasingly leery of seeing the past through the present. "There's a problem with judging just from what survives," David Webb told me. "Every epoch we see a whole new suite of big animals, because we have a good solid record of mammals in every continent. But what we can't say is what their ecological needs were in terms of living species." Theoretical cladism didn't quite relegate Simpson's work to the children's library, however. It would have had to dig very deep to root out all paleontological traditions of drawing connections between past and present, starting with the names of most fossil organisms, which liken them to living ones in various ways.

"Whether we like it or not," writes Philip Gingerich, the paleontologist who traced whale macroevolution, "we have inherited a basic approach to animal classification from our earliest forebears, and we have been schooled in this approach from the time we first began to speak. Thus history and past experience are important components of the somewhat complicated logic dictating the structure of an effective classification." Although he acknowledges that Simpson's way of classifying mammals needed revision in light of increased knowledge of their phylogeny, Gingerich maintains that "the wide use his classification received is evidence of the wisdom of a conservative eclectic approach to classification."

Even cladists found it hard to think about prehistoric beasts without recourse to adaptive scenarios. It was one such scenario, cited by Henry Gee in his book about cladistics, *In Search of Deep Time,* that posited glyptodon-killing ground sloths. The paper proposing the idea, published in the venerable *Proceedings of the Royal Society of London,* and provocatively entitled "*Megatherium,* the Stabber," based it on an impressive technical argument that the huge ground sloth's foreclaws could have delivered a blow powerful enough to kill large neighbors. "The forearms of *Megatherium* suggest that they were designed for sudden bursts," Gee summarized. "Armed with those claws, these arms could have slashed open the belly of an adversary or overturned a 1200 kilogram truck." Some other paleontologists found a killer *Megatherium* possible. The BBC's 2001 computer animation program *Walking with Prehistoric Beasts* shows one, although its victim is not a *Glyptodon* but a saber-tooth struck down as the huge sloth, a thief as well as a killer, steals its prey.

Gee evidently cited the article because it decoupled the sloth past and present so startlingly. "It makes an odd picture; some might say an unlikely one," he wrote. "But it is unlikely only in light of our conventional assumptions about interpreting the diverse past in the light of the limited selection of animals we have around us today." Yet the example fell short from one cladistic standpoint, because cladistics tries to use as many derived characters as possible to elucidate prehistoric relationships. In making a case for extinct ground sloths' dissimilarity to living tree sloths, the article's authors failed to consider all the knowledge of ground sloth anatomy that fossils have provided.

Startling theories decoupling sloth history and prehistory aren't new. An eighteenth-century Danish naturalist, Peter Lund, thought *Megatherium* had used its powerful claws to climb around in trees. "In truth, what ideas must we form of a scale of creation where instead of our squirrels, creatures of the size and bulk of the Rhinoceros and Hippopotamus climbed up trees?" he wrote. "It is very certain that the forests in which these huge monsters gambolled could not be such as now clothe the Brazilian mountains."

Darwin called Lund's idea "preposterous," however, and Owen set out to demonstrate that it was precisely the evidence of *Megatherium's* bones that supported its role as a ground-dwelling herbivore. "Owen very properly ridiculed these fanciful theories," W. D. Matthew wrote in a 1911 article about a sloth exhibit.

> In his brilliant and masterly argument, the great English anatomist showed how the teeth were adapted to the bruising and crushing of leaves and twigs; how the structure of the jaws and skull and arrangement of the nerve channels indicated loose, flexible lips and long, prehensile tongue adapted to browsing; how the long, loose-jointed forearms would enable it to lay hold of branches and small trees and drag them down within reach; how the powerful claws would enable it to dig around the roots of larger trees and loosen them, and the massive hind quarters and tail would give the necessary weight and fulcrum to pull down these trees when loosened in order to feed upon the upper foliage thus brought within its reach.

In 1994, David Webb interpreted ground sloth habits similarly, adding further evidence of herbivory. "With no need for speed," he wrote, "sloths had the advantage of low metabolism, and they easily converted their powerful digging claws and feet into leaf and branch stripping devices. . . . By making a few modifications in their unimpressive teeth, sloths were able to chew vast quantities of leaves. . . . Despite their lack of enamel, sloths developed tall-crowned, elaborately folded teeth with tracts of a hard substance called vitrodentine to supplement the soft dentine. Although sloth teeth wore down faster than enameled teeth, they compensated by growing continuously through the animals' life."

In 1996, I encountered another bit of evidence for ground sloth herbivory at a seminar at the U.C. Berkeley Museum of Vertebrate Zoology, at which the ecologist Paul Martin handed around a clot of dung found with sloth bones in a Grand Canyon cave. As far as I could see, the dung consisted entirely of strawlike material, which had been identified as globe mallow, a plant still common in the Southwest. I did not, of course, see what had produced the dung, and even if it had come from a ground sloth, that didn't prove that all such creatures were herbivores. But it did very tangibly bring history and prehistory closer together.

"*Megatherium,* the Stabber" would have been more convincing if it had offered some evidence beside powerful foreclaws for sloth carnivory. In any case, the article is as imaginative as neo-Darwinian adaptive scenarios, and Gee's skepticism about Simpson's synthesis seems more of a critique than an alternative. "Cladistic analysis is based on the study of homologous characters in organisms," Philip Gingerich writes. "Homology itself implies that the characters in question can be traced back to the same feature in a common ancestor, again presuming a knowledge of phylogenetic relationships."

Radical cladism contained a paradoxical element of reaction against the very thing it invoked to attack Simpson's synthesis—the vast welter of biological change. Henry Gee expresses a nostalgia for pre-evolutionist thinking that Cuvier would have understood:

> Given such confusion, you can only long for the eighteenth century, when Linnaeus could classify animals and plants without worrying about evolution. The archetypes of the classical world indicated a state of pure *being,* untroubled by process or change. To a palaeontologist, this view of life has a great deal to be said in its favor. Fossils do not *do* anything, they just *are.* It makes sense to look at them in that light—for what they are—without feeling obliged to fit them into a preconceived idea of adaptive scenarios or forcing them to play the part of missing links in a sequence of ancestors and descendants created after the fact.

This truncated vision of geological time seems a direct challenge to the continuous one in the Peabody murals. "As we know, Deep Time is not a movie, but a box of miscellaneous, unlabelled snapshots." Gee writes. A cladist might prefer Charles Knight's murals to Zallinger's, since Knight restored faunas without presuming to show the epochs flowing together. An even more cladistic art form might be an abstract prehistoric landscape inhabited by bones instead of living creatures. I have seen one such tableau.

Early Miocene droughts killed so many horses, camels, and other beasts at what is now Agate Fossil Beds National Monument in Nebraska that W. D. Matthew called it "one of the greatest fossil quarries ever found in America." He estimated in the 1920s that Agate Fossil Beds had yielded over 17,000 skeletons of three species alone—a rhino, *Diceratherium,* and two other ungulates, *Dinohyus* and *Moropus.* The last two were so different from anything living that he had trouble describing them. "These extinct animals are commonly called giant pigs," he wrote of *Dinohyus,* "although they are not very piglike in appearance and were not related to the pigs any more closely than to the ruminants. They were rather tall, but compactly proportioned, with two-toed feet like a bison's, very large heads with long muzzles and large, powerful tusks. . . . These formidable beasts were probably omnivorous like the pigs and bears, but better equipped than either to pursue and attack animal prey." Matthew struggled even harder with *Moropus:* "The name 'clawed ungulate' sounds like a contradiction in terms. The *Moropus,* however, belongs unmistakably to the ungulate division. It is related, although distantly, to the horses, tapirs, and rhinoceroses, but in its case the hoofs have been changed into large compressed claws on the forefoot and smaller claws on the hind foot. The animal is as large as a modern camel."

Zallinger's early Miocene section shows *Diceratherium, Moropus,* and a *Dinohyus* relative, *Daeodon,* beside a prairie river, and they do look strange. The Park Service used a similar scene by Jay Matternes to illustrate its 1980 *National Monument Handbook,* which describes the fossil mammals by similarities, ancestral or not, to living ones. "Let's watch *Moropus* as it ambles slowly across the plain, its strange stilted walk a little like that of the modern giraffe. . . . Like his cousins the rhinos, he isn't at all bright, and he has a very short temper. . . . He walks by himself and everything else detours around him. Look down toward the river, and we may see an exception. The two meter (six-foot) high 'pig' walking away from the river, covered with mud is heading right toward *Moropus.* His name is *Dinohyus . . .* and he's just as short-tempered and stupid as *Moropus.* He looks like a giant peccary."

When I visited the Agate Fossil Beds National Monument in 1997, however, the *Handbook* contained an "errata and comments" insert that disclaimed the use of terms "such as rhinoceros-like" as "more confusing than helpful . . . focusing more on appearance than science." And a new diorama in the visitor center conspicuously avoided drawing similarities with living animals. It showed a pair of *Dinohyus* confronting a carnivore named *Daphaenodon* over a rhino carcass while an alarmed group of *Moropus* looked on, a confrontation Zallinger might have painted. (The first paleontologist to work at Agate Fossil Beds, Erwin Barbour, was a disgruntled Marsh assistant and Cope spy.) The diorama's figures were skeletons, however. Even the stylized vegetation around the water-hole setting was leafless. Vultures painted circling in a hectic sky were the only nonskeletal creatures.

The dramatically posed bones made a vivid display and gave a strong sense of the beasts' strangeness. Certainly, the 1980 handbook had been vague in likening *Moropus* to a giraffe, *Dinohyus* to a peccary, and *Daphaenodon* to a dog. Still, the skeletons hardly gave a clearer idea than the paintings of what the creatures' living appearance might have been. Some visitors might have thought they were dinosaurs, since that assumption is common about any big fossil skeleton. The diorama wasn't even as decoupled from the present as it seemed, because it labeled the beasts with scientific names that, although few people could translate them, were as historically descriptive as the handbook's text. *Morupus* means "sloth foot," *Dinohyus,* "terrible pig." And despite its bare bones accuracy, the display's dominant impression was as much of macabre wonder as of scientific parsimony, trading the paintings' edenic ambience for a plutonic one.

Of course, a bare bones view of prehistory is basic to paleontology. From Cuvier on, anatomists had to make skeletal reconstructions before they could try "living" restorations. Indeed, the display Matthew was describing when he invoked Owen against Lund's arboreal giant sloths was of bones. "The skeletons are grouped around a tree trunk, in poses indicating the supposed habits and adaptions of the living animals," he wrote enthusiastically. "The Ground Sloth Group is the most realistic that has yet been attempted in the mounting of fossil skeletons, and the method of mounting, eliminating the upright steel rods ordinarily used, adds much to its effectiveness." David Webb described a more recent American Museum exhibit with equal enthusiasm, saying:

> They have a wonderful array of whole skeletons of all kinds of shelled and hairy edentates, and it's done sort of purist. They're just skeletons. They're more or less in lifelike poses, though by most people's standards, edentates aren't very lively, they just sort of sit there. But they've done a nice job—some of the bigger sloths are really reaching up in interesting ways. It's a very dark exhibit, so you have to kind of peer in and let your eyes adapt. Maybe the darkness is a good thing—it opens up your imagination.

According to Matthew's publicity-minded boss, however, skeletal restorations just weren't as effective as living ones. "Few persons are able to form an adequate idea of an animal from its skeleton," Henry Fairfield Osborn complained.

> It was soon found that very few visitors to the Hall of Vertebrate Paleontology appreciated the wonderful story told by the fossil skeletons of the past life of this continent, and in order to increase the educational value and the attractiveness of the hall in this respect, Mr. Charles R. Knight, the well-known animal painter, was invited to undertake the restoration of some of these animals . . . these restorations are regarded as working hypotheses which are of scientific value only in conveying a general idea of the external form and appearance, but they are of very great popular educational value since they serve to interest and attract the public.

Despite its undeniable storybook aspects, I think, many paleontologists sympathize with the popular desire to see the dead restored to a semblance of life. An interesting thing about Zallinger's mural is that, lifelike as it is, it doesn't sin against cladistics by presenting imaginative fossil genealogies, or even much in the way of adaptive scenarios. *Megatherium* isn't eating anything, animal or vegetable, although *Glyptodon* may be feeding on lichens. *Hyracotherium* isn't browsing, and *Equus* isn't grazing, although some distant *Mesohippus* and *Merychippus* have their muzzles in the grass.

The mural mainly just shows a kind of four-dimensional landscape with figures, and although one may surmise that *Hyracotherium* eventually evolved into *Equus* as tropical forests evolved into glacial grasslands, it doesn't insist on the idea. It simply joins the Eocene's landscape and fauna to the Pleistocene's in a seamless way that is hard to analyze. Although a product of art rather than nature, the mysterious seamlessness seems to me a deeper image of geological time than snapshots, living or skeletal. Movies are successions of snapshots. Zallinger's *Age of Mammals* is something else.

SIXTEEN

Exploding Faunas

A THIRD CHALLENGE TO SIMPSON'S SYNTHESIS came closer than continental drift to reprising the unexpected obstacle that Lord Kelvin threw in Darwin's path in 1866. Once again, an eminent physicist blindsided evolutionary theory, and in a monolithic way similar to Kelvin's brusque calculations of planetary cooling. Indeed, the new assault was literally monolithic. It flung a Mount Everest–sized interplanetary rock at neo-Darwinism's finespun web of environmental change, mutation, and adaptation.

As Simpson's science fiction novel showed, neo-Darwinism in the mid-twentieth century was not that much more confident than paleo-Darwinism had been about explaining the apparently abrupt break between the Mesozoic and Cenozoic. Trying to match the end of Zallinger's reptile mural with the beginning of his mammal one shows the difficulties. Despite continuities between the eras—swamps, palms, flowering trees—the jump from the Cretaceous horned dinosaur *Triceratops* to a dog-sized Paleocene mammal named *Psittacotherium* ("parrot beast") seems unnatural. As if to emphasize the mystery, Zallinger painted the mammal (named by Cope for its beaklike jaws) half-hidden in forest shade, and made nearby, even smaller ones hard to tell from roots or clods. He did include an element of continuity in *Cimolestes* ("fetus thief"), the snouty mammal at the reptile mural's end, but it was an afterthought, added in the 1970s, and, despite the suggestive name, he didn't show it doing anything related to dinosaur extinction.

Both murals hint at environmental change as a cause for the break. Cretaceous volcanoes erupt in the reptile mural, and the subtropical vegetation is sparse, as though growing on a lava flow. Similar peaks rise in the mammal mural's Paleocene background. Still, the volcanoes alone hardly explain the faunal jump, and the only suggestion of climate change between the periods is that the Cretaceous flora's sparseness makes it look drier. The murals seem to justify Simpson's fictional gloom about finding a solution to the problem.

There was a ray of light when Zallinger painted little *Cimolestes*, however. Although sequences of strata spanning the transition were "either scrappy or nonexistent" in most of the world, they had turned up in one place. "Western North America is one of the regions in which relatively continuous geological sections of fossiliferous upper Cretaceous and Paleocene sediments are preserved," wrote two vertebrate paleontologists who had been digging there, William Clemens of U.C. Berkeley, and David Archibald, then of Yale. (Archibald advised Zallinger about *Cimolestes*.) They described "a broad complex of flood plains" that had extended from southern Alberta to western New Mexico in the Late Cretaceous, bordering a receding inland sea to the east. "Deposition of new sediments on these flood plains was intermittent, as is the case with terrestrial strata. However, in many parts of the region, particularly the northern, at the Cretaceous-Tertiary boundary there is no evidence of exceptionally long periods of non-deposition, or other kinds of unconformities, comparable to those typical of sections of marine sediments."

Researchers had found a great diversity of vertebrate fossils in the western floodplain sediments—fish, salamanders, lizards, turtles, snakes, and birds, as well as dinosaurs and mammals. Beginning in the 1950s, they had used a new technique—screen-washing—to study in detail how faunas had changed. It was arduous, requiring the processing of vast amounts of sediment through fine-mesh screens to find small fossils. When I briefly tried it during my visit to the Hagerman horse beds in Idaho, I found no fossils, equine or otherwise, and I have seldom felt so worn out after a few hours of apparently moderate work. But screening gave a much more complete picture of animals' relative abundance through the Cretaceous-Tertiary transition's millions of years.

In a remote part of northeast Montana, described by Clemens as "miles and miles of miles and miles," paleontologists found some particularly encouraging fossils at Hell Creek, a badlands named for its summer temperatures. Barnum Brown had made its dinosaurs famous when he dug up the first, partial *Tyrannosaurus* skeleton there in 1902, hauling it 124 miles across the plains to the railroad. But it was a less spectacular aspect that made the area precious for Simpson's synthesis. Hell Creek had the world's most continuous known late Cretaceous to early Tertiary land strata. The region had been a swampy lowland bordering the receding sea during that time, and streams and ponds had left a virtually unbroken sequence of sediments. Its unique fossil record offered just the detailed picture of vertebrate population shifts that paleontologists were seeking.

One thing that the fossils in the area's Cretaceous strata, called the Hell Creek Formation, suggested was that dinosaurs had been decreasing gradually. Thirty dinosaur species were known to have lived ten million years before the formation's deposition, but Hell Creek itself apparently contained fewer than twenty. Within the formation, moreover, dinosaurs seemed to decrease in abundance and diversity from the lower to the upper beds. *Triceratops,* for example, seemed to be ten times more numerous near the formation's bottom than its top. No dinosaur fossils at all were known in the formation's top ten feet.

Dwindling quantities of dinosaur fossils weren't unique—other western formations revealed the same phenomenon. But something even more suggestive emerged from Hell Creek's vertebrate fauna. The mammal fossils found on the plains had hitherto been the typical late Cretaceous kinds—multituberculates, marsupials not unlike today's opossums, and eutherians like little *Cimolestes.* In 1965, however, Robert Sloan, a University of Minnesota paleontologist, and Leigh Van Valen, then of the American Museum, reported in *Science* that they had found unexpected beasts at a place in the Hell Creek formation picturesquely called the Bug Creek Anthills. It was, they wrote, "the earliest and lowest of three newly discovered distinct Cretaceous mammal faunas . . . that are transitional in community structure from characteristic, previously known, Late Cretaceous faunas to those of the early Paleocene. They contain the earliest North American species of four families previously thought to be restricted to the early Tertiary."

The Bug Creek fossils suggested that mammals had been getting more diverse and advanced while dinosaurs dwindled, a departure from the widely held notion, implicit in Simpson's novel, that they had remained archaic until dinosaurs vanished. Two of the new families found at Bug

Creek and the other sites were advanced kinds of multituberculates, and two were surprisingly modern eutherians. Sloan and Van Valen identified one of those as a specialized group of insectivores called leptictids, and the other was something new—an apparent ancestor of living hoofed mammals that they named *Protungulatum* ("first hoofed"). Of one species, they wrote: "*Protungulatum donnae* is presently the oldest genus and the oldest known species of ungulate. Five additional species, that can be closely related to this genus or to closely related genera, are now known from the Cretaceous. . . . Nothing of the known anatomy of *P. donnae,* dental or otherwise, precludes it from being an ultimate ancestor in the real rather than structural sense for the various orders of ungulates."

Something not unlike Osborn's legendary five-toed horse had apparently lived right alongside *Tyrannosaurus rex*. And more traces of unexpectedly advanced late Cretaceous mammals emerged. In 1969, Jason Lillegraven, a paleontologist then at the University of Kansas, reported finding "a definite taxonomic radiation of rapidly evolving placental mammals" in Alberta's late Cretaceous Edmonton Formation. Lillegraven didn't find *Protungulatum,* and he questioned some of Sloan's and Van Valen's classifications, but he thought his Edmonton fossils included ancestors of modern placentals such as carnivores and rodents. He drew a Simpsonian inference about this apparently sudden appearance of advanced mammals in otherwise conservative North American Cretaceous fauna. They might have arrived over a filter bridge opened by receding seas. "The two basal stocks appear to have been independently derived from Asiatic ancestors," he wrote. "Geological evidence suggests the possibility of faunal exchange across [the] Bering strait in Late Cretaceous time."

Screen-washing the sediments at Hell Creek and Edmonton was like putting the Late Cretaceous under a microscope, and finding a pattern that had been invisible. Instead of changing abruptly, Cretaceous and Paleocene faunas seemed to intergrade, and in a way consonant with the neo-Darwinian prediction that environmental change caused faunal change. As the North American inland sea receded, weather would have become more seasonal, vegetation less subtropical, and dinosaurs less common and diverse. The cooling also would have affected mammals, as more cold-adapted groups immigrated from the north Asian landmass, competing with natives. Some archaic North American mammals survived to leave fossils in overlying Paleocene strata, called the Tullock Formation, but most did not. Asian multituberculates largely replaced endemic ones, and marsupials nearly disappeared. The new mammals even might have competed with the already dwindling dinosaurs by using plant foods more efficiently or preying on their young.

"How did the Cretaceous world end?" Clemens and Archibald asked in 1981, summing up the scenario that had emerged from western fossil beds.

> During the latest Cretaceous most of the world's seas were being restricted in area while new regions and routes of dispersal were open to terrestrial organisms. The general cooling of climate . . . could well have been the product of changes in area and circulation patterns of the oceans and increase in continental areas. . . . These paleobiological data suggest the Cretaceous-Tertiary Transition was a period of several tens of thousands if not hundreds of thousands of years in duration, characterized by interaction of a complex of physical and biological factors producing a high net rate of decrease in biotic diversity.

It was a thoroughgoing scenario, supported by three decades of work, and also co-authored by a paleobotanist then at Yale, Leo Hickey. It was not unchallenged in 1981, however. Indeed, it was presented in response to a new theory that proposed to solve the dinosaur extinction mystery while sidestepping Simpson's synthesis.

A geological colleague of Clemens's at Berkeley had started it unintentionally. Walter Alvarez had been trying to date the Cretaceous-Tertiary boundary in marine sediments in the 1970s, a study that included measuring residues of iridium and other rare elements. Iridium mainly falls from outer space to the earth's surface in minute traces, and Alvarez and his co-workers hoped to use them to measure how long the sediments had taken to accumulate. What they found surprised them. Above and below the Cretaceous-Tertiary boundary, the sediments contained the expected minute traces. In sites in both Italy and Denmark, however, iridium concentrations at the boundary were far greater than expected, and this iridium "spike" suggested that much more of the element than normal somehow had arrived from space at that time. Since heavenly bodies such as asteroids and comets are rich in iridium, the team inferred that the spike was the result of the earth's collision with one at the end of the Cretaceous.

Walter Alvarez's father, Luis, a 1966 Nobel laureate in physics who liked to explore other disciplines, contributed greatly to the conclusion. He also put his scientific prestige behind a further inference. Reporting their results in a famous 1980 paper in *Science,* the Alvarezes theorized that the collision could have caused the extinction of dinosaurs and other organisms by catastrophically, if briefly, changing the earth's environment. A thick pall of toxic dust from an explosion thousands of times more powerful than a hydrogen bomb might have enveloped the globe, caused months of icy darkness, and prevented algae and plants from photosynthesizing. Without plant growth, many food webs would have collapsed, with particularly disastrous effects on large, active animals such as dinosaurs and marine reptiles, the biggest ones to disappear. Effects would have been less lethal on smaller creatures, which could take refuge, migrate quickly, go dormant, live on stored food, or otherwise dodge the catastrophe and the ensuing months of "global winter"—creatures like the mammals, birds, turtles, lizards, crocodilians, frogs, salamanders, and fish that survived.

The Alvarez theory sidestepped not only Simpsonian ideas of gradual, environment-driven extinction but historic paleontological assumptions of planetary integrity. It is hard to imagine even the "catastrophist" Cuvier regarding mass extinction by cosmic blast with equanimity, and the idea of a wayward comet blithely erasing eons of diligent adaptation would have appalled the industrious Victorians. "[S]o profound is our ignorance, and so high our presumption," Darwin wrote in the *Origin*, "that we marvel when we hear of the extinction of organic beings; and as we do not see the cause, we invoke cataclysms to desolate the world." For all his agnosticism, Darwin retained a sense that mammals are a higher life form than dinosaurs were. "The inhabitants of each successive period in the world's history have beaten their predecessors in the race of life," he wrote, "and are, insofar, higher on the scale of nature." If mammal succession had merely hinged on a falling mountain, even that sense of progress—a last vestige of unhappily relinquished Christianity—might have foundered.

Although neo-Darwinism didn't retain faith in inevitable progress, adaptation remained central, and Darwin's sense that adaptation was "going somewhere" lingered. Simpson's novel manifests this not only in its narrator's hopes for the future of Cretaceous mammals, but in its disenchantment with dinosaurs. The only real discovery Magruder makes about them is that they are *not* warm-blooded, contradicting what younger paleontologists were saying in the 1970s. "Now that I am living among dinosaurs, I know that they are all cold-blooded in the usual sense of the word," he concludes. "The idea that they might be . . . warm-blooded (homiothermal) had . . . a lot of publicity . . . just because it was touted as sensational news. There really was no evidence for it, and when it ceased to be news it was quietly dropped."

Simpson evidently found the asteroid theory's catastrophist implications no more credible than dinosaur warm-bloodedness. He dismissed them as summarily, writing in 1982 that the evidence for gradual extinction made it

> highly unlikely that the mass extinction resulted from one sudden extra-terrestrial event, such as the flare of a distant nova. . . . Still less likely is a hypothesis that is the latest at the time of this writing: that a large asteroid or meteorite crashed into the earth and raised such a dust cloud that the sun was obscured and photosynthesis was stopped for an unstated length of time (but one so short that it must appear as instantaneous in the fossil record). This has been forcefully advanced by Luis W. Alvarez, a physicist, but is viewed with doubt by most paleontologists.

A public used to the idea (albeit not the reality) of nuclear holocausts and star wars found the impact theory sensationally credible, however. *Science* published the Alvarezes' 1980 paper with minor cuts and some quibbles by a paleontologist referee, David Raup of the University of Chicago. Raup became a supporter, as did many paleontologists, particularly invertebrate specialists like him. The theory also explained a Late Cretaceous disappearance of myriad small marine organisms, since a die-off of photosynthesizing algae during months of global darkness would have destroyed their food source. Indeed, some scientists seemed to have been waiting for just such a theory to emerge. "The remarkable aspect of the Alvarezes's work—the part that has produced budding excitement among my colleagues rather than the ho-hum that generally accompanies yet another vain speculation—lies in their raw data on enhanced iridium at the top of the Cretaceous," Stephen Jay Gould, another invertebrate paleontologist, wrote. "For the first time we now have the hope (indeed the expectation) that evidence for extraterrestrial causes of mass extinction might exist in the geological record. The old paradox—that we must root against such plausible theories because we know no way to obtain evidence for them—has disappeared."

Beside being based on evidence, the theory was elegantly parsimonious, killing any number of Cretaceous birds with one huge stone. Dramatic simplicity made it popular. "The excitement was extraordinary," Richard Fortey, a British expert on trilobites, recalls. "All around the world evidence could be drafted in to support the meteorite theory as section after section revealed an iridium anomaly. . . . The possibilities of jobs and grants seemed to bob to the top of this ferment like dumplings in a rich stew." Skeptics found themselves on the defensive, outflanked in a blitzkrieg of theory. "The ingredients for a juicy controversy were in place," Fortey observes. "A radical new interpretation versus an onslaught from what could be labeled as a phalanx of fuddy-duddies and traditionalists."

Vertebrate paleontologists who worked at Hell Creek like Clemens and Sloan were the most prominent doubters, and they put up enough resistance to make the exchange a heated one. While not challenging the evidence for an asteroid impact in their 1981 summation, Clemens, Archibald, and Hickey attacked the Alvarezes' hypothesis that it had caused the Late Cretaceous mass extinctions. The hypothesis, they claimed, rested on three questionable assumptions: that the extinctions had been simultaneous; that they had affected successful, expanding groups of organisms; and that they had exterminated organisms because of shared characteristics, for example, the large size common to dinosaurs and marine reptiles. The three authors used the uniquely continuous strata at Hell Creek to attack those assumptions.

"The pattern is difficult to explain in terms of a sudden, all-encompassing catastrophe," they maintained. The Hell Creek dinosaurs had apparently disappeared gradually before the impact, while a mass plant extinction indicated by a change in fossil pollen seemed to have occurred well after the dinosaurs' disappearance, although before the impact. Some small mammals such as marsupials and multituberculates predicted by the asteroid theory to have survived an impact had also undergone widespread extinction in the late Cretaceous, apparently after the dinosaur die-off. "The multiplicity of patterns of extinction strongly argues against any hypotheses invoking some kind of catastrophic, short, sharp shock as the causal factor of the terminal Cretaceous extinctions," Clemens et al. concluded. Following the pattern of other students of extinction by butchering T. S. Eliot's lines, the outcome of these evaluations is clear:

This is the way Cretaceous life ended . . .
Not abruptly but extended.

Asteroid advocates quickly counterattacked—the scientific stakes were high. As a geologist, James Lawrence Powell, observed, "the Alverez theory would remain merely a scientific curiosity" if a Late Cretaceous impact had not killed off 70 percent of species. But if it had killed that many or more, the life sciences would never be the same. "Our conception of the role of chance in the cosmos, our view of life and its evolution, our understanding of our own place—each would be irrevocably altered."

In October 1982, Luis Alvarez devoted much of a lengthy talk at the National Academy of Sciences to arguing against Clemens, with whom he had been conferring. Clemens had helped the Alvarez team to find iridium at Hell Creek, and they had found it in a stratum, the "Basal Z Coal," that generally was considered the Cretaceous-Tertiary boundary. But Clemens thought that the dinosaur and plant extinctions had occurred before the Basal Z Coal was deposited. Alvarez denied this in several pages of technical math and testy rhetoric. "[O]ur preferred scenario is tied solidly to a well documented catastrophe that is the most severe event of which we have any record," he said. "I really cannot conceal my amazement that some paleontologists prefer to think that the dinosaurs, which had survived all sorts of severe environmental changes and flourished for 140 million years, would suddenly, and for no specified reason, disappear."

Alvarez ended with arguments that he called "overwhelmingly convincing" as to the "precise synchronicity" of the various extinctions, and deemed it unlikely that such simultaneity had been a coincidence unrelated to an asteroid impact. "In physics," he concluded, "we do not treat seriously theories with such low *a priori* probabilities. (But if you look closely at the writings of Archibald, Clemens, and Hickey, you find that they do not really have a viable competing theory—one that explains some reasonable fraction of the observational data. I think it is correct to say that their theory is that our theory is wrong!)." He then tried to palliate this breathtaking snub of Simpson's neo-Darwinism by deploring his 1866 precursor in the Darwinism-snubbing business. "And finally, if you feel that I have been too hard on my paleontologist friends and have given the impression that physicists always wear white hats," he said, "let me remind you of a time when our greatest physicist, Lord Kelvin, wore a black hat and seriously impeded progress in the earth sciences."

Hell Creek and other western sites became ground zero for impact researchers after this intellectual blast. In 1981, geologists found an iridium "spike" at the Cretaceous-Tertiary boundary of the Raton Basin in New Mexico, and noted that fern spores temporarily replaced angiosperm pollen at the boundary. Since ferns replace higher plants after historic volcanic explosions, they saw this as evidence for a major plant extinction at the boundary instead of before it, as Leo Hickey had maintained. Such evidence eventually led Hickey to abandon the gradualist position. In a 1988 paper, he and another paleobotanist, Kirk Johnson, reported studying 2,500 plant fossils at two hundred Rocky Mountain locations and finding that about 79 percent of Cretaceous plants had disappeared at the Tertiary boundary, simultaneously with a temporary predominance of fern pollen. "I became a believer," said Hickey in 1991. "This evidence is incontrovertible: there was a catastrophe. We really hadn't been looking at the record in enough detail to pick this extinction up, and we weren't disposed to look at it as a catastrophe. I think maybe that mind set persisted a little too long."

In 1984, two paleontologists, Jan Smit and S. Van Der Kaars, claimed in *Science* that *Protungulatum* and other advanced mammals that Sloan and Van Valen had attributed to the late Cretaceous had really lived in the Paleocene, and not alongside *Tyrannosaurus.* They concluded that the mammal sites at Bug Creek Anthills and similar locations only seemed to be of Cretaceous age because erosion had "reworked" dinosaur teeth and other genuine Cretaceous fossils into them. Rains had washed such dinosaur fossils out of older formations and carried them into the Paleocene stream courses that actually formed the Bug Creek sediments. *Protungulatum* and its associates were really Tertiary mammals, unimplicated in the extinction of either Cretaceous marsupials or dinosaurs.

Some vertebrate paleontologists resisted the claim. In 1986, Sloan, Van Valen, and two other paleontologists, Keith Rigby and Diane Gabriel, published a *Science* article upholding a gradualist scenario at Hell Creek. "Data supporting the beginning of the ungulate radiation well in advance of dinosaur extinction, the end of the Cretaceous, and the impact 'extinction' event are substantial and conflict with the account of Smit and van der Kaars," they maintained. Others were less sure. In 1996, David Archibald wrote that although he had "accepted the conclusion that the Bug Creek localities were Cretaceous in age," he had not found any such localities when he did his doctoral work at Hell Creek in the 1970s. "Rather, the localities even up to within about ten feet of the formational contact and the K/T boundary yielded only mammals that were typical of the late Cretaceous." He concluded that "a Paleocene age for all the Bug Creek sites is the majority opinion," although "definite evidence remains elusive, and shall remain elusive."

Luis Alvarez's National Academy talk already had challenged the other gradualist evidence at Hell Creek, the dinosaurs' apparent disappearance ten feet below the Cretaceous-Tertiary boundary. Increased fossil-hunting offered a further chance to test what was called the "ghastly gap," and Peter Sheehan, an invertebrate paleontologist at the Milwaukee Public Museum, took it. He and his colleagues organized volunteers who paid $800 apiece to spend a summer combing Hell Creek for fossils, leading

Sloan et al. to comment hopefully in their 1986 paper that the "Dig-a-Dinosaur project" would supply "a large quantitative base" with which to verify the ghastly gap's existence.

The volunteers collected 2,500 dinosaur fossils in the formation from 1987 to 1989, but their results disappointed the gradualists. According to Sheehan, they found dinosaur fossils within 60 centimeters of the Cretaceous-Tertiary boundary, and no real difference in dinosaur density from the lowest Hell Creek strata they searched to the highest. "Because there was no significant change between the lower, middle, and upper thirds of the formation," he wrote in a 1991 *Science* article, "we reject the hypothesis that the dinosaurian part of the ecosystem was deteriorating during the latest Cretaceous." A 1992 PBS documentary showed another volunteer team, led by Sheehan's associate David Fastovsky, finding a dinosaur claw near the K-T boundary in southeast Montana. "There doesn't seem to be any evidence for a gradual decline in their diversity," Fastovsky said.

Gradualists persisted, however, and got a plug in some 1989 *National Geographic* Dig-a-Dinosaur coverage, which described a team finding a marsupial jawbone as "the most excitement that day." The article quoted Diane Gabriel saying that the creature had been "slightly larger than a chihuahua—a giant for its time," and it paraphrased Robert Sloan's view of Late Cretaceous mammal significance:

> Sloan believes that 200,000 years before the K-T boundary a receding sea level had created a land bridge between North America and the long-isolated Asian continent. A plague of little Asian mammals invaded North America and began eating the same flowering plants that most dinosaurs ate.
>
> "The mammals ate much less food per animal," says Sloan. "But there were so many of them. They ate the last of the dinosaurs out of house and home."

A *New York Times* article on the Society of Vertebrate Paleontology's 1985 annual meeting gave the impression that gradualists felt oppressed, however. "Even among scientists," Robert Sloan told the reporter, "bias toward the impact hypothesis is often strong because catastrophe theories are very seductive." Robert Bakker expressed his frustration more vehemently: "The arrogance of these people is simply unbelieveable. They know next to nothing about how real animals evolve, live, and become extinct. But despite their ignorance, the geochemists feel that all you have to do is crank up some fancy machine and you've revolutionized science. . . ."

In effect, they're saying this: 'We high tech people have all the answers and you paleontologists are just primitive rock hounds.'"

An informal survey of meeting attendees found that "very few" accepted the impact extinction theory, although most believed there had been an impact at the Cretaceous-Tertiary boundary. The article focused on William Clemens's interpretation of dinosaur bones he had collected on Alaska's north slope. Although the Cretaceous climate there had been relatively warm, without freezing temperatures, the site had nonetheless been far north, with four months of winter darkness. The *Times* quoted Clemens:

> The challenge in this is to discover how dinosaurs adapted to the rigorous seasonal regime of daylight and darkness in the polar region. I don't believe they survived by migrating twice a year, because the distances would have been too great for them. With the photosynthetic growth of plants cut off by darkness, herbivorous dinosaurs may just have shut down for the winter, or found some other way to reduce their food requirements. But survive they did, as we see in the fossil record. What does this say about the impact theory of extinction? It's codswallop.

Some time later, Luis Alvarez responded with a vehemence that rivaled Bakker's, and oddly echoed the *New York Herald*'s Cope-Marsh invective. The *Times* reported that he called Clemens "inept at interpreting sedimentary rock strata" and paleontologists generally "not very good scientists . . . more like stamp collectors." Gad! The *Times* reporter seemed less gratified by such sensational fulminations than his 1890 *Herald* counterparts, however, and voiced mild impatience with the ongoing scientific failure to agree on a journalistically satisfactory solution. "The debate has crystallized," he concluded, "into a conflict between opposing camps whose partisans rarely seem to change their minds or soften their positions, whatever the objective evidence may be."

After his father's death, Walter Alvarez tried to be judicious when interviewed for the 1992 PBS documentary in which David Fastovsky denied gradual dinosaur decline. "There was a little more to evolution than Darwin realized," Alvarez said. "Darwin had it right, but he didn't have the whole story. Darwin said that a species has to be . . . well adapted, and we're finding that that's true, but there's an occasional really bad day when a huge rock falls out of the sky and makes terrible disturbances." But there was a little more to the impact controversy than that.

SEVENTEEN

The Revenge of the Shell Hunters

INVERTEBRATE PALEONTOLOGISTS LIKE STEPHEN JAY GOULD had good reason to welcome the Alvarezes' extinction theory. It supported a new evolutionary paradigm that some of them had initiated, one that questioned not only Darwinian ideas about progress and competition but many other assumptions about life that paleontology had accumulated. The paradigm had already gained wide acceptance when the impact extinction theory emerged. Mammals proved a peculiar stumbling block for it, however.

Vertebrate paleontologists had tended to dominate evolutionary theory ever since Cuvier's time. Darwin, who spent years becoming the world's fossil barnacle expert, was a great exception to this, but Darwin was a great generalist, using any handy organism to support his ideas. Specialization prevailed after 1870, and while invertebrate experts like Charles Schuchert and C. D. Walcott contributed to evolutionary thought, the likes of Marsh and Osborn eclipsed them. Cope's associate, Alpheus Hyatt, was equally important in developing neo-Lamarckism, but Hyatt was a mollusk specialist, and Cope's amblypods and creodonts got more attention. Simpson, with typical self-confidence, thought vertebrate paleontologists prevailed in theorizing because they were "more biologically oriented" and had "particularly good bodies of pertinent and well analyzed data."

Invertebrates have always been paleontology's bread and butter, however. Omnipresent fossil shells became the main way of identifying and dating rock strata when a civil engineer, William Smith, began using them for that purpose in the 1790s, and they increasingly had economic applications, not least in the fossil fuel industry. Such things translate into jobs, and by the mid twentieth century, most paleontologists, industrial or academic, specialized in invertebrates. Indeed, Simpson estimated in 1960 that there were only about 130 professional vertebrate paleontologists in the world. As computers allowed fossil-counting on an unprecedented scale, dizzyingly abundant shells and carapaces increasingly attracted young researchers with theoretical ambitions. Gould's colleague Niles Eldredge, an American Museum curator, recalls:

> In the 1960s, I was one of a number of graduate students interested in evolutionary paleontological research. . . . it was invertebrates that caught the eyes of most of us, since invertebrate fossils are found in great profusion. . . . So naturally we thought it would be far easier to study evolution if we had large samples of brachiopods, or clams, or snails—or in my case trilobites. . . . Basically, the evolutionary pattern we expected to find was Darwin's originally predicted pattern; slow, steady gradual changes within species as we looked up the geological column.

Eldredge found, however, that trilobite species remained largely the same over millions of years, although new, slightly different species occasionally appeared as a result of geographic isolation. Gould perceived similar patterns in snail evolution, and in 1972 they co-authored a famous paper proposing the theory of punctuated equilibria, which maintains that the fossil record's apparent pattern of long stable periods occasionally broken by abrupt appearances of new species is not an artifact of missing rock strata, as Darwin thought, but actually the way evolution really works. According to the theory's "equilibria" part, well-adapted species tend not to change unless their environment does. According to its "punctuation" part, when environmental changes do disturb adaptation, new species evolve so quickly that transitional forms tend not to appear in the fossil record. Richard Fortey, a trilobite expert, observes:

> [Eldredge and Gould's] new theory was set up in opposition to the notion of "gradualism," a slow and more or less continuous change or shift that nudged whole populations toward the new species. This was considered to be the dominant model for evolution in the aftermath of the "modern synthesis" of evolution in the 1930s—and a rather supine acceptance of

this creed ensured that the "punctuated" view, when it appeared, was heralded as startlingly novel.

At first, Eldredge and Gould applied the theory mainly to individual species, but they later decided that entire ecosystems remain in stasis until environmental changes cause episodes of rapid evolution. Eldredge notes:

> Stability is the norm until a physical environmental event (such as a spurt of global cooling) disrupts regional ecosystems sufficiently that a threshold is reached, and relatively suddenly many species become extinct. Only after such spasms of extinction do we find much evidence of speciation— i.e. actual evolutionary change. What then drives evolution? To my mind, and to many of my colleagues as well, it seems clear that nothing much happens in evolutionary history *unless and until physical environmental effects disturb ecosystems and species.* Competition among genes and organisms for reproductive success is important, but in and of itself insufficient to create the dominant patterns we encounter when we examine the history of life.

Punctuated equilibria accepted the basic neo-Darwinian mechanism of evolutionary change—natural selection of adaptive genetic mutations. But it said that the mechanism operates mainly at times of environmental change rather than continually. The theory thus deemphasized the explicit competition in both paleo-Darwinism and neo-Darwinism, as well as the implicit progress. Both older paradigms saw natural selection as a process continually at work seeking new avenues of increased fitness, not one that switches off when an organism successfully adapts to its environment.

Simpson seemed not to grasp punctuated equilibria's radical implications, and he called it "only a restatement. . . . of points made, in essence, long before"—points made by Simpson, in fact, when he had theorized that accelerated "quantum evolution" could occur in some organisms during times of environmental change. The implications became evident when the asteroid theory hit, however.

"I used to argue," Gould wrote in 1981, " that mass extinctions might be rip-roaring fun to discuss but were relatively unimportant for the ultimate disposition of life and its history. I was then caught up in some common prejudices about inherent, stately progress as a hallmark of life's history. . . . I can't think of any other idea I once held (after age five) that I now regard as so foolish. . . . The history of life has some weak empirical tendencies, but it is not going anywhere intrinsically. Mass extinctions do not merely reset the clock; they create the pattern. They wipe out groups that might have prevailed for countless milleniums to come and create ecological opportunities for others that might never have gained a footing. And they do their damage largely without regard to perfection of adaptation."

Punctuated equilibria and the asteroid theory put a new spin on extinction. According to both paleo- and neo-Darwinism, extinction is a process that, like natural selection, operates continuously as fitter species edge out the less fit. Darwin imagined the living world as a huge slab into which, like splitting wedges, natural selection had driven thousands of species. As selection drove any particular wedge deeper, competition would loosen adjacent ones, and some would drop out—become extinct. But this metaphor didn't prevail in Gould's and Eldredge's model, wherein adapted species interlocked to form stable ecosystems until environmental change, not competition, dislodged them.

David Raup, the University of Chicago paleontologist who had refereed the Alvarezes' 1980 impact paper, took the new extinction paradigm a step further. Whereas Eldredge and Gould had seen earthly disruptions such as climatic or sea level changes as the main equilibria punctuators, Raup pointed out that most common, widespread species are resilient to such things, having evolved in a dynamic biosphere. Many tropical species won't go extinct when the climate cools, for example, because they can retreat to persisting tropical areas. Raup concluded that "for geographically widespread species, extinction is likely only if the killing stress is one so rare as to be beyond the experience of the species, and thus outside the reach of natural selection." In effect, extinction was not an intrinsic part of the evolutionary pattern, but something imposed from outside. And what could be more extraneous than asteroid impacts? By this logic, indeed, the *only* thing that could cause a mass extinction like the dinosaurs' was something like an asteroid.

Raup and his colleagues at the University of Chicago extended their analysis to mass extinctions other than the Cretaceous one. David Jablonski found no significant correlations between them and earthly disruptions such as sea level changes or rising mountains, but he did find evidence— high iridium levels, craters, shocked quartz—of correlation between them and extraterrestrial impacts. Jack Sepkoski produced a data base suggesting that mass extinctions might even occur on a regular basis as the earth's orbit passes through asteroid belts or other clusters of heavenly bodies. The analysts posited a 26-million-year cycle of extinction peaks.

The fortuitous fit between punctuated equilibria and the Chicago mass extinction studies seemed almost as elegantly parsimonious as the Alva-rezes' theory itself. "A statistical analysis of all the data that Sepkoski has amassed on families in the fossil record really does seem to substantiate the cyclical repetition of mass extinction events—some bigger than others, of course," Niles Eldredge wrote. "The correlations are not perfect, but they seem to be good enough to have inspired the search for a cause of the ex-tinction." Stephen Jay Gould was more effusive. "In short," he wrote, "if mass extinctions are so frequent, so profound in their effects, and caused fundamentally by an extraterrestrial agency so catastrophic in impact and so utterly beyond the power of organisms to anticipate, then life's history either has an irreducible randomness or operates by new and undiscovered rules for perturbations, not (as we always thought) by laws that regulate predictable competition during normal times."

The concept of heavenly projectiles as life's driving force would have in-trigued an oenophilic Persian astronomer of nine centuries ago:

> The ball no question makes of ayes and noes,
> But right or left as strikes the player goes;
> And He that tossed thee down into the field,
> *He* knows about it all—He knows—HE knows!
> Edward FitzGerald, *The Rubáiyát*
> *of Omar Khayyam,* 50

Of course, it was one thing to expound the paradigm, which does have its Rube Goldberg side, and another to find conclusive evidence for it. As-tronomers weren't able to discover an extraterrestrial cause for the Sep-koski's 26-million-year mass extinction cycle. And even if trilobites and snails did not progress through competition, but simply readapted them-selves after periodic impacts, did that mean that all organisms have evolved that way?

Some vertebrate paleontologists thought not. Philip Gingerich soon published several papers challenging Eldredge's and Gould's ideas. Statisti-cally studying Paleocene and Eocene mammal lineages, he found that ap-parent examples of punctuated equilibria in the microevolution of genera like the Peabody mural's little primate, *Pelycodus,* resulted from a relatively small number of specimens used in earlier studies. Analyzing 255 speci-mens of *Pelycodus* jaws and teeth collected from throughout several thou-sand feet of strata, Gingerich concluded that the genus exhibited "a rela-tively long period of continuous gradual phyletic change." He found five

Pelycodus species "connected by a continuous, gradually evolving chain of related populations" in the strata.

On a macroevolutionary scale, the "cosmic punctuation" paradigm went against the long-held assumption that mammals had progressed through competition from small-brained, egg-laying groups to large-brained, live-bearing ones, an assumption not without foundation. The known fossil record, at least, suggested that placentals had replaced other groups except where geographical isolation intervened, and that larger-brained placentals had replaced smaller-brained ones. Moreover, the record contained little fossil evidence that any of the main three mass extinctions during mammals' tenure had terminated a major vanished group, such as triconodonts or multituberculates. If such extinctions were the main force behind evolutionary change, but had not affected mammals, then why had mammals changed? Why hadn't they simply stayed small and nocturnal, while the feathered dinosaurs that survived the Cretaceous extinction maintained the dinosaurs' 140-million-year dominance?

On the other hand, punctuationists could ask whether other vertebrate groups manifested the mammals' apparent competitive progressiveness. Dinosaurs seemed not to show a sustained trend toward larger brains, although some were smarter than others. And major groups of dinosaurs and other vertebrates had succumbed to mass extinctions, not only in the Cretaceous but at the end of every period in which vertebrates had lived. Did this mean that mammals were an exception to the evolutionary rule? Or did it mean that mammalian evolution really might be less competitive and progressive than it seemed?

Zallinger's murals seem oddly prescient of such questions. Monumentally tranquil, the *Age of Reptiles* looks like a world that only comet impacts could change. It proceeds leisurely from big saurians in Paleozoic swamps to big saurians in Mesozoic ones, and while there are hints of gradual progressive shifts, as with the advent of flowering plants, there is no visible competition. The "stately reptilian demigods" waddle or wallow untroubled by confrontations. The only hint of violence is an *Allosaurus* feeding gravely on a flayed and flattened carcass. As Vincent Scully remarks, "There are no horrors here. It is a pastoral painting, idyllic." The *Age of Mammals* is a mob scene by comparison. The figures seem to jostle as the tableau rushes from small beasts in tropical swamps through large ones in temperate savannas to huge ones in autumnal tundra. Conflict is ubiquitous, from the Eocene's *Oxyaena*-tormenting coryphodons to an Ice Age bison charging a dire wolf. The central Oligocene section is particularly frantic. A hyenalike creodont gnaws a ribcage, while saber-toothed *Hoplophoneus* claws a terrified *Protoceras,* and the ashen titanothere seems to bellow at the cosmos.

In response to the questions, punctuationists sometimes oversimplified things. "Mass extinctions change the rules of evolution," David Jablonski told *National Geographic.*

> When one strikes, it's not necessarily the most fit that survive; often it's the most fortunate. When their environment is disrupted, groups that have been healthy can suddenly find themselves at a disadvantage. Other species that have been barely hanging on squeak through and inherit the earth. The best example is the mammals. Dinosaurs and mammals originated within ten million years of each other about 220 million years ago. But for 140 million years dinosaurs ruled, while mammals stayed small and scrambled around hiding in the underbrush. Mammals all basically looked alike—squirrelly or shrewish and no bigger than a badger—until the dinosaurs disappeared. Then they took off. Within ten million years there were mammals of all shapes and life-styles: whales and bats, carnivores and grazers. Mammals just couldn't do anything interesting until the dinosaurs were out of the way.

Stephen Jay Gould approached the questions more subtly. He acknowledged that studies like Gingerich's seemed to show a furry propensity toward continuous gradual evolution. He maintained, however, that such studies followed "traditions of work and expectation" more than they recorded "actual relative frequencies of nature." And he launched a broad, sustained critique of traditional simplifications about mammal progress and competition.

Gould attacked the basic idea that monotremes represent a step from egg-laying reptiles to live-bearing eutherians. "If evolution were a ladder toward progress, with reptiles on a rung below mammals, then I suppose that eggs and an interclavicle would identify platypuses as intrinsically wanting," he wrote. "But . . . evolution proceeds by branching and not (usually) by wholesale transformation and replacement. . . . The presence of premammalian characters in platypuses does not brand them as inferior or inefficient." Platypus survival demonstrates, he maintained, that successful adaptation, not antiquated anatomy, is what counts: "If anything, this very antiquity might give the platypus more scope (that is, more time) to become what it really is, in opposition to the myth of primitivity; a superbly engineered creature for a particular, and unusual, mode of life. The platypus is an elegant solution for mammalian life in streams—not a primitive relic of a bygone world."

Gould noted, furthermore, that the other living monotreme, the echidna, has a larger, more convoluted brain than many placentals and performs better on some intelligence tests than cats and rats. "The solution to the paradox of such adequate intelligence in such a primitive mammal is stunningly simple," he concluded. "The premise—the myth of primitivity itself—is dead wrong. . . . The reptilian features of monotremes only record their early branching from the ancestry of placental mammals—the time of branching is no measure of anatomical complexity or mental status."

Gould also found oversimplification in neo-Darwinian scenarios of placental evolution, such as Simpson's horse lineage. While praising the great synthesizer for seeing a branching bush instead of an orthogenetic ladder, he faulted him for retaining "biases imposed by the metaphor of the ladder." According to Gould, Simpson had "restricted his bushiness as much as possible and retained linearity wherever he could avoid an inference of branching." In his genealogy's earlier phases, from *Hyracotherium* to *Miohippus,* Simpson erroneously had told "a story of linear descent, only later interrupted by copious branching among three-toed browsers." Gould thought he particularly had erred in emphasizing "the supposedly gradual and continuous transformation" between *Mesohippus,* the mid-Oligocene horse in the Peabody mural's background, and the late Oligocene genus, *Miohippus,* a transition so gradual, according to Simpson, that experts had trouble distinguishing the two.

The "enormous expansion of collections since Simpson had proposed this hypothesis" had allowed the paleontologists Donald Prothero and Neil Shubin to "falsify Simpson's gradual and linear sequence for the early stages of horse evolution and introduce extensive bushiness into this last stronghold of the ladder," Gould said. Far from having trouble distinguishing the two genera, Prothero and Shubin saw reliable differences between *Mesohippus* and *Miohippus* foot bones, thus determining that *Miohippus* had branched from a *Mesohippus* stock that then survived for another four million more years, during which the genera lived side by side. Moreover, each genus contained a bush of species, which also often coexisted. One Wyoming formation contained the bones of three *Mesohippus* species and two *Miohippus* ones.

The Oligocene equid genera and species tended to arise with geological suddenness and persist for long periods with little change. "Such stasis is apparent in most Neogene [later Cenozoic] horses as well, and in *Hyraco-therium,*" Prothero and Shubin concluded. "This is contrary to the widely held myth about horse species as gradualistically-varying parts of a continuum, with no real distinctions between species. Throughout the history of horses, the species are well-marked and static over millions of years. At high resolution the gradualistic picture of horse evolution becomes a complex bush of overlapping, closely related species."

On a larger scale, Gould leveled his critique at the hoary notion that North American beasts had "conquered" South America after the Panama land bridge formed. He noted that the inexorable statisticians Raup and Sepkoski estimated that a similar number of South American families had actually migrated north across the land bridge as North American ones migrated south, and that a similar percentage of genera from both continents had died out after the land bridge formed. It was only because most of the South American mammals that moved north were tropical and had gotten no farther than central Mexico, that the North American mammals seemed to have triumphed: "[T]he old story of 'hail the conquering hero comes'—waves of differential migration and subsequent carnage—can no longer be maintained."

Gould even found a punctuational use for A. S. Romer's idea that the Eocene flightless bird *Diatryma* had challenged mammalian dominance. Gould did not question *Diatryma*'s valor. "The gigantic head and short, powerful neck identified *Diatryma* as a fierce carnivore," he wrote in an essay cinematically entitled "Play It Again, Life." "Like *Tyrannosaurus,* with its diminutive forelimbs, but massive head and powerful hindlimbs, *Diatryma* must have kicked, clawed, and bitten its prey into submission." Far from regarding *Diatryma*'s reign as a brief hitch in mammal progress, however, Gould saw it as evidence that the beasts' post-Eocene prevalence over the birds was just luck, and he buttressed his argument with another group of predacious birds. Although not closely related to *Diatryma,* the phorusrhacids, which inhabited South America through the Cenozoic, also had huge heads and beaks, and they may have competed with the native marsupial carnivores. They outlived them, anyway, and later migrated over the land bridge as far as Florida. Gould saw them as a "strike two" against ideas of inevitable mammalian hegemony. "Birds had a second and separate try as dominant carnivores in South America," he concluded, "and this time they won."

Gould left "Play It Again, Life" out of his next essay collection, perhaps because, soon after its magazine publication, an ornithologist announced a new interpretation of *Diatryma*. Allison V. Andors of the American Museum thought it was a waddling relative of ducks and chickens, whose massive beak was adapted to browsing, as with birds such as rails today. "Far from posing a threat to animals high in the Eocene food chain, *Diatryma* appears to have occupied a lower trophic level than hitherto believed and to have assumed the role of a primary consumer," Andors concluded. If Gould dropped the essay for that reason, however, he may have been too scrupulous. Other paleontologists ignored Andors's analysis, as though a "fierce carnivore" bird was just too good to lose. An Eocene bird very like Zallinger's *Diatryma* stalks through the BBC's *Walking with Prehistoric Beasts,* persecuting mammals right and left.

Anyway, a ducklike *Diatryma* hardly punctured punctationism. Whether it was a fierce carnivore or a waddling browser, impact theorists were ready with another reason than beastly competition for its late Eocene demise. One of the times that Raup and Sepkoski had proposed for their 26-million-year cycle of asteroid or comet bombardment fell in the late Eocene, just about twenty-six million years after the Late Cretaceous extinction, and roughly when *Diatryma* vanished. In 1982, geologists had found evidence of a late Eocene impact, an iridium spike and tektites, in Caribbean sediments. And *Diatryma*'s family had not been the only major group to disappear around then. Most of the other groups that figure in the *Age of Mammals*'s Paleocene and Eocene sections did too. Alligatorlike champsosaurs disappeared, and the first big mammals that had grown into the dinosaurs' emptied niche—Cope's and Marsh's mesonychids, uintatheres, heavy-tailed pantodonts, and "beaver-hog" tillodonts—died out. Impact theorists began referring to a "Terminal Eocene Event."

If an Eocene impact had killed most of the early large mammals, might not other impacts account for later mammal extinctions? The entelodonts had disappeared in the early Miocene along with creodonts and condylarths, and this had occurred roughly twenty-six million years after the Eocene. No sign of an impact crater was known from then, but that didn't mean that none existed. Since the early Miocene was almost twenty-six million years ago, moreover, might another yet-undiscovered impact have caused the mass extinction of mammoths, ground sloths, phorusrhacid birds, and other megafauna that ushered in the present Holocene epoch? Might humans themselves be impact survivors, punctuated out of hunting and gathering equilibria by a cosmic blast, a trauma so severe that, like childhood abuse, it had erased all memory of it?

EIGHTEEN

Simpson Redivivus

ONE ASPECT OF ZALLINGER'S *AGE OF MAMMALS* does seem more applicable to punctuated equilibria than its stately neighbor. Because the smaller mural's landscapes change so much, each epoch could stand as a picture on its own, giving a sense of the Gould-Eldredge theory's abrupt shifts. Zallinger literally punctuated his epochs by placing foreground trees between them—an epiphyte-draped palm closes the Paleocene, a stout sycamore the Eocene, a spindly pine the Miocene, a small aspen the Pliocene. Trees also punctuate the *Age of Reptiles*—a fan palm closes the Jurassic, a primitive conifer the Triassic. But the bigger mural evokes less sense of environmental shifts because the landscape remains much the same, from Paleozoic swamps to Mesozoic ones.

Punctuationism can certainly apply to mammals, as Gould's essay on fossil horses shows. Indeed, according to his posthumously published magnum opus on evolutionary theory, the mammoth, so progressive-looking in the mural, was exemplary of stasis in one classic fossil analysis. Gould noted that when a paleontologist friend of Darwin's, Hugh Falconer, published an omnium-gatherum on the genus in 1862, he emphasized its stability. "One section is devoted to the persistence in time of the specific characters of the mammoth," Falconer wrote in a letter to Darwin, "I trace him from before the Glacial period, through and after it, unchangeable and unchanged as far as the organs of digestion (teeth) and locomotion are concerned." To be sure, Falconer believed that mammoths demonstrated the permanence of species, not evolution, punctuated or otherwise. But Gould thought his conclusion—that mammoth species only seemed to change because immigrant species replaced them—anticipated "a primary inference of punctuated equilibrium—that a local pattern of abrupt replacement does not signify macromutational transformation in situ, but an origin of the later species from an ancestral population living elsewhere, followed by migration into the local region."

Gould's friend Elisabeth Vrba, a South African expert on antelopes, became mammal punctuation's leading advocate. "Vrba's classic studies of African antelopes stand out," he wrote, "for detailed data on one of the must successfully speciose of vertebrate higher taxa." Vrba found that South African antelopes living before about 2.5 million years ago had teeth and other anatomical features typical of forest-dwelling browsers, whereas ones living afterward had the teeth of savanna grazers. She thought a drastic shift in global temperature and precipitation then had caused a "turnover-pulse" not only in antelopes, but in plants, hominids, rodents, and marine invertebrates. "According to the classic Darwinian view," she wrote, "the 'engine' that drives the evolution of species is competition between organisms; extinctions occur when species outcompete others for the same resources. . . . But the fossil record appears to tell a different tale. Species seem to arise and vanish in pulses of varying intensities, with many appearing and others disappearing at the same time."

Many other paleontologists applied the theory to mammals in varying degrees. Donald Prothero and T. H. Heaton found more punctuation than gradual evolution in the classic White River fauna, the mammals that Joseph Leidy had first described in the 1850s. "Most species are static for 2–4 million years on average, and some persist much longer," Prothero and Heaton wrote. "Only three examples of gradualism can be documented in the entire fauna, and these are mostly size changes." Outlining North American's Cenozoic in general, David Webb might have been invoking Zallinger's tree-punctuated imagery. "Thanks to improved land mammal chronologies," he wrote, "the tempo of faunal turnover is now seen to be strikingly syncopated. Stately chronofaunas, made up of stable sets of slowly evolving taxa, persist in the order of 10 to the 7th power yr. Chronofaunas are terminated by rapid turnover episodes in which the native taxa experience rapid evolutionary rates and other taxa appear in intercontinental migrations."

Punctuationism's mammal applications were just one facet of its broad and lasting appeal. A 2002 newspaper science page article quoted David Jablonski reciting "an impressive array of examples in the fossil record, from snails to horses . . . a core of solid analyses that are very convincing," in support of the paradigm. "One scientist analyzed the evolutionary 'trees' of 34 different types of scallops," the article paraphrased Jablonski, "and found only one that displayed gradual evolution over time. The remaining 33 stayed pretty much the same from generation to generation. . . . Another scientist found gradualism in only 8 of 88 lineages of trilobites." The same science page carried a sidebar citing possible evidence that cosmic impacts might have affected dinosaur evolution in the Triassic and Jurassic extinctions as well as terminating nonavian dinosaurs in the Cretaceous one.

When combined with dinosaurs, sudden and catastrophic evolutionary change made irresistible copy, especially after the discovery of a 100-mile-wide crater on the Yucatan coast provided the "smoking gun" for a Late Cretaceous impact. It was front-page news when Peter Sheehan published his Dig-a-Dinosaur data analysis in the journal *Geology* in 2000. "What we found suggests that the dinosaurs were thriving, that they were doing extremely well during that time," Sheehan told the Associated Press. "The asteroid impact brought a sudden and very abrupt demise to species that were healthy and doing well." Sheehan said his data cast doubt on "the theory that the dinosaurs died out slowly and that the asteroid impact was simply an end-the-misery trauma for the vanishing species."

Paleontologists kept disagreeing about the evidence at Hell Creek, however. The Associated Press article also reported William Clemens's response: that Sheehan's study had not considered how dinosaur diversity had declined there from that of five or six million years earlier, or how an asteroid impact had affected other animals. "You need to consider the whole fauna," Clemens concluded. "Why did amphibians go through this period unaffected? There was a diversity of birds, and they go through the period unaffected." When I asked David Archibald about Sheehan's study in 2001, he said: "The data don't agree with Sheehan. What you have to go on is the published record, and very few people buy Sheehan's results. The data aren't there."

Plenty of other data existed. Archibald's own studies at Hell Creek and environs had indicated a dinosaur diversity of fourteen to twenty species, fewer than in earlier Cretaceous formations. "[T]he evidence from Garfield County supports Van Valen and Sloan's contention that the decrease was real and not an artifact of sampling," Archibald had written of the Bug

Creek Anthills site in 1982. Close stratigraphic study later convinced another researcher, Donald Lofgren, that the presence of both dinosaurs and advanced mammals like *Protungulatum* in strata at Bug Creek was the result of "reworking" and showed neither that *Protungulatum* had lived in the Cretaceous nor, as some researchers thought, that dinosaurs had survived into the Paleocene. "If this interpretation is correct," he wrote, "the survival rate of mammal species is approximately 10% and of dinosaurs 0%, making catastrophic mass-extinction scenarios more plausible." But Lofgren still did not assume a sudden dinosaur extinction, observing: "[S]uch scenarios are still not compatible with the high specific survival rates (55–71%) of non-dinosaurian vertebrates or the entire vertebrate fauna (exceeding 50%)."

At a Cretaceous-Paleocene site in Wyoming, Jason Lillegraven and J. J. Eberle concluded that diversity of nonavian dinosaurs had remained "high within upper levels" of the late Cretaceous fossil-bearing strata they were sampling, and saw no strong evidence that advanced mammals like *Protungulatum* had coexisted with them. The "seventeen varieties of non-avian dinosaurs" they found were about the same number as at Hell Creek, however, and they did not interpret the disappearance of dinosaur fossils above the Cretaceous boundary as proving a sudden extinction. "Although the late history of diversity of local non-avian dinosaurs seemed to have had a 'sudden' biostratigraphic termination," they wrote, "we do not necessarily suggest that dinosaurian extinction was of a 'catastrophic' nature within a biologically relevant interval. Based upon existing uncertainties in interpreting the local biostratigraphic record, their demise could have extended for thousands, or perhaps tens of thousands, of years."

Even one of punctuationism's co-authors had expressed some unease about its ability to explain extinctions. "We have been witnessing, I think, an understandably enthusiastic but still off-scale, over-zealous jump into a bandwagon," Niles Eldredge wrote in 1987. "The arena has been the plausibility of this or that astrophysical scenario—with not enough attention to the basic question: what really is the pattern of that monstrous extinction event?"

Addressing that basic question, David Archibald and Donald Prothero published books in the 1990s analyzing the pattern of two mass extinctions involving mammals, the Cretaceous and Eocene. Both authors accepted that impacts had probably occurred, but concluded that, as Prothero put it, "their effect on climate or extinction seems to have been minimal." Both saw evidence of a number of extinction events clustered toward the ends of those times, rather than of two large events coincident with extraterrestrial impacts. Both saw a combination of climatic and geological changes as the major causes of extinction. "Thus far, marine regression as a possible cause of major extinction looks robust," Archibald wrote of the Cretaceous. "By this analysis, its predicted effects show a tighter fit with the paleontological data than do the impact or volcanic theories." Prothero attributed the Eocene extinctions to a major cooling of global climate beginning in the mid-Eocene and lasting to the early Oligocene. "[T]he biggest extinctions of the Cenozoic seem to coincide with the initial formation of glaciers on Antarctica," he wrote. "Rapid rifting between Antarctica and Australia may have allowed moisture to precipitate as snow over the South Pole."

At the same time, some catastrophist claims turned out to be overstated. As Stephen Jay Gould acknowledged in his magnum opus, evidence of impacts coincident with mass extinctions other than the Cretaceous proved inconclusive. (There is no evidence that an impact caused the Pleistocene one, so we can't use that to excuse our neuroses.) Simpson's synthesis underwent a media resurgence of sorts as gradualist interpretations of mass extinctions persisted. The 1992 PBS dinosaur documentary had favored impact theory, for example, but the BBC's 1999 *Walking with Dinosaurs* computer animation series took a more gradualist stance, emphasizing the effects of Cretaceous volcanism, although still taking artistic advantage of impact fireworks.

Punctuationism accordingly took some knocks. People noticed in 1998 when Simon Conway Morris, an English paleontologist whose 1970s work on early Paleozoic invertebrates had influenced Gould's ideas, attacked him. "Gould sees contingency—evolutionary history based on the luck of the draw—as the major lesson," Conway Morris wrote. "Such a view, with its emphasis on chance and accident, obscures the reality of evolutionary convergence. Given certain environmental forces, life will shape itself to adapt. History is constrained, and not all things are possible. . . . If such a quality as intelligence can arise both in human beings and in the octopus—an eight-armed sea animal without a bone in its body—then perhaps there is a course and a direction to evolution that would be achieved despite diverse anatomical starting points."

In the surprisingly fashionable venue of a 1999 *New Yorker* article, Robert Wright, a science journalist, lambasted Gould for "his championing" of "punctuated equilibrium—the idea that evolution proceeds in fits and starts, and spends much of its time moving nowhere in particular." Wright reiterated Simpson's 1978 observation that punctuationism was simply a "new label" for old ideas, claiming that other scientists had "widely criticized" Gould for "pronouncing Darwinism dead." He cited examples of evolutionary progress through Darwinian competition, including the "strong tendency" of North American mammal brains to increase in size during the Cenozoic, and North American mammals' apparent decimation of smaller-brained South American ones after the Panama land bridge formed. He suggested that Gould's dislike of some early evolutionists' "social Darwinist" ideology had motivated his attacks on competition and progress, but warned that he was playing into the hands of creationists by challenging a neo-Darwinian mainstream.

Wright's attack was more political than scientific, and his paleontological examples were dated. Yet, although unmentioned in the *New Yorker,* evidence had been accumulating that challenged one of the punctuation paradigm's commonplace assumptions about the Cretaceous extinctions in particular and mammalian evolution in general. It was becoming increasingly evident that Jablonski had been oversimplifying in 1989 when he told *National Geographic* that early mammals "all basically looked alike," and that they "just couldn't do anything interesting until the dinosaurs were out of the way." Of course, paleontologists from Marsh to Simpson had already shown that dinosaur age mammals *did* interesting things and were *not* alike, but those facts had gone over most people's heads, as Simpson ruefully discovered. Ever-accelerating research was beginning to change that, however, and to put a new spin on mammal-dinosaur relations.

Perhaps the main thing Jablonski meant when he described Mesozoic mammals as uniformly uninteresting was that they were all small, which reflects a human bias. We are large animals ourselves, and a thundering brute gets our attention, as their predominance, whether placid stegosaur or frenzied titanothere, in Zallinger's murals demonstrates. This is a scientific flaw, since most prehistoric vertebrates were, as most vertebrates are today, no bigger than a cat, although it is understandable given the artistic obstacles to juxtaposing midgets with titans. To show the real small-big ratio, the *Age of Mammals* would have had to show hundreds of little beasts scurrying around its titanotheres and mammoths, which might seem more Hieronymus Bosch fantasy than paleontological restoration.

Zallinger compensated for our myopia by rendering just a few small creatures as ironic asides, an example of what Vincent Scully called his "cunning." The irony is muted in the *Age of Reptiles,* where he painted a few dainty dinosaurs like *Compsognathus* skittering under the "stately reptilian demigods" and later added the tiny mammal *Cimolestes*. It is explicit in the mammal mural, wherein a Lilliputian indifference to Brobdignagian posturings pops up throughout. The late Paleocene opossum, *Peradectes,* goofily ignores dinosaurlike *Barylambda;* an Eocene rodent named *Paramys* blithely washes its face as coryphodonts harass *Oxyaena;* a Pliocene ground squirrel, *Epigaulus,* sports tiny horns as though to burlesque the antlers and tusks paraded above. Rabbits play a traditional role as puncturers of megafaunal pretension. One named *Palaeolagus* scratches its ear underfoot of frenzied *Brontops,* and, thirty million years later, another long-eared lagomorph named *Panolax* hops nonchalantly away from a confrontation between ursine *Agriotherium* and lupine *Amphicyon*—Br'er Bear and Br'er Wolf.

A trip to the woods will reflect Zallinger's ironies. As hikers scan the horizon for charismatic megafauna, chipmunks ransack tents, marmots gnaw car engine hoses, and deer mice build nests in glove compartments. At night, flying squirrels raid suspended food sacks, and woodrats steal wristwatches. The ironies can be scary as well as annoying. When we camped along Yellowstone's Hellroaring Creek one moonless night, my wife was awakened by what she thought was a large animal, possibly a grizzly, just outside our tent. In the morning, when she reached down to tie her shoes, she fell over backward. The "grizzly" had been a mouse which, for some unfathomable microfaunal reason, had spent the night skillfully gnawing through her shoelaces at each of the eyeholes, leaving the pieces deceptively in place.

Zallinger's mammalian microfauna challenge the assumption that small animals are less important then big ones, and, by implication, that mammals "lost" to dinosaurs by occupying a microfaunal niche. A shrew's world is lesser than a lion's only in a quantitative sense, although it is very different, physically and biologically. Small animals use energy less efficiently than large ones, so their lives and memories are short, but gravity impacts them less, so they are stronger and tougher. They can fall hundreds of feet without crippling injury and perform muscular feats beyond our imagination. I once watched a lactating weasel dart down a burrow on the California coast and quickly emerge carrying a pocket gopher that probably weighed more than she did. She took it to her kits in a series of four-foot leaps through knee-high grass, as if a wolf were to jump through brush with an elk in its mouth.

As their prevalence since the Triassic demonstrates, small mammals have an importance proportionate to their diversity, abundance, and adaptability. It is a little surprising that any field paleontologist could find them inconsequential, since they have the same impacts on bone hunters as on other campers. The diary of Arthur Lakes, one of O. C. Marsh's Como Bluff collectors, features a litany of complaints about microfauna, which stole food, clothing, and tools, including a shovel. One of their "nightly amusements" was to fill his boots with beans and rice. The abundance of Lakes's tormenters may seem surprising to a casual visitor at Como Bluff, among the West's more barren-looking places, but micromammals throng even greater desolations. Walter Granger's 1907 Fayum notes describe a place far from water or vegetation, but fennecs, cat-sized desert foxes, plundered supplies. When George Olsen tried to trap them, they stole the bait, then walked off with the trap.

An adventuresome American Museum paleontologist, Michael Novacek, seemed to echo Lakes's and Olsen's pique when he wrote that small mammals "might be more aptly called vermin than beasts." Yet Novacek recalled that his teachers, Jason Lillegraven and David Vaughn, regarded microfauna as "the new frontier of research," offering unique insights into prehistoric physiology and ecology. "In fact," Novacek wrote, "both these early mentors shared a widespread disdain for much of the research on dinosaurs and big fossil mammals, which they characterized as superficial, driven more by a desire to make a splash on the public scene than by a wish to solve evolutionary questions."

The more we learn about prehistory, indeed, the more prevalent and ingenious mammalian microfauna seem. The Chinese discovery in 2000 of the smallest Mesozoic mammaliaform yet known, a creature named *Hadrocodium* ("heavy or full head"), which weighed the same as a paper clip when alive, changed ideas about the rate of early mammal evolution. Although it lived about 195 million years ago, *Hadrocodium*'s brain, jaw, and inner ear were much more advanced for that time than previously known. "It was a little smart cookie with an extended brain," said one of its describers, the prolific Zhe-xi Luo. Its brain size and other features seem more typical of beasts that lived forty-five million years later, suggesting that dinosaur dominance arrested mammal development less than previously thought.

One might even say that dinosaurs lost out to mammals by not seizing the microfaunal niche in the Triassic. "Could it be," wondered Lillegraven, "that the spectacular adaptations seen among the dinosaurs were of a somewhat more 'superficial' nature, having been related primarily to adaptations for predation (or protection from it), sexual displays, and large body size? Perhaps large body size itself provided physiological or mechanical constraints to evolutionary experimentation within more fundamental tissue-, organ-, or system-level adaptations that did not exist for the smaller mammals." If any single trait allowed mammals to survive the Mesozoic, anyway, it probably was their smallness and consequent powers of survival and reproduction. And they kept on being small. One living species, Kitti's hog-nosed bat, is *Hadrocodium*-sized, and at least one Cenozoic fossil insectivore was smaller, an estimated 1.3 grams compared to *Hadrocodium*'s two grams.

Even dinosaurs demonstrate smallness's advantages. Although their avian branch—represented in the reptile mural by *Archaeopteryx*—didn't evolve until the Jurassic, some birds were robin-sized or smaller in the Late Cretaceous. As with mammals, avian smallness probably helped them outlast their big relatives. And they, too, kept on being small—the Cuban bee hummingbird may be the tiniest warm-blooded vertebrate ever. Smallness is not an evolutionary cul-de-sac but a dynamic realm of primary potential, perhaps more than bigness, which, judging from the rise and fall of "stately chronofaunas," can have a certain dead-endedness. As Zallinger's mural shows, most big mammals disappeared without descendants. And smallness was a realm in which, it began to seem in the 1990s, the little Cretaceous eutherians had done even more evolving than previously suspected.

NINETEEN

Wind Thieves of the Kyzylkum

SIMPSON'S RETREAT FROM MOSCOW IN 1934 was only a temporary hiatus in Central Asian evolutionary grail quests. After World War II, Soviet bloc scientists took over the search for early mammals in Mongolia, assisted by local collaborators. The Soviet Union fielded well-equipped Gobi expeditions in the late 1940s, which, beside making impressive dinosaur discoveries, brought the Cold War into bone-hunting by questioning the Cretaceous date of the Americans' 1920s Flaming Cliffs mammal fossils. Then, from 1963 to 1971, eight Polish-Mongolian expeditions led by Zofia Kielan-Jaworowska, an early mammal authority, explored the Gobi.

Inspired during Nazi occupation underground schooling by one of Roy Chapman Andrews's books, Kielan-Jaworowska combined Andrews's logistical skills with Granger's and Matthew's paleontological ones to find the best Mesozoic mammal fossils yet, along with many Cenozoic mammals and dinosaurs. Crawling around with magnifying glasses in 104°F heat, she and her teams discovered fifty mammal skulls at the Flaming Cliffs, and a hundred more in various other locations, along with enough postcranial bones to generate new ideas about how Mesozoic Mongolia's multituberculates, deltatheres, and early eutherians had looked and behaved. Indeed, they collected so many new fossils that they "became upset at not finding any specimens of the genera represented in the New York collections" and were relieved when they finally located three well-preserved skulls of *Zalambdalestes,* the Americans' major eutherian find. (They found a single relic of the Andrews expedition itself at the Flaming Cliffs, a bottlecap. *Sic transit gloria.*) They also proved that the Flaming Cliffs mammals really were Cretaceous, to the relief of the American Museum, which had sent Malcolm McKenna to Warsaw with copies of the Andrews expedition's contour maps to make a case for their age.

Mammal knowledge grew further after Mongolia became independent, when McKenna and Michael Novacek led a new series of American Museum expeditions there. In 1993, they found what Novacek called "the choicest spot for Cretaceous fossils in the Gobi, if not the entire world," at a place called Ukhaa Tolgod. The site eventually yielded "nearly 1,000 mammals," which, like the contemporary Gobi's jerboas and mice, had lived and died around Cretaceous desert oases. Exquisitely preserved deltathere skeletons provided convincing proof that the formerly mysterious group was more closely related to marsupials than to eutherians. Another skeleton, of the famous *Zalambdalestes,* was so complete that it included "epipubic" bones, the small pelvic bones that Cuvier had found when he demonstrated that his little clawed mammal from the Paris gympsum was a marsupial. The bones suggested that, although a eutherian, *Zalambdalestes* might have given birth to young in a less developed state than modern placentals. (Doubt has arisen, however, as to whether epipubics support the marsupial pouch—they may be just muscle supports.)

"Undoubtedly the best preserved mammal fossils of this age come from the Gobi desert, where in some cases complete skulls and skeletons have been collected," observed David Archibald. Still, the late-twentieth-century Gobi expeditions did little more than the 1920s ones to find Osborn's and Simpson's dreamt-of grail, to find, that is, anything remotely resembling a five-toed horse. Although different from North American late Cretaceous mammals, the Mongolian ones were known Mesozoic kinds—multituberculates, deltatheres, early eutherians. They revealed little more than American fossils as to the origins of modern placentals.

It was as though the grail was being withheld for one "parfit, gentil knight," and in the 1970s, something like one appeared. By then, an impoverished East Bloc mired in revolt and military disaster no longer was sending well-funded expeditions to Asian badlands. That didn't stop a Russian paleontologist named Lev Nessov from exploring a region that attracted him—the Kyzylkum Desert of western Uzbekistan. Located southeast of the Aral Sea, it had not attracted many scientists, and its paleontological potential was obscure when Nessov first went there in 1974.

Only one western Asian Cretaceous mammal fossil was known before 1978, from neighboring Kazakhstan. Nevertheless, as the Soviet Union collapsed around him, Nessov went to the Kyzylkum eleven times from 1974 to 1994. He took third-class trains as far as they went, then hitchhiked or carried his minimal supplies the rest of the way. A site called Dzharakuduk, "the well of the escarpment," where a 200-foot chunk of Cretaceous sediments cropped out of the desert, proved particularly rewarding.

The Dzharakuduk sediments, the Bissekty Formation, date from the mid-to-late Cretaceous, about eighty-five to ninety million years ago. Nessov found many dinosaur fossils in them—sauropods, hadrosaurs, ceratopsians, tyrannosaurs. They had inhabited a swampy coastal plain covered with sycamore, laurel, viburnum, and other temperate-to-subtropical plants. Nessov also got down on his hands and knees and found many teeth and jawbone fragments of eutherian mammals. Their jaws were tiny, and it was unclear just how they had looked or behaved, since Nessov found no complete skeletons. Yet their molars were not the sharp-cusped ones typical of bats, shrews, and other insect eaters. They were more rounded, like those of omnivorous creatures such as raccoons. Because most known eutherian mammals that long ago had sharp-cusped molars, this suggested that the Kyzylkum fossils ate more plant material than eutherians had before, and that they heralded a change in the group. Nessov had not found the first human ancestor in Asia, but he may have found something more important for evolution overall—the first known Mesozoic relative of modern placental ungulates.

Nessov's personality fell a bit shy of Galahad's. David Archibald, who first met him at a conference in Kazakhstan in 1990, said he "could be irascible, and sometimes bordered on being eccentric, but at the same time could be brilliant." Indeed, L. A. Nessov seems to have been a modern Russian version of O. C. Marsh and E. D. Cope rolled into one. He was an established academic like Marsh, and spent his career rising through the ranks at Leningrad University, from laboratory assistant to senior scientific researcher. Yet an energy like Cope's drove him to pursue a bewildering variety of subjects and travels during a strenuous life. He died even younger than Cope, at the age of forty-eight, in 1995, but his publications included papers ranging from Precambrian ascidians and Devonian fishes to Mesozoic ecosystems, and his fieldwork included much of Asia. He personally found eighty dinosaur fossil sites in the former Soviet Union.

"Nessov was very, very widely read," David Archibald told me, "and this was at a time when in the Soviet Union you weren't supposed to read stuff from outside. But he really knew the material. 'David, look at this paper,' he'd say to me, 'that tooth looks like here, and there.' He liked to work at home, and sometimes he'd work all night. So he was working on the Kyzylkum mammals long before the Soviet Union's collapse. He began publishing on them in 1983 or '84."

Nessov's energy propelled him on tangents at times. "Lev named everything he found, and we've had to prune that a bit," Archibald said, echoing Osborn's complaint about Cope and Marsh. "He also got some strange ideas." Nessov had first decided that the Kyzylkum mammals were condylarths, the primitive ungulates that Simpson found nondescript, but that Zallinger painted as looking more diverse. (Cope classified condylarths as an order, but later paleontologists have used the name more loosely to designate early ungulates that don't fit into better-understood groups.) All known condylarths lived in the Cenozoic, however, so their presence twenty-five million years earlier seemed improbable, and Archibald convinced Nessov that they weren't condylarths. Figuring out what they *were* was more difficult, but Nessov and Archibald eventually agreed that they seemed to be very early ancestors of beasts like the Bug Creek Anthills' mysterious *Protungulatum*. The Kyzylkum fossils were so early, however, that they didn't place them with any known group, but used a name that Nessov had coined. They called them "zhelestids," a Kazakh-Greek composite meaning "wind thieves," which they put in quotes because, as Archibald put it, "some 'zhelestids' have a more recent common ancestry with Ungulata than other 'zhelestids.'"

They weren't certain that zhelestids *were* linked to, or "monophyletic" with, early ungulates like condylarths. "Archaic ungulates retain so many characters primitive for eutherians and lack so many of the characters found in later ungulates that the monophyly of Ungulata, including archaic ungulates, is difficult to test," Nessov wrote in a paper co-authored with Archibald and Zofia Kielan-Jaworowska. Nondentologist readers will have to take on faith the analysis by which they tested the link. "In 'zhelestids' and Late Cretaceous eutherians," they wrote, "the postmetaconule crista forms the metacingulum, while in ungulates the metacingulum is formed by the postmetaconule crista continuing into the metastylar lobe." Despite the difficulties, however, they concluded: "The discovery by the first author of new ungulatomorphs (the 'zhelestids') from the Late Cretaceous of western Asia . . . helps to clarify characters seen in earliest Paleocene archaic ungulates in North America and permits a more testable hypothesis of ungulate monophyly."

It was unclear why the many Mongolian expeditions hadn't found zhelestids. Nessov thought Central Asia might have been too cool and dry for them. But they had definitely lived "cheek by jowl (or more accurately nose to toe)" with western Asian dinosaurs for over twenty million years. And they had apparently begun to move into other areas with similar ecological conditions by the Late Cretaceous. Archibald told me that zhelestids have been found in North America, from about seventy-five million years ago, and in Europe. At U.C. Berkeley, William Clemens showed me a jaw fragment cast from a little creature named *Gallolestes* that inhabited what is now Baja California in the Late Cretaceous. Its molars were so tiny that I needed a magnifying glass to examine them, but Clemens gave a detailed explanation of why they resembled the Kyzylkum zhelestid's molars. *Gallolestes* seems to have inhabited a moist coastal habitat similar to that of southwest Asia at that time. Clemens thought that their ancestors might have immigrated from Asia during a warm period, then have been confined to the southern half of North America when climate cooled later.

Nessov's fossils seemed evidence that a sister taxon of ungulates, at least, had started branching in the mid-to-late Cretaceous, and paleontologists such as Archibald and Kielan-Jaworowska found them convincing. Archibald spent a field season at Dzharakuduk with Nessov in 1994, and created a new clade, the Ungulatomorpha, for the fossils in 1996. He has returned with other Russian colleagues five more times since Lessov's death, and found many more fossils. "Lev did mostly surface collection, but we've gone to extensive screenwashing," he said. "We have lots of postcranial material now. We don't yet have whole skeletons like they do from Mongolia, but we have nearly whole skulls, and limbs and vertebrae. The problem is knowing who they go to."

Nessov's finds at Dzharakuduk also included a few teeth of a creature he named *Kulbeckia* in 1993. When Archibald worked the site from 1997 to 2000, he found enough additional teeth and bones to suggest that *Kulbeckia* was related to *Zalambdalestes,* the long-snouted hopping Mongolian eutherian. Malcolm McKenna and Michael Novacek had studied *Zalambdalestes* skulls and noted similarities to those of living rodents and rabbits, particularly in that "the main pathway of the carotid artery ran in two branches on either side of the middle of the skull . . . a striking departure from the usual situation in placental mammals, in which the carotid crosses the base of the skull away from the midline and through the middle ear cavity." *Zalambdalestes* also had skeletal similarities to *Barunlestes,* a late Cretaceous Mongolian genus; to a possibly rabbitlike Paleocene group from China called mimotonids; and to a definitely rabbitlike early Eocene Mongolian group. This suggested that Nessov had found not only the oldest known relatives of the largest mammals (assuming that artiodactyl ungulates like hippos are indeed whales' closest living relatives), but the oldest known relatives of some of the smallest. When Archibald and his colleagues did cladistic analyses of zalambdalestids and zhelestids in 2001, zalambdalestids seemed even more likely to be related to rodents and rabbits than zhelestids did to ungulates.

The fragmentary, technically challenging Asian micromammals seldom reached popular media versions of the extinction debate, yet their implications for it were profound. If some mammals had begun to evolve modern characteristics like ungulate dentition and lagomorph circulation twenty-five million years before the dinosaurs' disappearance, then the idea that an abrupt mass extinction had been the only factor freeing them from an "uninteresting" primitive equilibrium made less sense. "Their descendants appeared in North America at the close of the Mesozoic," Archibald wrote of zhelestids, "where I believe they wreaked havoc on the hapless native marsupials." Such scenarios implied a more gradual and progressive picture of mammal evolution than the impact punctuation paradigm's.

Although this support for their "Asian Eden" ideas would have pleased the likes of Matthew and Simpson, living paleontologists didn't all hail zhelestids. In 2000, Novacek, McKenna, and others reported the discovery of zhelestid-like teeth with zalambdalestid skulls in upper Cretaceous Mongolian strata, and speculated that zhelestids might simply have been blunt-molared zalambdalestids, eliminating "all evidence for the presence of ungulates or other members of the crown group Placentalia in the Mesozoic of Asia." David Archibald told me, however, that zhelestid skulls are quite different from zalambdalestid ones, although it would a few years before researchers could cast some light on their appearance and behavior by reconstructing a skeleton. Perhaps they will prove to have had five toes—with a tiny hoof at each end.

Other paleontologists seemed not to notice the Kyzylkum's tiny bones. Neither of punctuated equilibria's architects mentioned zhelestids or zalambdalestids in theoretical works at the turn of the century. Niles Eldredge wrote in a 1999 book that dinosaurs had kept mammals "fairly generalized, smallish, and undifferentiated physically" until "the violent scene that put an end to the Mesozoic world." In his 2002 opus *The Structure of Evolutionary Theory,* Stephen Jay Gould stuck to the idea that "mammals would still be rat-size creatures living in the ecological interstices" of a dinosaur world if a Late Cretaceous bolide hadn't hit. Of course, if Gould and Eldredge had noticed zhelestids, they could have tucked them into a punctuationist context. They could have said that since these microfauna entered the fossil record abruptly, their appearance must indicate that environmental change had caused a turnover pulse then, and probably not a very big one, because zhelestids were still small and relatively unspecialized.

Nobody has questioned zhelestids' Cretaceous age, anyway, and evidence of Mesozoic mammal diversity continues to grow. In May 2002, a group of paleontologists, again including Zhe-xi Luo, announced discovery of a eutherian that had lived in the early Cretaceous, about 125 million years ago, and that had already evolved at least one seemingly advanced trait. So perfectly preserved in northeastern Chinese lake shales that a black mass of carbonized hair surrounded it, the skeletal creature was named *Eomaia scansoria,* "climbing dawn mother," because its claws suggested that it was a nimble scrambler on tree trunks and branches. It differed in this from terrestrial mammaliaforms found in the same formation, although, like them, it had teeth and jaws adapted to eating animals. "*Eomaia* may have reached food sources that were not available to other mammals or have scampered overhead to avoid predation," speculated one biologist, "or it may have combined both activities."

"For studying early evolution of mammals, this is a dream come true," Zhe-xi Luo said of *Eomaia.* "All this early evolution appears to have come about in the shadow of the dinosaurs." He then bowed to convention, however, by adding, "but only after the dinosaurs disappeared did the placental mammals begin to radiate far and wide."

One hot new discipline produced even stronger advocates of Mesozoic mammal progress than neo-Darwinian vertebrate paleontology. In the 1980s, the development of DNA analysis revived the possibility, thought likely by Marsh and Osborn, that early members of modern mammal groups would be found in the Cretaceous. By comparing the DNA of living species, technicians could estimate not only how closely related they are (providing more evidence, for example, of a link between artiodactyls and whales), but how long ago the various modern groups branched off from the extinct Mesozoic ones. And many estimates put that branching somewhere in the mid-to-late Cretaceous.

Some estimates were startling, like those of Mark Springer, a geneticist at the University of California, Riverside, and his colleagues. According to their analysis, modern placentals originated at least 100 million years ago,

Figure 16. *Eomaia scansoria.* Art by Mark A. Klingler / Carnegie Museum of Natural History.

and successively branched into four superorders. The first, called "Afrotheria," includes elephants, hyraxes, aardvarks, and manatees; the next, "Xenarthra," sloths, anteaters, and armadillos; the next, "Euarchontoglires," primates, rodents, and rabbits; and the last, "Laurasiatheria," insectivores, bats, carnivores, ungulates, and whales. (According to Springer, the many similarities between groups from different superorders—between manatees and whales, for example—result from convergent evolution.) This succession raises a problem like that of the tribosphenic molar's origin, because some researchers think it shows that placentals originated in Gondwana and spread north, while others still think they originated in Laurasia and spread south. The four superorders themselves seem on their way to being generally accepted, however.

Such a "primitive" origin for the mighty mammoth would have caused Stephen Jay Gould both joy and sorrow, because, while undercutting conventions of a mammalian progress ladder with elephants near the top, it also undercut conventions of Mesozoic mammal stasis. Two DNA analysts, Sidhir Kumar and S. Blair Hedges, tipped the punctuationist applecart a bit when they wrote: "The origin of most mammalian orders seems not to be tied to the filling of niches left vacant by dinosaurs but is more likely to be related to events in earth history."

Ironically, modern vertebrate paleontologists tended to resist such assertions by geneticists, since the known fossil record doesn't support them. (The earliest known elephant relative, for example, is a 55-million-year-old Moroccan fossil called *Phosphatherium escuilliei*.) "On one hand, what they write about higher relationships seems perfectly reasonable," said Philip Gingerich of the DNA researchers, "But on the other hand, what they write about the timing of divergences seems completely unreasonable." David Archibald found "no support for a Late Cretaceous diversification of extant placental orders" in his cladistic analysis of zhelestid and zalambdalestid fossils. He told me he thought competition from dinosaurs had made Mesozoic mammalian evolution "pretty ho-hum" compared to the Cenozoic.

DNA analysts have pointed out, however, that the fossil records of other modern vertebrate groups such as bony fish do support molecular clocks ticking backward well into the Mesozoic. DNA-based Cretaceous origin estimates seem an at least possible further challenge to the traditional "interesting huge dinosaur–boring tiny mammal" paradigm of Mesozoic life.

If dinosaur extinction wasn't the only factor causing eutherian mammals to evolve "anything interesting," however, what else was? Nobody believed anymore that they were orthogenetically inclined to evolve hooves, high-crowned molars, and bigger brains. On the other hand, the environmental changes, whether sudden or gradual, that contributed to late Cretaceous dinosaur extinction were anachronistic to the appearance of zhelestids and zalambdalestids twenty to thirty million years earlier, and even more so to *Eomaia scansoria*'s appearance fifty million years before that. According to Darwinism, mammals were changing to adapt. If they were adapting to something beside dinosaurs or drastic environmental changes, what was it?

TWENTY

The Serpent's Offering

THE GIANT SNAKE DANGLING FROM a fruit tree at the beginning of the *Age of Mammals* is one of the mural's most arresting figures, and its most enigmatic. The Edenic reference is clear, but exactly why it should come at the start of a picture of Cenozoic evolution is less so. Zallinger probably wasn't suggesting that a serpent offered some kind of malign temptation to our biological class as they emerged from the Mesozoic, although the contrast between the placid dinosaur mural and the turbulent mammal one does seem to raise the issue. Yet the biblical serpent evolved from older myths in which it was not an evil tempter but an embodiment of generative natural powers, particularly those of plants. In the latter sense, Zallinger's serpent may have a symbolic significance to modern mammal origins.

Vincent Scully noted a basic difference between the mammal mural and its famous neighbor. "Everywhere," he wrote of the *Age of Reptiles,* "the space is deep, punctuated with figures in echelon and diminishing to treed mesas far away. It suggests an earth young and uncrowded, open to many wonders. That deeply receding space is very important to the fresco." Then, in what may be art criticism's sole response to it, Scully evaluated the other picture: "In contrast, Zallinger's much smaller *Age of Mammals* of 1961– 67, also in the Peabody, is choked with greenery, and, with the exception of the trumpeting mammoth who concludes it, the animal forms are much less isolated and arresting than those in the *Age of Reptiles.*"

Figure 17. Serpent in tree (Paleocene) from Zallinger's *Age of Mammals* mural. Courtesy Peabody Museum of Natural History, Yale University, New Haven, Conn.

The *Age of Mammals* that Scully described was indeed "much smaller" than the reptile mural. The "trumpeting mammoth" is in the Time/Life cartoon—the Peabody's merely raises its trunk inquiringly. Scully's mistake doesn't invalidate his observation that the Great Hall's life-sized dinosaurs are more "isolated and arresting" than the annex's scaled-down beasts. He might have allowed, however, that the saurians benefit from a high ceiling as well as "deeply receding space." And there was another, less estimable, reason for their spaciousness, of which he perhaps was unaware.

"I had felt it unfortunate that the dinosaur mural portrayed these great beasts in a modern landscape where their remains now occur," Paul O. Dunbar, the museum's director, recalled. "The great cliffs of Navajo sandstone make a beautiful and impressive backdrop . . . but the environment is totally unlike that in which their evolution occurred." The reptile mural's sparse plants provide a "young and uncrowded impression" because they are props on an anachronistic stage that tells relatively little about the Mesozoic's many environments.

Dunbar accordingly planned the second picture to reflect western North America's real, shifting Cenozoic landscapes. Whether "choked with greenery" or not—and I think it does evoke far horizons despite its cramped venue—the *Age of Mammals* integrates art with scientific authenticity. And while geology and climate were the neo-Darwinian forces driving the shifts, plants were its main indicators. The mural's only nonliving signs of climate change are thinning clouds, dwindling waterways, and ice fields. It is the plants that tell the story—rainforest palms, savanna sycamores, prairie cottonwoods, tundra birches.

The mural's paleobotany has dated somewhat. (Dunbar recalled that Zallinger found his plant adviser less helpful than the others.) The tree from which the serpent dangles was probably an herb; the Paleocene forest generally is too tropical for its time and place; and the Miocene savannas' mountain laurels and hydrangeas seem more like garden club displays than prehistoric wilderness. Yet Zallinger's rendering of the evolving continent's overall plant diversity is vivid and sometimes subtle. Grass modulates from ferny green in the Oligocene to pale umber in the Miocene and reddish gold in the Pliocene, evoking the shift first from a moist warm climate to a dry warm one, and then to a cool one in which frost reddens leaves.

The reptile mural's plants seem almost incidental to its animals. The mammal mural's floral enfoldment of its fauna shows a deep interpenetration, and it thus implies one explanation other than dinosaur extinction for the appearance of modern placentals—that plants influenced it. Whether Dunbar or Zallinger intended any such implication is unclear. Flowering plants, angiosperms, coincide with mammals in both murals, however, and, as I have said, the dangling serpent that marks our era's dawn carries strong generative associations. While perhaps seeming to bar the little Paleocene mammals from "the fruit of the tree of knowledge," it also might be inviting them to taste it. And, oddly enough, a Peabody Museum paleontologist was one of the first to articulate such an explanation, although before Zallinger's or Dunbar's time.

The explanation has always been scientifically obscure, perhaps partly because that rare mammal in the nineteenth century, a French Darwinian, first conceived it. In 1863, Darwin remarked in a letter to Lyell that he had heard of a botanist who, unlike most of his Gallic compatriots, considered "descent with modification" a good idea. Five years later, he wrote the man, Gaston de Saporta, saying he had read with great interest his papers on fossil plants, and was gratified to learn that he was "a believer in the gradual evolution of species." Saporta was honored, and they began a correspondence. A scion of Provençal nobility, Saporta had taken up paleobotany in 1855 after his first wife's death in childbirth had shocked him out of youthful dilettantism. By the mid 1860s, he had become one of the first to study the chronology of plant fossils systematically, drawing on Tertiary lake bed deposits in the country around Avignon. Darwin hoped his lake strata leaves, seeds, and flowers would do for plant evolution what horses had done for animal evolution, and he said so in an 1872 letter: "[T]he close gradation of such forms seems to me a fact of *paramount* importance for the principle of evolution," he wrote. "Your cases are like those of the gradation of the genus *Equus* recently discovered by Marsh in North America."

In 1877, Darwin got an unexpected letter from Saporta in response to his recent book on the importance of insect fertilization to flowering plant reproduction. Temporarily forgetting "close gradation," Saporta took up Darwin's fertilization ideas and applied them to the origin of angiosperms, suggesting that they "could not have diversified without insects," and vice versa. Insects and angiosperms, he wrote, "had to follow an evolutionary path side by side . . . from the moment when the effects of crossing and the mode of fertilization (entomophilous fertilization) became manifest." Because of their relationship, "the vegetable kingdom, and the angiosperms in particular, took flight, ramifying in all directions and offering from all sides the most varied, unexpected, and ingenious combinations," while insects also "took flight" into the "Diptera, Hymenoptera, Lepidoptera, Hemiptera, and many Coleoptera" that feed at and fertilize flowers.

Darwin's reaction was enthusiastic, perhaps envious. "I am surprised that the idea never occurred to me," he replied, "but this is always the case when one hears a new and simple explanation for some mysterious phenomenon. . . . I formerly showed that we might fairly assume that the beauty of flowers, their sweet odor and copious nectar, may be attributed to the existence of flower-haunting insects, but your idea, which I hope you will publish, goes much further and is much more important."

Darwin reacted differently, however, to a related idea in Saporta's letter, which continued: "I admit another series even more remarkable for its consequences. I wish to speak of the development of mammals, linked with that of plants as the latter is to that of insects. If plants were unable to develop except through the influence of insects . . . the mammals, for their part, could not have begun their upward march unless that of the vegetable kingdom had already been under way." Mammals, Saporta thought, had remained "feeble" and "subordinate" until the Tertiary's beginning, "just when the vegetable kingdom, fully completed, contained almost the same elements, and largely the same genera, as presently." Only with the help of new foods produced by flowering plants could mammals have diversified and gained predominance over the animal kingdom.

Darwin's response to this second "new and simple explanation" was guarded and puzzling. "With respect to the great development of mammifers in the later geological periods following from the development of dicotyledons," he wrote, "I think it ought to be proved that such animals as deer, cows, horses &c. could not flourish if fed exclusively on the gramineae and other anemophilous monocotyledons, and I do not suppose that any evidence on this head exists." Alleging with his usual exaggerated modesty, that he was not "properly grounded in botany, " Darwin seemed to confuse grasses (Gramineae)—which are flowering plants, although their wind-pollinated (anemophilous) flowers are tiny—with the more primitive, nonflowering ones that predominated in the Mesozoic. His confusion may have been genuine. Many botantists at the time thought monocot angiosperms like grasses and palms more primitive than dicots like oaks and apples. Yet Darwin surely knew that cows and horses, at least, can be "fed exclusively on the gramineae" (that is, on oats, hay, bran, etc.), and he might better have suggested that the experiment be conducted by feeding them on non-angiosperms like cycads, conifers, ginkgos, and ferns.

Darwin's reply is so baffling as to seem evasive, and he said nothing further about the mammal-angiosperm theory. Their correspondence languished, and he didn't keep Saporta's letters. The 1877 one only survived because the paleobotanist had made a copy of it. Perhaps Darwin felt that Saporta, with his unusual and enviable ideas, was proving less of a disciple and informant than he had hoped. More concerned with organisms diverging competitively than "marching together," Darwin may have found the idea that mammalian evolution was tethered to that of trees and bushes confusing, perhaps subtly inimical to his tenuously held belief that mammals were "higher on the scale of nature" than their "predecessors."

Saporta published his idea in 1879, although he seemed a bit defensive about it. He acknowledged that many large herbivores had existed in the Mesozoic, when plants remained primitive. Still, he wrote, they were merely "enormous reptiles" like Gideon Mantell's iguanadons, with jaws able to grind the hardest, least nutritious vegetable substances. "Plants provided animals with richer foods only as they became more developed," he insisted.

> [L]and animals, after quickly attaining a remarkable degree of organic complexity, found themselves unable to progress further on their own; they were forced to await the other kingdom's progress. That explains why, when one encounters the mammals before the Triassic's end, and again toward the middle and end of the Jurassic, they always are rare, scrawny, imperfect, in fact, stymied. The vegetation of those same epochs was impoverished, its forms coarse and monotonous. It didn't advance until long after, toward the end of Cretaceous times, and only then did a similar development of mammals occur . . . the lack of tender and succulent plant parts long opposed an increase in herbivorous mammals, and, by necessary consequence, of carnivores. As long as that state of affairs persisted, the entire class could not multiply or perfect itself.

Darwin's evasive skepticism may have discouraged Saporta. He didn't publish on the idea again, and there is no evidence that he conducted experiments such as feeding cows and horses only on primitive plants. (Darwin, although an avid experimenter, apparently didn't try it either, but foddering a Victorian household's horses with pine needles and ferns might have seemed like fueling a modern one's cars with paraffin.) Saporta's idea fell into scientific limbo, and not without reason, since he had provided no evidence for it except the circumstance that Cenozoic mammals evolved after angiosperms did. Even in 1879, dinosaur extinction seemed a more likely explanation for modern mammals' appearance than angiosperm

Figure 18. Jacob Wortman with dinosaur bones. Courtesy American Museum of Natural History Library.

evolution. They apparently had evolved soon after dinosaurs disappeared, but some time after angiosperms became common. Notwithstanding their collective obsession with mammalian origins, Owen, Huxley, Marsh, and Cope ignored Saporta's idea, if they even heard of it.

Soon after their deaths, however, another anomalous evolutionist resurrected the idea. Cope's and Osborn's restless collector Jacob L. Wortman wanted to be a full-fledged scientist like them, and had the ability. Dapper and gregarious, he had published a brilliant comparative study of vertebrate teeth at age thirty, and sometimes filled in for Joseph Leidy as a Philadelphia Academy lecturer. Wortman lacked the funds and influence for scientific preeminence, however, and circumstances forced him, like John Bell Hatcher, to roam from one job as a museum assistant to another.

After rivalry with Barnum Brown drove him from the American Museum in 1899, Wortman joined Andrew Carnegie's new Pittsburgh Institute, just endowed with over a million dollars by its steel magnate benefactor. An admirer of Marsh's "astonishing" dinosaurs, Carnegie wanted his institute to collect even bigger ones, but Wortman had done more than enough of that backbreaking work at Como Bluff, and he still dreamed of taking over the largely undescribed Tertiary mammal fossils of Marsh's Peabody collection. He proved a less compliant employee than his boss at the institute, a scientific bureaucrat named William J. Holland, had anticipated. Holland forced him to resign in 1900 after a squabble about sauropod bones. Wortman, Holland complained, "became very angry; told me to 'go to hell' [and] covered me with uncomplimentary epithets."

Osborn then finally helped his disgruntled former assistant get a job at the Peabody Museum cataloguing mammal collections, which Wortman doubtless hoped would lead to the Yale paleontology chair he coveted. At least, it finally gave him time to develop a "theory of mammal descent" that his fellow bone hunters had probably heard over badlands campfires. "The facts of mammalian distribution in the Northern Hemisphere may be briefly stated as follows," he wrote in the second of a Peabody monograph series on Marsh's treasures.

> Early in the Mesozoic, there appeared small, mammal-like forms, which were widely distributed over both the Northern and Southern hemispheres. Representatives of these species continued throughout the Cretaceous, and finally disappeared in the early stages of the Tertiary. . . . One fact in connection with these Mesozoic forms stands out clearly and distinctly, and that is that as far as we are able to judge from their fragmentary remains, the progress of their evolution toward any of the higher mammals was very slow indeed.

On the other hand, the oldest Tertiary strata, Cope's famous Puerco beds in New Mexico, contained what Wortman described as "the remains of a rich mammalian fauna, thirty-one species in all, composed largely of the representatives of the higher, or Eutherian, subclass." Noting that

Saporta and other paleobotanists believed that angiosperms had evolved in a then subtropical Arctic region during the Cretaceous, Wortman speculated that modern placentals had evolved there along with them, then spread southward into North America and Eurasia as climate cooled during the Tertiary. "The existence of the higher types of Mammalia was manifestly impossible before the appearance of the necessary plants upon which they so largely depend for food," he wrote, echoing Saporta, "and I shall therefore assume the existence of a close relationship between the development of the one and the origin of the other."

Despite its prestigious venue, however, Wortman's idea got no more attention than Saporta's, and for like reasons. Both men were outsiders, and Wortman lacked even Saporta's vestigial *noblesse.* He complained to Charles Schuchert, the shell hunter who did get Marsh's paleontology chair, that he had "failed to receive any word of encouragement or appreciation from the Yale authorities." After cataloguing only carnivores and primates, he left the Peabody abruptly in 1903, the year it published his angiosperm theory, and drifted away from science, sick of taking orders from men who knew less about paleontology than he did. He ranched a few years in Nebraska, bone-hunting in his spare time, then spent his last two decades running a drugstore in Brownsville, Texas. Osborn glossed over this waste of talent in a 1926 obituary, saying that Wortman had "retired" to Texas.

Another former boss didn't even gloss. "It is to be regretted that his impulsive temperament led him to abandon his paleontological studies in 1900," William Holland blandly misinformed readers of the Carnegie Institute's journal. "He literally 'shook the dust' of paleontology from his feet at that time and absolutely refused to read anything in the science in which he had already achieved for himself an enduring reputation. It was a curious act of renunciation, the psychology of which is hard to explain."

Wortman's idea probably wouldn't have caught on even if he had been more influential. Like Saporta's, it was based on little more than the circumstance that angiosperms and placentals both became predominant in the Tertiary. Like Saporta, Wortman offered no evidence, or even speculation, as to how eating angiosperms might have influenced Cretaceous mammals enough to foster the Puerco formation's "rich mammalian fauna." Since then, paleontologists have approached the two mens' idea cautiously, if at all. It occupies a paragraph in Peter J. Bowler's 525-page history of evolutionary biology from 1860 to 1940; not a word in Gould's 1,433-page evolutionary theory treatise.

Simpson skirted the idea skeptically in one of his later books. He thought the likelihood that many Cretaceous and Paleocene mammals had been arboreal or browsers implied a relationship between their radiation and "a spread of forests and shrubby veldts." But he saw no correlation between that and the "basic radiation" of flowering trees and shrubs. "The radical changes in faunas in the Mesozoic-Cenozoic turnover do seem to demand correlation with some worldwide ecological change," he concluded, "but more definitive designation of its nature is one of the problems on which paleontologists must still work." Jason Lillegraven skirted it more optimistically when he speculated that plant-mammal coadaptations in the mid-to-late Cretaceous might have been "even more direct than we now suspect" and that "the roots of many Cenozoic orders of mammals" would eventually be found there. Michael Novacek skirted it more cautiously when he wrote that "the Paleocene saw the radiation of many small mammals that specialized in fruit and other diets. These were the source of tree-dwelling groups like the primates. . . . Both the mammals and the plants 'took flower.'"

One reason the idea languished was that angiosperm origins kept receding in time. Simpson saw no correlation between the "basic radiation" of angiosperms and Paleocene mammals because angiosperm origins had been dated back from the Late Cretaceous, as Saporta and Wortman envisioned them, to the Early Cretaceous, or even the Jurassic. (The earliest known fossils now thought to be of flowering plants date from the Early Cretaceous, about 130 million years ago, and DNA analysis seems to support this.) In a 1971 paper, William Clemens saw a correlation between angiosperms and Mesozoic, not Cenozoic, vertebrate evolution. "Reflecting the rise of the angiosperms," he wrote, "a pattern of marked evolutionary radiation with the extinction of some older lineages during the Middle Cretaceous characterizes the history of most terrestrial vertebrates." Clemens thought the mid-Cretaceous predominance of low-browsing duckbilled and horned dinosaurs over long-necked sauropods was a response to "new sources of food provided by the diversification and increase in geographic range of the angiosperms." He thought that the herbivorous multituberculates also had evolved new groups, ptilodonts and taeniolabids, in response to the new food sources. Writing before Nessov discovered zhelestids, he couldn't see a direct relationship between angiosperms and mid-Cretaceous metatherians or eutherians, because they had "dentitions in which shearing was the primary or only function, suggesting that they were small carnivores or omnivores." But he thought that the coevolution with angiosperms of spiders, beetles, flies, butterflies, and bees had "provided an expanding source of food for small vertebrate carnivores and omnivores."

Recently, invertebrate paleontologists have raised further problems by challenging the temporal evidence of insect-angiosperm coevolution that prompted both Saporta's and Clemens's ideas about it. A 1995 study suggested that the insect families usually associated with angiosperms—bees, flies, beetles—had reached a high diversity almost a hundred million years before flowering plants appeared. Apparent fossil bee nests in trees of Arizona's 220-million-year-old Petrified Forest, for example, imply that insects evolved features like sticky tongues and sucking mouthparts for feeding, not on flowering plants, but on earlier kinds. Jack Sepkoski, the study's co-author, speculated that flowers might have appeared in response to insect evolution instead of vice versa. Such evidence did not refute Saporta's idea of insect-angiosperm interaction, since he also had thought insects affected angiosperm evolution, but complicated it by so widely separating their origins.

Such a separation would similarly complicate Saporta's idea about mammals. If insects influenced angiosperm evolution more than the other way around, then mammals, emerging in the Late Triassic, might have done so too. The herbivorous multituberculates appeared in the Jurassic, and other early groups probably ate soft plant parts when available. As living species such as tayras show, many predators are not choosy about their diet, despite their specialized teeth. Mesozoic mammals may have been as important to pollination, seed broadcasting, and other plant functions as squirrels, monkeys, possums, and mice are today. The apparent climbing ability of the first known eutherian, *Eomaia scansoria,* suggests that it was in the forest canopy, perhaps exploiting pollen, fruit, and nectar along with small animal prey.

Dinosaurs may also have affected angiosperms, of course, and Robert Bakker accused mammal experts of ignoring them when they insisted that Cretaceous micromammals, "however tiny and unimportant," had had "a major impact on the evolution of angiosperm fruits, nuts and leaves." Bakker thought that the horned and duckbilled dinosaurs that appeared in the Early Cretaceous might have furthered, even caused, angiosperm evolution by mowing down gymnosperm competitors and forcing flowering plants to reproduce more efficiently. (Their teeth and jaws did allow more grinding and chewing than those of most earlier dinosaurs.) "In their way, dinosaurs invented flowers," he wrote. "Without them, perhaps, our modern world would yet be as dull green and monotonous as was the Jurassic flora." It seems equally possible, as Clemens proposed, that ceratopsians and hadrosaurs evolved in adaptation to feeding on early angiosperms, but Bakker had a point in maintaining that dinosaurs must have used plants in complex and intensive ways.

Mammals may have interacted even more intensively with vegetation, however. The enhanced sense of smell that helped them at night would have been an advantage in finding and evaluating plant foods, particularly those from aromatic, insect-attracting species. Smelling and tasting food allow for a more selective approach than just swallowing it. As Saporta noted, large animals can crop huge amounts of rough vegetation; small ones have to be more discriminating, going for the most nutritious items. Mesozoic mammals' smallness may have given them greater influence over some plant resources than big dinosaurs. Some may have collected and stored seeds and other plant foods, for example.

Early mammals' olfactory powers could have allowed them to exploit a peculiarly important plant phenomenon. This struck me one day when I was walking in the California coast range. Wild pigs had dug up entire grassy hillsides in search of edible roots and tubers, and I noticed that they also were rooting in the bare shade under large oaks. They probably were sniffing out truffles, the spore-bearing bodies of the many fungal species that live symbiotically on plant roots, absorbing water and nutrients from the soil and getting food in return. The fungi, called mycorrhizae, "mushroom-roots," are indispensable to the survival of most higher plant species and played a major, and early, role in their evolution. Many mammal species eat truffles, spreading the spores in the process, and this indirect relationship with plants may be as old as the direct one.

Of course, the possibility that some Mesozoic mammals coevolved with flowering plants doesn't necessarily mean that mid-to-late Cretaceous eutherians did so. When I mentioned the idea to David Archibald, he said it remained unprovable for lack of evidence. A scarcity of transitional fossils continues to sideline Saporta's idea. There is still no fossil proof that modern herbivores—rodents, rabbits, ungulates—evolved in the Cretaceous, much less that eating angiosperms influenced their evolution.

Still, it is suggestive that Nessov's Kyzylkum fossils, the "oldest well documented eutherian-dominated fauna in the world," as Archibald put it, have even slight similarities to modern herbivores. An Early Cretaceous origin for angiosperms doesn't necessarily mean that flowering plants reached anything like their present ecological dominance in the Mesozoic.

One Late Cretaceous fossil flora, at least, suggests that zhelestids could have been adapting to relatively new vegetation ninety million years ago. The flora, preserved in volcanic ash at Big Cedar Ridge in Wyoming, was rich in dicot angiosperms, but most were herbs or vines. Nonflowering plants—tree ferns, cycads, and conifers—formed the predominant forest cover, although there were far fewer species of them. Only one angiosperm tree, a palm, was common. The paleobotanists who studied the Big Cedar Ridge flora thought it might have been typical of upland regions, and that the angiosperm trees and shrubs that grew then—like sycamore and viburnum—might have been confined to moist stream corridors and floodplains. Such a pattern might have had something to do with zhelestids' presence among the sycamores and viburnums of southwest Asia, and with their apparent absence from Mongolia's uplands. According to Nessov,

> [T]he environmental conditions (temperature, moisture, and dynamics of these factors) which reigned on the coastal plains of southwest Asia, in comparison to those prevailing on more central areas of Asia, led to a greater diversity of mammals . . . this resulted in more intense competition between ecologically similar forms and in a faster evolutionary rate, especially in the Late Early Cretaceous. . . . It is possible that the rarity of multituberculates in humid and semi-humid coastal plains of southwest Asia led to the faster evolution of therians toward plant consumption; all the more so in that the insectivorous niche there must have been close to saturation, and hence subject to fierce biological competition.

If Cretaceous eutherians were fierce biological competitors, it might have affected more than other mammals. One thing that puzzles me about prevailing dinosaur extinction theories is that they don't explain the disappearance of chicken-sized dinosaurs like the ones in Zallinger's reptile mural. Something about the Cretaceous extinction was universally lethal to large land animals, but not to small ones, so why didn't some small nonavian dinosaurs survive? One possibility, at least, is that Late Cretaceous ones had to compete with mammal groups that hadn't existed in the Late Triassic or Jurassic, when nonavian dinosaurs did recover from mass extinctions.

However plants affected Mesozoic mammals, and whether or not angiosperms allowed eutherians to "flower," there's no doubt that the vegetable kingdom influenced Cenozoic mammal evolution, as Zallinger's "greenery choked" chronofaunas imply. Plant-mammal interactions have been central to our era, according to the paleobotanist Scott L. Wing, who observes:

> It may well be that the limited areas of open vegetation during the early Tertiary precluded the radiation of large browsing or grazing herbivores and favored the evolution of small, frugivorous terrestrial, arboreal, and volant types. These small frugivores may in turn have favored the success of the animal dispersed plants characteristic of closed forests. . . . Under the combined influences of drying and cooling in the mid-Tertiary, these closed-forest ecosystems were fragmented. Plants with shorter life cycles and less wood tissue diversified in this climatic regime, and larger grazing and browsing herbivores evolved to take advantage of this expanded resource. These large herbivores in turn probably favored the success of such herbaceous and grassland vegetation by physically disturbing it through feeding.

The English biologist Tom Wakeford called large mammal herbivores "the early architects of the savanna landscape . . . each with its own internal microbial garden, travers[ing] every habitat that could support them." Indeed, the hoofed abundance and diversity of Miocene savannas supports this idea. Up to twenty genera of grazing and browsing ungulates simultaneously inhabited western North America, from elephantine gomphotheres and giraffe-necked camels, to gazelle-sized horses and pronghorns. An important factor in this diversity, no doubt, was the fact that many trees and shrubs, like the oaks, chestnuts, and ginkgos in Zallinger's mural, bore fruits or seeds edible by large herbivores.

As Daniel Janzen's studies of contemporary Central America savanna and dry forest show, many tree species there with large, tasty fruits—jicaro, guapinol, guanacaste—depend largely on ungulates to spread their seeds. Some are dwindling because the big herbivores they evolved with are gone, although reintroduced horses, which consume the fruits as though their 7,000-year New World extinction had never happened, are helping restore some. The Miocene western plains resembled a more temperate version of those habitats, with large-fruited trees like honey locust, osage orange, prickly pear, and wild plum as well as oaks and chestnuts, and the same gomphothere, camel, rhino, and horse genera as in Zallinger's mural have been found in a Honduran fossil deposit. David Webb, who collected there in the 1960s, showed me their fossils at the Florida Museum of Natural History. "It's probably going to be dreadfully disappointing," he warned, "everything's so fragmentary." But the very fragmentation was significant, because most of what remained were the parts for interacting with plants—massive jaws, high cusped rhino molars, and high-crowned camel and horse teeth.

Webb defined the mid-Miocene's "stately chronofauna" as "a coevolving set of primary consumer species which also had regular coevolving relationships with the producer species of the savanna flora" and attributed its longevity to a "scheme of coevolutionary interactions against a background of relatively stable environments." Animal diversity literally fed on plant diversity as long as climate remained equable. As the North American climate grew cooler and drier in the late Miocene, however, "the vast herds of ungulates themselves probably accelerated the decline of savanna and the expansion of steppe. . . . A positive feedback loop between faunal and floral evolution may have accelerated the process that established steppe conditions in the early Pliocene."

A strange thing happened then. It is well known that the high-crowned teeth of horses, elephants, and other modern grazers are an adaptation to the abrasiveness of silica-rich grasses. It is less well known that, when modern grazers were evolving, many grasses underwent a major biochemical shift, metabolizing a different carbon isotope, C_4, than most plants, which use the isotope C_3. "You can tell from the chemistry of fossil teeth whether an animal eats mainly C_3 or C_4 vegetation," Bruce MacFadden told me.

> So grazers on modern grasses will have a distinct C_4 carbon signal in their teeth and browsers and other animals eating more primitive grasses will have a C_3 signal. The isotopes also tell something about the climate because C_4 grasses are more arid-adapted and C_3 are more cold-adapted, so in a sense you also look at climate change as well as vegetation change as well as paleo diets of ancient herbivores. There seems to have been a major global climate change about five to seven million years ago. Prior to that time, all ecosystems were dominantly C_3. The dominant grass ecosystems today are C_4, but they're a relatively recent arrival on the ecological scene.

The reason for the shift from C_3 to C_4 in grasses is unclear, as are its implications. It may have been related to changes in atmospheric carbon dioxide concentration, and thus to the cooling, drying climate of the late Miocene epoch. Traces of C_4 vegetation as much as twenty-five million years old have been found since MacFadden talked to me in 1993, however, complicating the matter. And, although less diverse, landscapes after the change still teemed with megafauna. In southern California's Pleistocene, for example, twenty-one large mammal and bird species preyed or scavenged on sixteen large mammal herbivore species, which were abundant enough to leave more fossil spores of a common dung fungus, *Sporomiella*, in the soil than in modern pastures.

Such shifts show the complexity and subtlety of plant evolution, anyway. Whether or not fossils reveal them, prehistoric plant-animal interactions must have been important. The best point that Robert Wright made in his 1999 *New Yorker* critique of Stephen Jay Gould was that another environmental force beside climate, geology, or extraterrestrial impacts drives evolution. "Natural selection talks only about 'adaptation to changing local environments," he quotes Gould as saying. "[T]he seas come in and the seas go out, the weather gets colder, then hotter etc. If organisms are tracking local environments by natural selection, then their evolutionary history should be effectively random as well." But, Wright maintains, Gould left something out of his description of changing local environments. "This would be good logic if environments consisted entirely of sea and air. But a living thing's environment consists largely of other living things: things it eats and things that eat it."

This other force—which drives organisms along a continuum from predation to parasitism to commensalism to mutualism—still has tenous names like "coevolution," or "symbiology." Competition is part in it, as is adaptation, but it occasionally seems to transcend both by producing things so novel that they change everything. Plant and animal life's main component, the eukaryotic cell, is now thought to have evolved from an association of several kinds of prokaryotic cells (bacteria), some still discernible in eukaryotic cells as organelles such as chloroplasts and mitochondria. Cells might have begun by devouring or infecting others, but eventually evolved an "accommodation" in which some came to live and function within others. In this sense, mammalian evolution began with symbiosis and has continued as such, because each mammal's body consists not only of its component eukaryotic cells, but of viruses, bacteria, fungi, and other organisms that live on and inside it. Those populations, such as the bacteria without which intestines could not digest food, have undoubtedly been major evolutionary factors, and some current theories about this are even more surprising than Mark Springer's DNA analysis of elephant origins.

One theory suggests, for example, that retroviruses, the kind of viruses that, among other things, cause AIDS, might have contributed to the evolution of the mammalian placenta. Some "endogenous" retroviruses occur in the genetic material of mammal cell nuclei, the "genome," and may play roles in physiology and development. According to the science writer Frank Ryan: "Certain viruses have the capacity to fuse mammalian cells into confluent sheets of cytoplasm with many nuclei and no cell membranes between them. . . . Fusion of cells is also a feature of the mammalian placenta, with the formation of a microscopically thin and confluent tissue layer, called the synctium, that is the final barrier between maternal and fetal circulation." Some microbiologists and biochemists think retroviruses contribute to synctium development in living mammals and may have played a role in the placenta's evolutionary origin. If such things do occur, then the relationship among heredity, the organism, and the environment could prove to be much more complex than Darwinians such as Weismann and Simpson have thought. Lamarck's vision of transmutation could take on renewed significance.

Elusive as fossil evidence of Cretaceous plant-eutherian coevolution may prove, it seems as likely to turn up as fossil proof of microscopic symbioses. There is no denying Saporta's basic claim, at least, that angiosperms produce richer foods than other plants. Whether or not proof ever emerges, the possibility remains that Cenozoic mammals' growth in diversity and intelligence was related to "greenery" as well as dinosaur extinction. So our distant ancestors indeed may have tasted of the "fruit of the tree of knowledge," although the ethical implications of that may depend on us.

TWENTY-ONE

Anthropoid Leapfrog

A SCRAP OF POSSIBLE EVIDENCE for Gaston de Saporta's idea came to light in the 1960s. Among the apparently Cretaceous mammal fossils that Sloan and Van Valen found at Bug Creek was a single lower molar that may have belonged in the same genus as teeth from Hell Creek's Paleocene Tullock formation, 100 miles to the west. That genus, named *Purgatorius*, after Purgatory Hill, where it was found (so called for the Dantean torments of hauling fossil matrix down it), was thought to be an early primate. As with zhelestids, an adaptation to plant as well as animal foods distinguishes early primate teeth from other primitive eutherians'. The tooth implied that a distant human relative had been eating fruit before the dinosaurs vanished.

Like the other supposedly Cretaceous Bug Creek mammal fossils, however, the single molar seems more likely to be Paleocene, and paleontologists aren't even sure that it belongs in the genus *Purgatorius*. Anyway, Cretaceous primates wouldn't necessarily prove that modern mammals co-evolved with flowering plants, any more than Cretaceous ungulates would. Far from seeing a Mesozoic *Purgatorius* as evidence of such progress, Stephen Jay Gould found it exemplary of randomness in his 1981 essay hailing the Alvarezes' comet theory. "It may not have been the only member of our order," he wrote, "but there probably weren't many of us back then. Suppose that *Purgatorius* hadn't pulled through—and remember, it probably owed its continuation to luck or to adaptations not related to features we value in primates because we have capitalized on them. Primates would not have reevolved."

Still, primates *are* noted for ingenious relationships with angiosperms. Jacob Wortman had our order particularly in mind when he revived Saporta's idea in 1903. He did so in the introduction to his monograph on Marsh's primate fossils, theorizing that we might have evolved from arboreal Cretaceous marsupials "with grasping hands and feet" that began to feed on the angiosperms appearing above the Arctic Circle. Implying that primates achieved a placental grade separately from other modern orders, this seems a suitably orthogentic idea for a former Osborn paladin, although King Henry might have liked a possum for an ancestor even less than an ape.

Wortman's idea of primate origins brings up an aspect of Zallinger's *Age of Mammals,* which may have seemed anomalous when I said that the mural is "about us." Human origins, the transition from apes to hominids in the past five million years, certainly is prehistory's most "about us" aspect, and is almost as popular a subject as dinosaurs. A row of books on it always abuts theirs in libraries and bookstores. During my Peabody visit in 2000, I found a bright spot in the mammal annex, a shiny human evolution exhibit, which, Mary Ann Turner told me, had replaced a musty titanothere one. But the mammal mural itself doesn't show hominids, or even apes and monkeys. There is good reason for the omission, because hominids, apes, and monkeys didn't evolve in North America; they only arrived here in the last millisecond of deep time. But it makes the mural's connection with the human present seem tenuous, at least superficially.

Another important stage of primate evolution does appear in the mural, however. At its far left, in the early Paleocene, a bushy-tailed, long-snouted creature named *Plesiadapis* scampers along a fallen branch below the giant serpent. *Plesiadapis* is regarded as an early primate and may have been arboreal, although it probably resembled a squirrel more than a monkey, since it had gnawing incisors, a gap, or "diastema," between canines and molars, and long claws instead of nails.

Plesiadapis raises a question that perplexes evolutionists more than that of how humans evolved from apes: how apes might have evolved from creatures like *Plesiadapis.* In a way, it is a bigger question, because the gap between *Plesiadapis* and a chimpanzee is much greater than that between a chimpanzee and a human. Chimpanzees share 98 percent of their DNA with us, and they use crude tools and live in complex social groups. *Plesia-*

dapis and its Paleocene relatives didn't resemble any living animal much, but the closest thing to them certainly isn't an ape or monkey.

The gap between *Plesiadapis* and anthropoid primates (monkeys, apes, and hominids) is so wide that it is surprising that creationists don't dwell on it. Zallinger's mural contains a hint of what might have bridged it, however—the two tiny Eocene creatures named *Pelycodus* that huddle near the *Oxyaena-Coryphodon* confrontation. Compared to *Plesiadapis,* they have marked anthropomorphic qualities—a binocular gaze, prehensile, nailed digits, and elongated hind legs. The cartoon in Time/Life's book *The World We Live In* also shows an even more gremlinesque homunculus clinging to a branch just below them. It is named *Tetonius,* and it likewise has binocular vision and prehensile paws. Why Zallinger left *Tetonius* out of the mural is unclear (his adviser, Elwyn Simons, a primate specialist, told me that he doesn't remember), but it did share the North American Eocene with *Pelycodus.* Cooling climate extirpated such creatures from this continent after the Oligocene. Ones like them survived in the Old World tropics, however, and their distant descendants, called prosimians, still do. *Pelycodus* probably was related to a group called adapiforms, which includes living Malagasy lemurs, Asian lorises, and African pottos and bushbabies. *Tetonius* probably was related to one called tarsiiforms, of which the only survivors are the Malay Archipelago's bug-eyed tarsiers.

Early paleontologists found fossil prosimians in Europe, although it took them a while to realize it. In 1822, Cuvier thought one from the Paris gypsum was a small pachyderm and named it *Adapis,* "rabbit" (*Plesiadapis* means "near *Adapis*"). Owen made an opposite mistake in 1838, first identifying the dawn horse *Hyracotherium* as a primate. Once the fossils' similarity to living prosimians became evident, transmutationists began to suspect that they might also be ancestral to the anthropoid primates. As with horses, however, prosimian evolution remained relatively little known until collectors began to explore the North American badlands.

In 1870, Joseph Leidy acquired a Bridger Basin fossil that he named *Notharctus,* "false bear," because, although he thought it was a "small, extinct pachyderm," it had anatomical characters suggesting it "was probably as carnivorous in habit as the raccoon and bear." Its jaw and teeth puzzled him, however, and provoked the diffident naturalist into uncharacteristic speculation. "In many respects the lower jaw of *Notharctus* resembles that of the existing American monkeys quite as much as it does that of any living pachyderm," he wrote. "The resemblance is so close that but little change would be necessary to evolve from the jaw and teeth of *Notharctus* to that of a modern monkey. . . . A further reduction of a single pre-molar would give rise to the condition of the jaw of the Old World apes and man."

As it happened, Leidy's rare plunge into theorizing was dead right. *Notharctus* turned out to be an adapiform prosimian, in fact, a close relative of Zallinger's *Pelycodus*. Cope and Marsh were hot on Leidy's heels, and in the following years, they coined six other names for fossils the same as *Notharctus*. They also found many genuinely new prosimian genera in the badlands, demonstrating that Eocene North America had supported a welter of lemurlike, lorislike, and tarsierlike beasts. Then, of course, they disagreed about them.

When Wortman sent him a bulbous-skulled little fossil from Wyoming, Cope named it *Anaptomorphus homunculus* and decided with typical brashness that it was the ancestor, not only of "the Malaysian genus of lemurs, *Tarsius*," but of all anthropoid primates. "In conclusion," he wrote after describing it in his "Bible," his bulky compendium of Tertiary fossils, "there is no doubt that the genus *Anaptomorphus* is the most simian lemur yet discovered, and probably represents the family from which the true monkeys and men were derived. Its discovery is an important addition to our knowledge of the phylogeny of men."

Marsh thought that *Anaptomorphus* was the same as *his* genus, *Antiocodon*, and that Cope's claims of possible anthropoid ancestry were presumptuous. "The energy of Cope has brought to notice many strange new forms, and greatly enlarged our literature," he sniffed in his 1877 keynote speech to the American Association for the Advancement of Science.

> The relations of the American Primates, extinct and recent, to those of the other hemisphere offer an inviting topic, but it is not in my present province to discuss them in their most suggestive phases. As we have here the oldest and most generalized members of the group, so far as now known, we may justly claim America for the birthplace of the order. That the development did not continue here until it culminated in Man, was due to causes which we can only surmise.

Cope's choice of the tarsierlike *Anaptomorphus* as an anthropoid ancestor showed his usual lucky prescience, however. Tarsiiforms do have qualities that seem to link them more closely to anthropoids than adapiforms. Living tarsiers' upper lips aren't cleft like most mammals', including lemurs, and their snouts are short like those of apes, while their brains' olfactory region is comparatively small. They have a well-developed visual cortex, and their reproductive anatomy is more anthropoid than that of other prosimians, with a menstrual cycle and a placenta resembling those of apes and monkeys. On the other hand, as Marsh implied, such living similarities don't prove that Eocene tarsiiforms evolved into anthropoids, and some anatomists saw more similarities between anthropoids and adapiforms.

The main trouble with Cope's audacious idea, as Marsh also implied, was that paleontologists hadn't found any early anthropoid primate fossils with which to test it. When such fossils did come to light, decades later, the question of tarsiiform or adapiform origins quickly became one of the main evolutionary controversies, although the fossils were so small, sparse, and diverse that the technical level of argument made hominid discussions seem elementary. Anthropoid origins occasioned a kind of bone hunters' leapfrogging through the twentieth century, and the debate remained so inconclusive that Kenneth D. Rose, a biologist at Johns Hopkins, lamented in 1984 that either the adapiforms or tarsiiforms "almost certainly gave rise to the anthropoid primates . . . in the late Eocene or early Oligocene, but the precise pedigree of higher primate groups is still hotly debated."

For the century's first three decades, the earliest known anthropoidlike fossils were the ones that the Austrian collector Richard Markgraf had found at Fayum. These consisted of parts of four lower jaws, and paleontologists surmised from their teeth that two small apelike genera and two small monkeylike genera had inhabited Oligocene Africa. Two of the jaws, which the German paleontologist Max Schlosser attributed to creatures he named *Parapithecus* and *Propliopithecus,* were nearly complete, and correspondingly influential. "These were widely figured in books on paleontology and human evolution and came to serve as the basis for almost all discussions of the origin of the higher primates," Elwyn Simons writes. The paucity of the material limited understanding, however. Schlosser thought that *Parapithecus* might have been an early human ancestor, but he misjudged the number of its teeth because a piece was missing from the jaw Markgraf sent him. *Parapithecus* eventually proved to have teeth more like a New World monkey's than an ape's.

Then an apparent early anthropoid fossil turned up in Asia, ironically, while Andrews and Granger were combing Mongolia for the "missing link." Perhaps to compensate Barnum Brown for his competitors' mounting celebrity (Brown had also just lost his wife), Osborn had the Frick family send him to look for southern Asian fossils in 1921. Brown was well aware of Osborn's and the media's missing-link fever, and, dinomania notwithstanding, he doubtless dreamt of a sensational primate find. He went to India, where he collected many bones as well as a new wife, a vacationing American co-ed he nicknamed "Pixie." The bones, from a famous fossil site, now in Pakistan, called the Siwalik hills, included primate fossils. They were of Miocene and Pliocene age, however, much later than Fayum's, and Osborn's curatorial Mordreds hastened to throw cold water on any hope that the Siwalik primates might be the "dawn man" King Henry craved. W. D. Matthew wrote that Brown's three anthropoid ape jaws indicated an evolution "in the direction leading more to the type of modern gorilla and chimpanzee than toward the human type."

Brown then got wind of an older site in the Burmese teak forests, an uninviting prospect by conventional standards, but not one to daunt Osborn's bloodhound. "Looking for bones in the jungle is one of the mortal sins of paleontology," observed "Pixie," whose real name was Lilian. "It simply *isn't done.* Which is why Barnum does it. He likes to find things where no one else will look for them." In 1923, the newlyweds trekked for two months over "indescribable" trails through northern Burma's Pondaung Hills, where geologists had found Eocene sandstone outcrops. Lilian, who published several books on their adventures, called it "rainbow-chasing, the rainbow being an elusive ribbon of varicolored rock—red, blue, and yellow—that unwound through woodlands, alternately appearing and disappearing beneath the heavy undergrowth." Arriving at a village named Mogaung, they made such a stir that in 1978, the current headman, who had been a boy of ten at the time, recalled them vividly fifty-five years later. "I remember a white man and woman coming on horseback with several oxcarts of supplies," he told two American paleontologists. "The man would ride off each morning and return late in the evening with his horse packed with odd-shaped rocks. The woman, who wore pants but was strikingly beautiful, would sit in camp and play with a small dog, whose hair she was constantly combing."

One day, Brown's bag of "odd shaped rocks" included a half-dollar-sized jaw fragment containing three teeth. It was not a find to excite the discoverer of *Tyrannosaurus,* and Brown was too busy getting more bones and struggling with southeast Asia's virulent malaria to care much what it was. "Something petrified filled every inch of cart space," Lilian recalled, "and at one village we had to jettison part of our supplies in order to stuff in more petrified bones. Our pockets bulged too. . . . Even my saddlebags carried a quota of prehistoric sundries—a rib here, a backbone there, teeth, toes, sections of tail. Small wonder that the natives fled in terror when we rode into Monywa." Brown modestly called his Pondaung Hills haul "small, but better than all combined collections made heretofore," but he was unaware that it contained a primate fossil, as he was unaware that an Englishman named G. D. P. Cotter had found similar jaw and tooth fragments in the Pondaungs in 1913. But then, Cotter also was unaware that he had found primate fossils, since his finds weren't identified until 1927.

Despite the overloaded carts and saddlebags, Brown's jaw fragment reached the American Museum safely after he and Lilian sent their hoard to New York. Apparently, nobody cared to look for a missing link in Barnum's "better than all combined" collection, however. The fragment sat in a drawer until 1937, when it caught the attention of Edwin Colbert, then a fairly recent employee. "It was associated with some other fossils, all of which were very fragmentary," he wrote, "and because of its small size and its rather unpromising appearance it seemingly had been overlooked."

Colbert tentatively identified it as the jaw of an early anthropoid. "It seemed to me that because of the form and development of the teeth and because of the deep, short jaw, this fossil might very well be an early forerunner of the anthropoid apes," he recalled, and he named it *Amphipithecus*, "near ape." Cotter's 1913 find had been identified as a primate by then, and named *Pondaungia*, but it was too fragmentary for detailed description. Colbert was able to say that a primate with apparently advanced characteristics had inhabited Eocene Asia, although the jaw was not complete enough to confirm whether its owner had been an advanced prosimian or a primitive anthropoid.

The discovery would have gratified Osborn, who died believing firmly that Asia would produce missing links and dawn men. But further irony beset his legacy in the 1950s, when *another* overlooked fossil turned up in the American Museum's collection, this time from Africa. Elwyn Simons "came across a small piece of forehead bone, identified as a 'possible primate'" that had been lying around since Markgraf had either given it to Granger during the 1907 Fayum expedition or sent it to Osborn afterward. William K. Gregory had noted the bone's monkeylike character briefly in a 1922 book on mammal teeth, but he had not identified it positively. Simons surmised that it belonged to *Apidium*, the genus that Osborn had named from a Fayum jaw fragment in 1908, but had considered a small artiodactyl.

The find encouraged Simons to lead a series of Peabody Museum expeditions to Fayum in search of more primate fossils in the 1960s and 1970s. Combing fine sand deposits neglected by the titan-hunting Osborn team, they found limb and cranium bones as well as numerous jaws, and added new genera to Fayum's primates. Four of these they classified, along with *Propliopithecus,* in the same family with apes, although there was evidence that, unlike living apes, some had tails. *Parapithecus* and *Apidium* they classed in an extinct family they named the Parapithecidae because the fossils had characteristics similar to New World as well as Old World anthropoids. (*Apidium,* the size of a South American squirrel monkey, turned out to be the commonest Fayum primate, or the commonest one fossilized, probably because crocodiles had caught many unwary juveniles and left their remains in bottom sands.) The oldest fossil genus they found, *Oligopithecus,* had characteristics of both anthropoids and prosimians, although it was unclear whether it resembled adapiform or tarsiiform prosimians. *Oligopithecus* implied that prosimianlike African creatures might have given rise to the anthropoid primates either in the late Eocene or early Oligocene.

Meanwhile, however, the two Americans to whom the Mogaung headman described Barnum Brown's 1923 expedition were encountering more Pondaung Hills primate fossils (which was curious, because one of them, Donald Savage, had discovered *Oligopithecus* in 1961 during Simons's first Fayum expedition). On a 1978 trip to Burma, Savage and his colleague, Russell Ciochon, saw a number of jaws found by local paleontologists. "Of the four primate jaws they had discovered," Ciochon wrote, "two were *Pondaungia* and one was *Amphipithecus.* The fourth may represent a type of primate previously unknown to science." The new fossils gave a much clearer picture of the Eocene creatures than Barnum Brown's jaw fragment did. *Amphipithecus* had been a gibbon-sized animal, larger than known prosimians, with characteristics linking it to New World as well as Old World anthropoids. Ciochon thought that the Burmese genera showed "a combination of lower and higher primate features, with the latter considerably more prominent, indicating that they were at or across the evolutionary transition from prosimian to anthropoid," and he concluded that such creatures had evolved in Asia, then migrated to Africa. "Once in Africa, these early higher primates continued to evolve," he added, "with some populations becoming ancestors of the 30 to 35-million-year-old Fayum primates of Egypt (and ultimately of all Old World monkeys, apes, and humans). Other populations crossed the then-narrow equatorial Atlantic Ocean by island-hopping along a series of volcanic islands. In this way they reached South America and became the ancestors of New World monkeys."

Meanwhile, Elwyn Simons and his teams kept finding more Fayum primates—there were at least twenty-one species by 1992. These included tarsiiforms and adapiforms as well as early anthropoids, and some of the early anthropoids were of late Eocene age. "This arguably represents more taxonomic diversity of primates, especially higher primates, than has been demonstrated before in one so spacially and temporally limited area," Simons wrote. "These facts argue that an important, perhaps primary, radiation of anthropoideans took place in the African Eocene." Simons speculated that anthropoids might have evolved from neither adapiform or tarsiiform prosimians, but from "a third Eocene ancestral line, very likely in Africa."

Ciochon thought anthropoids had evolved from adapiforms, however, and Philip Gingerich, who agreed, observed that when he compared early tarsiiform prosimians, early adapiform ones, and early anthropoids, the anthropoids shared more characteristics with adapiforms. Most of these were technical dental characteristics. But one convincingly simple trait was that early anthropoids' body masses resembled those of cat-sized lemurs and lorises more than those of rat-sized tarsiers, implying that the first anthropoids had resembled the former. "The most likely area of origin of higher primates, based on present evidence, appears to be Africa and/or South Asia," Gingerich diplomatically concluded. "This is the region . . . which lies between the known distribution of *Amphipithecus* and *Pondaungia* in Burma and *Oligopithecus* in Africa. All three of these genera have the distinction of being ambiguous adapid-simiiforms at a time when simiiform primates [anthropoids] were first differentiating."

A problem with tarsiiforms as anthropoid ancestors was that no fossil creatures resembling them were known from the Old World Eocene, but that didn't discourage the tarsiiform faction. "A balance of cranial, dental, soft tissue, and molecular evidence," wrote two of them, "support the hypothesis that tarsiiforms and anthropoids are sister taxa. . . . This implies that higher primates are descendants of some species that would perhaps be classified as tarsiiform." And early tarsier fossils eventually appeared. Some isolated teeth from the late Paleocene of North Africa, with the jaw-breaking name of *Altiatlasius,* were judged to have tarsiiform affinities in

1990. Then, in 1994, definite Old World tarsier fossils finally turned up in 45-million-year-old Eocene rocks at a limestone quarry near the village of Shanghuang, China. Like Fayum, the site was rich in primates, including adapiform as well as tarsiiform prosimians, making it seem yet another candidate for an anthropoid Eden.

The Shanghuang site had something else that was new. With the prosimians were remains of another little creature (it had probably weighed about three ounces in life), which its discoverers, a team led by the Carnegie's Chris Beard and Tao Qi of the Beijing Academy, named *Eosimias,* "dawn ape," because it had characteristics of both tarsiiforms and anthropoids. "Within Anthropoidea, *Eosimias* appears to occupy a very basal phylogenetic position, certainly before the diversification of Parapithecidae, Oligopithecinae, Platyrrhini [New World monkeys] and Catarrhini [Old World anthropoids]," they wrote. "Given this position, *Eosimias* yields new insights into the phylogenetic relationships of simians with respect to other primates." Beard and his colleagues thought similarities between *Eosimias* and the Burmese fossils suggested that *Amphipithecus* and *Pondaungia* might be "younger, more derived members of the basal simian radiation sampled at Shanghuang." Turning one of the adapiform faction's arguments on its head, they surmised that *Eosimias's* tininess, along with other characteristics, made it a likelier relative of tarsiiforms than of adapiforms. "*Eosimias* is smaller than all known adapiforms except the specialized and temporally late *Anchomomys,*" they wrote. "Hence, one must postulate an episode of phyletic dwarfing, which is rare among mammals, to derive *Eosimias* from adapiforms."

Eosimias was striking enough to get front-page news coverage, although it took a back seat to another Shanghuang fossil, which was the smallest primate yet found, a tarsiiform prosimian weighing half an ounce. "They were furry little things, with long tails, express-train metabolism and hearts that probably beat like jackhammers," gushed one article. Then, more soberly, it quoted a scientist as saying, of *Eosimias:* "We have the first unambiguous evidence that is able to bridge the anatomical gap between lower and higher primates."

Simons and Ciochon both might have asked what was so ambiguous about *their* respective evidence, however, and the anthropoid leapfrog continued. In the mid-1990s, more fossils emerged from Myanmar (formerly Burma), including not only skull and leg bones of the known *Amphipithecus* and *Pondaungia* but jaws of two new, smaller genera named *Bahinia* and *Myanmarpithecus.* Some skull features of the new *Amphipithecus* and

Pondaungia fossils suggested that they may have been much less anthropoidlike than previously thought, and by the early 2000s, Ciochon and Gingerich had decided that the two genera had really been adapiform prosimians not unlike Leidy's *Notharctus.*

"Dental and cranial resemblances between pondaungines and anthropoids," they judged, "are probably functional convergences resulting from a diet consisting of hard and tough food objects utilized by these Southeast Asian primates." The scantier fossils of *Bahinia* and *Myanmarpithecus* remained ambiguous, and *Bahinia* seemed to have similarities to *Eosimias.* But in a 2002 paper, Ciochon and a colleague echoed Simons's earlier speculation that anthropoids had not evolved from either adapiforms or tarsiiforms, but from "a third Eocene ancestral line, probably in Africa." A sidebar speculated that anthropoids had "originated in Africa in the late Mesozoic or early Cenozoic and remained an endemic African radiation until the later Oligocene, when they began to spread into Asia and Europe."

In the late 1990s, some teeth from early Eocene North Africa suggested that a tiny extinct primate named *Algeripithecus* might have been an earlier version of Fayum creatures like *Oligopithecus* and *Apidium,* further pushing back their pedigree. But "late Mesozoic or early Cenozoic" anthropoid ancestors have yet to turn up in Africa, and even the Fayum generas' pedigree has dubious aspects. They have the anthropoid skull bona fides of fused frontal bones and bony cups behind the eye sockets, which prosimians lack, but their legbones suggest that they might have been part of a different anthropoid lineage from living ones. They seem adapted to leaping, like lemurs and tarsiers, rather than to walking or brachiating like monkeys, apes, or hominds. So the first true ape ancestors may still be undiscovered somewhere in Africa, although it would seem strange that they should have stayed there placidly throughout the Eocene while adapiforms and tarsiiforms swarmed around the globe.

Anthropoid leapfrog will continue as more fossils turn up. Still, the game has pretty well vindicated *some* kind of prosimianlike beast as a link between the Peabody mural's *Plesiadapis* and apes, or at least between some kind of African *Plesiadapis* counterpart and apes. (The squirrellike Paleocene primate has been found in Eurasia as well as North America, but not in Africa.) It has strengthened another macroevolutionary thread in Zallinger's tapestry.

The mural's *Pelycodus* and the inexplicably omitted *Tetonius* also point to a weakness in the links connecting primate groups, however. Learning what apes evolved from won't necessarily show how they evolved. The reasons for the shifts from early hoofed mammals to whales, and from apes to hominids, seem crystal clear by comparison to a prosimian to anthropoid shift. Both whales and hominids evidently evolved new anatomical features as adaptations to habitat changes. Whales evolved a streamlined shape and flippers as they moved from land to sea; hominids evolved an upright stance as they moved from trees to the ground. With their abrupt but progressive appearances in the fossil record, the stages of whale and hominid evolution fit both the gradualist and the punctuationist paradigms.

The prosimian to ape transition features no obvious scenarios. Prosimians and nonhominid anthropoids (except some baboons, macaques, and other Old World monkeys, which, like hominids, are relatively recent) live in tropical forest, and apparently always have. The length of their survival in that habitat shows their adaptation to it, and their disappearance from cooling or drying regions shows their dependence on it. Why, then, did apes diverge from prosimians? Apparently, they weren't moving into a new habitat, and the environment wasn't changing around them. Neither gradualist nor punctuationist neo-Darwinism has ready explanations for evolutionary changes that don't involve adaptations to environmental changes.

On the other hand, habitat changes can be subtle. The fact that most prosimians are small and nocturnal, and most anthropoids larger and diurnal, is suggestive. Even in the forest's shadow, the move from night to day can be as momentous as that from land to sea. A prosimian lineage might have shifted from night to day as it began to spend more time seeking plant foods than animal ones. Such a lineage could have displaced older occupants of the diurnal herbivore niche, multituberculates and marsupials, as those groups' late Eocene disappearance from the Old World suggests. Ample plant foods in the tropical forest canopy would have encouraged increased size and other typical adaptations. As Elwyn Simons speculated in one article, apes may have evolved because of "the Oligocene primates' arboreal way of life, involving feeding on leaf buds and fruits near the ends of branches," which meant that "certain kinds of dexterity" contributed to survival.

It is nonetheless a big jump from forest prosimian to forest ape, which continues to hint that there might have been more to it than natural selection. Geoffroy Saint-Hilaire's idea that change might occur by sudden mutations has continued in various forms. Simpson considered it the most likely alternative to neo-Darwinism. "The most formidable fairly recent development that is possibly not clearly consonant with the synthetic theory is the hypothesis of non-Darwinian evolution, incorrectly so-called," he wrote. "Advanced by biochemists, not organismal biologists, this holds, in simplest and most general terms, that many, most, or possibly even all genetic changes in populations (or taxa) are not affected by natural selection and occur at random and at constant rates." Simpson was skeptical that any mutation could long remain unaffected by selection. But Stephen Jay Gould raised one possible example, again involving the horse, that Darwinian sacred cow, in a 1980s essay.

Gould first attacked the "traditional" neo-Darwinian scenario of a geographical barrier splitting an ancestral population, with the two descendant populations then adapting by natural selection to their new habitats:

During the past decade, the predominance of this mode has been challenged by a variety of new proposals advocating an interesting twist or reversal of perspective. They all argue that reproductive isolation can arise rapidly as a result of historical accidents with no selective significance at all. In this case, reproductive isolation comes first. By establishing new and discrete units, it provides an opportunity for selection to work. The ultimate success of such a species may depend upon the later development of selected traits, but the act of speciation itself may be a random event.

Gould then brought up chromosomal speciation, an old hypothesis whereby spontaneous change in the number and form of chromosomes could cause a new species to arise. A problem with it was that an individual with changed chromosomes would be at a selective disadvantage, but Gould saw one way around that. If the species had a harem structure, a mutant male would engender chromosonal changes in large numbers of offspring, and mutant chromosomes could quickly become predominant. He concluded:

My colleague Guy Bush, of the University of Texas, tells me that horses provide, circumstantially at least, a strong case for chromosomal speciation. They all maintain the harem structure of kin breeding. Their seven living species (two horses, two asses, and three zebras) all look and act pretty much alike despite some outstanding differences in external color and pattern. But their chromosome numbers differ greatly and surprisingly—from thirty-two in one of the zebras to sixty-six in that paradigm of the unpronounceable, Przewalski's wild horse.

Whether this could apply to primates is another question, of course. Among living ones, some genera like tarsiers and gibbons, the most primitive apes, are monogamous, obviating speciation through harem reproduction. Other prosimian and anthropoid genera are more polygamous, and some do approach a harem structure, particularly the more herbivorous kinds, like gorillas. Closely related primate species can have very different chromosome structures, so chromosonal speciation is not outside the realm of possibility, although exactly how it might help to transform a small, nocturnal prosimian into a large, diurnal anthropoid remains unclear.

Given how little is known for certain about anthropoid evolution, nothing with any scientific validity is outside the realm of possibility. As I've said, new ideas such as the theory that retroviruses may have effects on the mammalian genome raise Lamarckian echoes, although not with any applications that I know of to anthropoid origins. Even a whisper of Osborn's orthogenesis lingers around one mysterious phenomenon, although it is more likely to be a result of neo-Darwinian convergent evolution. As I said earlier, the almost simultaneous, apparently unrelated appearance of civilization in the Old and New worlds can seem orthogenetic. What we now consider anthropoid primates also appeared almost simultaneously (by geological time standards, at least) in the Old and New worlds. New World monkeys, isolated in South America since the late Oligocene, are not as big and smart as Old World monkeys and apes, but they are much more "anthropoid" than either adapiform or tarsiiform prosimians. Yet nobody knows how their ancestors reached South America.

If they got there from Africa across the Atlantic, as most paleontologists now think, they must have been related to African creatures like *Parapithecus* and *Apidium.* The fact that those two genera had teeth like them supports this. The Atlantic was already wide in the Oligocene, however, and primates aren't good swimmers. Sweepstakes distribution via volcanic islands is a possibility, as Ciochon suggested, but there is no evidence of it. If, on the other hand, they got there from North America via a Central American island chain, as some paleontologists suspect, the matter becomes more complicated. It simply may mean that Old World anthropoids had reached North America earlier, when tropical forest grew across a northern route. No North American anthropoid fossils are known, however, and if anthropoids stayed in Africa until the Oligocene, as Ciochon and others suspect, it might have been too late for them for them to cross the cooling higher latitudes.

If anthropoids didn't reach South America via an Atlantic sweepstakes

Figure 19. Simpson with a bush baby. Courtesy of Joan Simpson Burns.

or northern filter route, then the ancestors of New World monkeys could have been, not African anthropoids, but the abundant North American fossil prosimians in which Leidy and Cope saw anthropoid possibilities. Indeed, an early Oligocene fossil prosimian from Texas named *Mahgarita* may have similarities to New World monkeys, although it also has similarities to Old World adapids. If a North American prosimian like *Mahgarita* was indeed their ancestor, then New World monkeys are not anthropoids, but a primate group that independently evolved similar traits like diurnal habits, color vision, and higher intelligence.

However they evolved, there is a haunting familiarity to New World monkey faces, despite thirty-five million years of isolation from other primates. I once spent a day in a Honduran river delta that might have been prehistoric Fayum. Howler and capuchin monkeys trooped through the mangroves as crocodiles lurked below, ready to fossilize unwary youngsters, and I felt as though I were seeing diminutive precursors of gorillas and chimpanzees. Howlers are placid herbivores that live in relatively simple societies centered around big males who roar and display to maintain dominance, but seldom fight. Capuchins, the classic organ-grinder monkeys, are clever, manipulative omnivores that live in complex, polygamous societies, which cooperate to fight rival bands and hunt squirrels.

I wish Simpson had featured some *Purgatorius*-like creature in his Cretaceous novel. It would have been fanciful, of course, although they say tarsiers steal into hunters' camps at night and play with fire embers. But it might have enlivened the story. There is a photograph of the elderly theorist with a distant *Purgatorius* relative perched on his shoulder, an African bush baby. Simpson's snapshot expressions are usually guarded, but his face wears a look of open, even giddy, delight as the living homunculus nestles under his ear.

A bush baby or howler monkey may be as close as a human can come to literally looking into the face of deep time. With its laboriously uncovered trail of tiny bones, anthropoid leapfrog eptitomizes how far we have come toward understanding mammal evolution. There is no Cinderella story more exciting to us, whether pleasurably or not, than that of Cretaceous gremlins evolving into the likes of George Gaylord Simpson. It also epitomizes how far we have to go. Like mammal evolution in general, it is still a mystery story. Although Darwin, Simpson, and their colleagues have given us a general idea of at least some of the processes involved, we still understand almost nothing about how prosimians evolved in the first place, and we still have only a vague idea of how some of them evolved into anthropoids, or, eventually, into the likes of Simpson. Beyond that, there remains the mystery of where, if anywhere, evolution—anthropoid and otherwise—is "going," a mystery that essentially will remain unsolved as long as evolution continues, of course, but that compels ongoing wonderment.

EPILOGUE

Cenozoic Parks

I began this book by comparing Zallinger's murals to Michelangelo's because, whatever their differences, history impels us to view them both as
mythological evocations of deep time. But Zallinger's *Age of Mammals* may
actually have more similarities to much older murals. People entered low,
dim chambers to see painted horses, mammoths, and bison in some famous other places. W. J. T. Mitchell recognized this when he wrote that
Zallinger's images look "as if they had always existed, like the anonymous
animal cave paintings at Lascaux."

The problem with this is that we know much less about Lascaux, Al-
tamira, Les Trois Frères and other great paleolithic cave paintings than we
do about the Sistine Chapel, so it is harder to relate their significance to
that of the Peabody mural. Archaeologists first speculated that the cave
murals functioned mainly as ritual hunting magic to increase or attract
prey, which would distance them from the Peabody annex's role. Analysis
of middens, however, has shown that the people who painted the caves
subsisted on reindeer, wild pigs, and rabbits, which they rarely painted,
more than on great beasts like mammoths, horses, or bison. This suggests
that the cave paintings were not, or at least not only, magical sites, but
mythological ones. They may have been, as Joseph Campbell thought, the
world's first "rendition in art of the mythological realm," the source of
"all temples since." If so, they do resemble Michelangelo's and Zallinger's

murals in function, and we can speculate, however tenuously, on their significance.

The cave artists' mythology was different from ours, of course. When they painted humans, it often was with beast or bird features, suggesting that they regarded themselves in ways strange to us. But I think their paintings show that they admired beasts much as we do, an admiration that links them to the Peabody annex. Zallinger could hardly have contemplated its dim, low-ceilinged space without thinking of the caves. They were much discussed in the 1950s; some, like Lascaux, had only recently been discovered. He must have welcomed the opportunity to learn not just from fellow restorationists like Charles Knight, who himself borrowed from cave art, but from the eyewitnesses. Zallinger's feeling for the beasts' foursquare presence, and his pleasure in rendering their meaty liveliness, certainly recall the cave artists'. His Ice Age *Equus* has the cave horses' wooly plumpness; his *Mammuthus* has the cave mammoths' shoulder fat deposits and trailing belly hairs. His *Megatherium* might be less like a stuffed toy if he had been able to model it on a Stone Age version.

Despite the many millennia that divide them, the murals have similar subtexts—the relationship of human life to an animal realm so vast and rich that humanity seems peripheral. The cave murals showed a present age of mammals, the Peabody mural shows a past one, but the difference may be less than it seems.

I suspect that Zallinger drew even further on cave imagery in rendering this subtext. His elfin prosimian *Pelycodus* has an odd resemblance to cave art's most haunting human image, the "sorcerer" of Les Trois Frères in Ariège, in southern France. The tailed, antlered human figure has a bug-eyed, pug-nosed, heart-shaped face like the tiny primate's—even its flexed forelimbs and ritually bent knees seem reflected in *Pelycodus*'s crouching pose in the *Age of Mammals.* If such cave figures represented mythic links between beasts and men, as anthropologists think, then *Pelycodus* plays a similar role. Its ancestral but uncertain connection to living humanity raises the "about us" side of the question more overtly posed by his mammoth and megatherium: whether life is "going somewhere."

The sense of recognition we find in other primates' eyes, even ones as distant as *Pelycodus*'s, encourages a sense that life is "going somewhere" with us—that humanity, or at least an anthropoid kind of intelligence, is somehow inherent in the evolutionary process. As Robert Wright wrote in his *New Yorker* critique of Stephen Jay Gould, "[I]f you look at the foundations of human intelligence—tool use, language, reciprocal altruism,

Figure 20. Rudolph F. Zallinger with his *Age of Mammals*. Copyright © 1966, 1975, 1989, 1991, 2000 Peabody Museum of Natural History, Yale University, New Haven, Conn.

coalitional contentions, and others . . . the eventual combination of these foundational properties in a single species was likely." Yet the 200 million years of mammal evolution during which humans did not appear lend weight to Gould's opposing view that "we may owe our evolution, in large part, to the great Cretaceous dying that cleared a path, yet spared our ancestors' lives to tread upon it," that "humans are here by the luck of the draw."

Gould once observed that when organisms seem most advanced, they may be closest to extinction—that horses seem to have progressed from small, stupid browsers to big, smart grazers, for example, because only one horse genus survives, an evolutionary has-been compared to the hundreds of living bat or rodent genera. "Many classic 'trends' of evolution are stories of such unsuccessful groups—trees pruned to single twigs, then falsely viewed as culminations rather than as lingering vestiges of former robustness," he concluded, and cited another twig: "I remind readers that one other prominent (or at least parochially beloved) mammalian lineage has an equally long and extensive history as a conventional depiction of a ladder of progress—yet also lives today as the single surviving species of a formerly more copious bush."

Created in the progress-mad 1950s and 1960s, the *Age of Mammals* might be expected to manifest Wright's viewpoint rather than Gould's. The text that explicated Zallinger's cartoon in *The World We Live In* certainly did so. The "main line of progression . . . has been an increase of perception—the development of more efficient sense organs and more complex and sensitive nervous systems . . . the great arsenal of the mammals," it proclaimed, echoing O. C. Marsh. Cartoon and mural both show such a "line of progression," from flat-skulled *Coryphodon,* which faces toward the past, to high-browed, forward-looking *Mammuthus.*

Other aspects of the *Age of Mammals* seem reflective of a less confident outlook, however. The two *Pelycodus* are not looking to the future—one peers into the past, the other, worriedly, at the beholder. The mural's only other primate, *Plesiadapis,* scrambles back toward the Cretaceous as though frankly daunted by the giant serpent's ambiguous offering. These are minor details, but another seems more significant. Traditional evolutionary narratives, Darwinian or otherwise, end with the advent of humans, and this often entails a sense that organic evolution's task is thus accomplished. Yet a close examination of the mammal mural reveals no suggestion of a human conclusion, no Clovis projectile point, for example, next to Zallinger's signature at the lower right corner. The mural doesn't even end with the "advanced" mammals that accompanied the Ice Age human invasion of North America—mammoth and bison—but with the megatherium and glyptodont that invaded from "evolutionary backwater" South America. Zallinger might be "rerunning the tape" of evolution, as in Gould's metaphor, and painting a world in which *Australopithecus* never evolved, or never evolved into world-conquering *Homo.*

Of course, Zallinger may have had no such sense of evolutionary contingency. He simply may have been rendering the traditional idea that the age of mammals ended when humans appeared—that humans aren't part of it, but of the "wholly recent," Holocene epoch, the age of man. If so, he would have been following the *The World We Live In*'s text: "[A]n upstart on the planet Earth, man nevertheless stands alone in the complex of nature, the master of vast and incalculable forces, and arbiter of his own fate."

In that sense, inserting even a spearpoint in the painting would have implied a false connection between the old world of wild beasts and the new one of civilization.

There is also another possibility. Zallinger's mural may express both evolutionary viewpoints, the forward-looking mammoth's and the backward-looking megatherium's. (As F. Scott Fitzgerald observed, "the test of a first-rate intelligence is the ability to hold two opposed ideas in the mind at the same time, and still retain the ability to function.") Overall, the painting can be read both ways. It certainly is going somewhere, in that the flora and fauna of its western North American locale change through sixty-five million years of climatic and geological events. It starts in jungle and ends on tundra. Yet there is also a sense in which it is going nowhere, because the habitats in the mural still survive, if not in western North America, then somewhere on the planet. The painting is "about us" in that it concerns the world we live in now as well as the kind of animal we evolved from. Almost all the mural's plants are recognizable living forms, from Paleocene palms to Pleistocene pines. Many animals in even the earliest epochs, like the opossum *Peradectes,* resemble living ones.

This other possibility has an interesting implication—that we, like visitors to Lascaux seventeen thousand years ago, are not divorced from the mythic realm of beasts, but step back into it when we return to the light of day. Or, at least, it implies that a kind of time travel to the age of mammals is possible, perhaps less compelling than Sam Magruder's return to the reptile age, but less terminal.

I attempted such a trip, and it took me to a surprising place. It began in Costa Rica years ago, when I went for a twilight stroll and almost stepped on a lesser anteater, or *Tamandua.* Like sloths and armadillos, their fellow members of the Xenarthra superorder, tamanduas *look* as though they branched off early from most other living mammals, with prehensile tails and powerful foreclaws for breaking into termite nests. They seem little changed from Eocene ancestors, and Zallinger painted a possible one named *Metacheiromys* ("partly hand mouse"), peering nearsightedly at *Mesonyx's* face-off with *Uintatherium.* On the other hand, *Metacheiromys* may have been more closely related to the pangolin, an Old World anteater of the genus *Manis,* which the geneticist Mark Springer classes in his Laurasiatheria superorder. But pangolins also are respectably antique. Indeed, their scalelike armor makes them look even more primitive than tamanduas.

Such creatures have acquired a peculiar resonance in recent years because of an extraordinary Eocene site, the Messel Shale, in the German state of Hesse. Dug up largely in the past three decades, Messel's 54-million-year-old fossil organisms are strikingly well preserved because they fell into a deep, steep-sided lake where anaerobic bottom conditions prevented decay. Eocene beetle wings still flash green, gold, and purple iridescence. Leaves, flowers, and fruits might have been kept in a plant press. Fish, frog, lizard, snake, bird, and mammal skeletons retain vestiges of scales, feathers, and fur, sometimes of internal organs and stomach contents.

Fossil life seems to have imitated art at Messel, because the site had a decided resemblance to the Peabody mural's Eocene landscape. Zallinger's uintatheres roam a clifftop over a swampy lake, and it is easy to imagine them tumbling off and sinking into the ooze. No uintatheres have emerged from the Messel Shale, but many smaller creatures found there are like the mural's. There is a species of *Hyrachyus,* the primitive tapir Zallinger painted trotting futureward from the *Mesonyx-Uintatherium* confrontation, as though seeking peace and quiet, ill-advisedly, in the Oligocene. There are relatives of the giant bird, *Diatryma,* and of the dawn horse, *Eohippus-Hyracotherium-Protorohippus.* There also are two possible relatives of Zallinger's little anteater, *Metacheiromys,* and, strangely, the 54-million-year-old fossils seem to resemble living anteaters more than they do *Metacheironomys.* Taxonomists thought them so like living tamanduas and pangolins, at least, that they simply gave them versions of the same names—*Eurotamandua* and *Eomanis,* "European tamandua" and "dawn pangolin." Other taxonomists later decided that *Eurotamandua* really was a pangolin relative, and that its striking likeness to the living tamandua was convergent. Still, the ancient beasts' similarity to living ones is startling.

In Costa Rica, I had felt as though I had almost stepped on a piece of the Eocene. Learning about the Messel Shale and its anteaters made me want somehow to travel even closer to fifty-four million years ago. I might have gone to the Messel site itself—with imagination, fossil deposits can be time travel destinations. I was seeking something more literally prehistoric than a Hessian oil shale mine, however. When I read that the Messel fossils had "close relationships . . . most of all to South-east Asia," I decided to go there and see if I could get a kind of stereoscopic view of the Eocene by bringing together what I knew of Messel and the Peabody mural with the living forest there.

It seemed an unlikely prospect when I arrived in roaring, fuming Bangkok. But I soon glimpsed something like it on a twilight visit to the hills of Khao Yai National Park, an afternoon's drive from the Thai capital. A blue-green forest very like Zallinger's overhung the road, and a large python undulating across could have been the mural's Edenic serpent—its diamond-patterned scales looked the same. Or it could have been one of Messel's skeletal pythons, whose bones are so perfectly articulated that they still might be crawling across the shale slabs they rest on. One had a small crocodile in its stomach.

The serpent was the first of many glimpses. Messel has produced fossils very like the nightjars that fluttered up from the Khao Yai roadside, and like the two million wrinkle-lipped bats which emerged from a nearby limestone cave just after sunset. Messel's hundreds of fossil bats are so well preserved that paleontologists can dissect their inner ears, study the cellular structure of their tissues, and learn their food preferences from insect prey in their stomachs. Other mammals I saw that night reminded me of the Peabody's Eocene fauna as well as Messel's—a hoofed, but also fanged, "muntjac" the size of *Eohippus* lying in the grass; a wolverinelike "binturong" the size of *Oxyaena* loping along the road. Herbivore and carnivore seemed to share something of the primal amorphousness that discouraged Simpson from popularizing his creodonts and condylarths.

At other Thai parks—Nam Nao, a pine-studded plateau; Khao Sok, a limestone-cliffed rain forest—I saw similar sights. Slender spotted civets and hoofed beasts smaller than muntjacs stole through the underbrush. Even the "age of reptiles" seemed to linger as it must have in the Eocene, when it was only ten million years in the past instead of sixty-five million. Muntjac-sized monitor lizards raced about, and spindly green *Draco* lizards glided among the treetops, so swiftly that it took me a few days to realize they were "flying." Of course, flying dinosaurs were everywhere, very old ones, like scarlet trogons, separated from neotropical counterparts since the Miocene, and emerald-plumaged broadbills with an *Archaeopteryx*-like air. At Khao Sok, little red chickens scratched on the floor of an outdoor restaurant; when I walked in the forest afterward, little red chickens—ancestral jungle fowl—scratched in the leaf litter.

These were only glimpses, however, never quite the 3-D Eocene-o-rama I had fantasized about. Animals vanished in a blink or manifested themselves invisibly, as with hoarse muntjac barks that issued startlingly from twilit underbrush. When I tried to ask rangers about primitive mammals—prosimian lorises, flying colugos—they shook their heads. They hadn't seen such things, or didn't understand my rudimentary Thai.

The forest resisted my attempts to look into its deep past as though everything that happened since was blocking the view. I saw many anthropoid primates—gibbons, macaques, langurs—but no lorises. I didn't see pangolins, colugos, or tapirs, and most of the mammals I did see, even the most primitive-looking ones, belonged to groups that evolved mainly after the Eocene. Muntjacs and their smaller relatives are musk deer; binturongs and civets are viverrids, carnivores related to cats and dogs. At least, musk deer and viverrids are fairly old members of the cervidae and carnivora. Other Thai mammals might have stepped out of the Peabody mural's Pleistocene section, like a sun bear that I saw dozing in a tree, elklike sambur grazing in a clearing, or elephants at the forest edge.

One of the elephants glanced at me, and inquiringly raised the tip of its trunk in an attitude not unlike the Peabody mammoth's. It was a smaller, more diffident beast, with short tusks and a gray, wrinkled hide, but its domed head and wise gaze were the same. It seemed to epitomize the difficulties of seeing into living animals' pasts. Asian elephants are related to mammoths, with high-crowned molars adapted to graze the silica-rich grasses of a cooling, drying planet. But the only grasslands in the Thai parks I visited were abandoned farms. Otherwise the elephants live in dense tropical forest that has not changed that much from what uintatheres inhabited fifty-five million years ago. And they seem to live comfortably enough. A guide told me that wild elephant dung is valued for medicinal purposes because they eat such a wide variety of plants.

It seems arbitrary that mammoth-toothed Asian elephants survive by browsing in forests when mastodons, whose cusped molars were adapted mainly to forest-browsing, are extinct. There the elephants are, however, although perhaps not for much longer, since proboscideans are another of Gould's pruned evolutionary trees. There have been initiatives to restore a branch by extracting DNA from frozen mammoths and cloning live ones. A French science historian might have been describing Zallinger's *Mammuthus* when she observed that resurrecting the genus, "would rediscover the circular image of eternal time, in which mammoths would gaze at us throughout eternity, at once worried and thoughtful, their deep eyes lost in the reddish mass of their hair." But so far high technology has proved little more practical than my Eocene-o-rama at time travel—even to the Ice Age that ended ten thousand years ago.

Deep time can be insidious as well as elusive, however. In a way, the most ancient beasts I encountered in Thailand's Cenozoic Parks were the most ordinary ones. To my surprise, the mammals that impressed me most there were tree squirrels. I had never seen such a diversity of them, from marmot-sized, black-and-white giant squirrels that crashed about the canopy at noon, to mouse-sized flying squirrels glimpsed at dawn, zipping along branches with lizardlike speed. New species kept appearing—brown, gray, or red—striped, spotted, or solid-colored. Southeast Asian tree squirrel diversity suggests that this is another group with very old origins on the greatest landmass.

In fact, tree squirrels and mammals like them have been around as long as the Cenozoic, probably much longer. Some of the ancient, possibly egg-laying multituberculates may have had a convergent resemblance to squirrels. Although Zallinger painted Cope's Paleocene genus, *Ptilodus*, on the ground, skeletons found since suggest that the genus was arboreal. Early rodents, like the Eocene *Paramys* that perches on a branch near Zallinger's *Coryphodon-Oxyaena* brawl, probably looked and acted much like living squirrels. Blithely washing its face, *Paramys* resembles the little red squirrels that now inhabit the Rockies. At Messel, a marmot-sized rodent named *Ailaravus* resembled today's southeast Asian giant squirrels. Creatures very like the present cosmopolitan genus, *Sciurus*, appeared in the Oligocene.

"They are still primitive, just beginning to specialize in various directions, and all are more or less squirrel-like" Simpson wrote of Eocene rodents, "for squirrels, despite a few specializations, are about the most primitive of rodents surviving today and therefore most like these ancestral forms." I had been just about as close to the Eocene trying to keep squirrels off my birdfeeder in Berkeley as in the Thai forest.

Squirrel evolution is even more mysterious than elephant, since they throng the largest of unpruned family trees. The big, orange-tailed fox squirrels in my backyard, for example, are not California natives, but a Midwestern species, *Sciurus niger*, which was introduced in city parks. First recorded here in 1935, it now is so abundant east of San Francisco Bay that Berkeley seems to have one per backyard, hanging upside-down on the Droll Yankee birdfeeder. When I lived in central Ohio in the 1970s, however, the species occurred in farm woodlots more than suburban areas, where the smaller gray squirrel, *Sciurus carolinensis*, predominated. It seems odd that a species that has trouble adapting to suburbs in its native range should adapt so well to suburbs outside it, although it may have succeeded here simply through lack of competition. There are no historical records of the native West Coast gray squirrel, *Sciurus griseus*, occurring in the East Bay, and more-recently-introduced eastern grays have replaced fox squirrels in parks west of the Bay. The fox squirrel occupation has been so speedy and pervasive, however, that it almost seems to transcend neo-Darwinian adaptation.

Fox squirrel mysteries don't end there. Visiting Ohio in 2001, I took a walk in a suburban woodland I had known in the 1970s. I don't recall seeing fox squirrels in it then, but I found them common in 2001. Gray squirrels also were common, as in the past, which seemed a change from the former situation wherein grays excluded fox squirrels. I didn't think much of this at first, but as I walked through the area, I began to have trouble distinguishing the species. Some individuals were gray squirrel–sized, but had orange bellies and tails. It was late spring, and they may have been juvenile fox squirrels. Yet the fact that both species thronged the place raised the possibility that, instead of competing, they had begun to interbreed, that I was seeing hybrids, conceivably a new species in the making.

Hybridization remains little understood, part of the "symbiogenesis" biologists are beginning to explore, wherein organisms evolve by converging. Odoardo Beccari, Alfred Wallace's contemporary in southeast Asia, was so puzzled by the anomalous fauna—"not only flying lizards, but flying squirrels, flying foxes, flying frogs, and, could we believe the Malays, flying snakes"—that he thought such creatures must have arisen during a "formative epoch" when hybridization had been more common. It is unclear what Beccari meant by a "formative epoch," and his neo-Lamarckian evolutionary ideas seem quaint now. His "formative epochs" might apply to mass extinctions like the late Cretaceous one, however, and it is possible to imagine a neo-Darwinian scenario wherein habitat shrinkage and fragmentation could disrupt populations to the extent that, as he wrote, "differing types could cross and produce offspring."

The place where I saw fox and gray squirrels together certainly resulted from habitat shrinkage and fragmentation. It was a "metro park," surrounded by farmland in the 1970s, but by freeways, factories, malls, and subdivisions in the 2000s. Changing habitat did contribute to at least one spectacular hybridization during that period. When I lived in the central Ohio countryside, I didn't hear coyotes howling, but they became common there by the mid 1990s, as in the east generally, and they evidently did so at least partly by hybridizing with remnant wolves and feral dogs. Eastern coyotes are larger than their western ancestors, and somehow better adapted to life in eastern forest and farmland. Biologists call them a "hybrid swarm," with "hybrid vigor." It would be interesting to see how this sudden change might show up in the fossil record. Maybe some of the primitive ungulates that exploded out of Asia just after the late Cretaceous extinction were hybrid swarms.

Tree squirrels may reprise the early Age of Mammals in more than their antiquity. They may be gearing up for the latest "formative epoch." A paleontologist once observed that mammals just after the Cretaceous extinction had marked similarities in form and behavior to those surviving in cities today—both groups being smallish, unspecialized and adaptable. My backyard's inhabitants indeed do have similarities to Zallinger's Paleocene ones. The fox squirrels resemble the primate, *Pleisadapis,* as well as the multituberculate, *Ptilodus.* The opossums, introduced along with the fox squirrels, more than resemble the opossum *Peradectes.* The raccoons that snooze under the sundeck are similar in size and shape to the omnivorous condylarth *Loxolophus,* which Zallinger painted with a black mask in the Time/Life cartoon. I can't see anything in the mural very like the skunks that dig up our flower bulbs, but defensive smelliness evidently goes back a long way, as I found with the opossum on the Ohio golf course. Maybe it was another adaptation to dinosaur dominance.

Seen from an F. Scott Fitzgerald perspective, Zallinger's mural suggests that Charles Lyell had a point in speculating that time might run backward as well as forward. I sometimes feel that even the Mesozoic is not so far off. One summer day, driving through big redwoods, I pulled off the road to look at some lilies. As I stepped out of the car, something scuttled underneath, so fast that I thought it was a lizard, but when I looked, I saw it was a shrew. The tiny beast seemed disoriented, sniffing the exhaust fumes, and I felt the uneasy pity that one does for wild animals lost in a machine world. But that was wasted. It zipped back into the redwoods, and I was left feeling like one of the dinosaurs looming above snouty *Cimolestes* in the *Age of Reptiles.*

Actually, shrews are no more closely related to *Cimolestes* than I am. Some Mesozoic mammals may have superficially resembled them, but shrews don't appear in the fossil record until after anthropoid primates do. Still, there's a basic durability about a small, furry animal in a forest, and also a basic novelty. As far as we know, nothing like it existed during the billions of years before the Triassic. Nothing more fundamentally new has come along in the 200 million years since.

I watched a fox squirrel gathering acorns from a young live oak today. It climbed the tree, picked an acorn, climbed down, hopped a few feet away, and buried the acorn. Then it did it again. And again. And again . . . as long as I watched.

NOTES

PROLOGUE. THE FRESCO AND THE FOSSIL

Page

xv "This is Rudolph Zallinger's *Age of Reptiles*": The reptile mural is usually likened to Giotto's frescoes because Zallinger based his technique on the *Libro dell'arte* of Cennino Cennini, a fifteenth-century admirer of Giotto's. Cennini lived closer to Michelangelo's time, however, and Giotto's work did not have the popular influence of the Sistine and Peabody frescoes.

"The Peabody hired him to paint the mural": Zallinger, "Making of the Age of Reptiles Mural" (http://www.peabody.yale.edu/mural/Rudy.html).

xvi "The art historian Vincent Scully was speaking outside the mainstream": Scully et al., *Age of Reptiles*, p. 10.

Scully called it "abundantly entitled": ibid.

"[T]hat wall is the most important one since the fifteenth century": Zallinger, "Making of the Age of Reptiles Mural."

"[A] modern monument": Mitchell, *Last Dinosaur Book*, p. 192.

"I was moved nearly to tears": Dodson, "Dodson on Dinosaurs," p. 5.

"Another dinosaur scientist, Robert Bakker, traced his vocation to seeing the picture in a *Life* magazine article": Bakker, *Dinosaur Heresies*, p. 9.

xvii   "Mitchell called them the 'totem' of industrial civilization": Mitchell, *Last Dinosaur Book*, p. 77.

"Zallinger 'never produced any work that came remotely close to his masterpiece'": ibid., p. 193.

xix   "[A] 1999 description of a visit by kindergartners to the Peabody": Jaffe, *Gilded Dinosaur*, p. 379.

"A lavishly illustrated 1910 book": Osborn, *Age of Mammals*, p. 150.

xx   "A few pages later, in describing the Eocene epoch": ibid., p. 165.

"The next animal one sees . . . standing over the skull of a uintathere": Osborn, "Prehistoric Quadrupeds of the Rockies," p. 711.

xxi   "Among Cope's finds were two fragmentary skeletons": I am grateful to Mary Ann Turner and other Peabody staff for this account of *Synoplotherium's* provenance.

"Osborn took Cope's side": Osborn, *Impressions of Great Naturalists*, p. 144.

xxii   "Gad! Gad! *Gad!*": Shor, *Fossil Feud*, p. 42.

"The flat claws are a unique peculiarity": Cope, *Vertebrata of the Tertiary Formations of the West: Book One*, 355.

xxiii   "In the 1960s, a paleontologist, Leigh Van Valen": Wong, "Mammals That Conquered the Seas," p. 73.

"Doubts later arose as to whether mesonychids were whale ancestors": Normile, "Whale-Ungulate Link Strengthens," p. 775.

"Evidence suggests that *Pakicetus* and other early Eocene cetaceans represent an amphibious stage": Gingerich, "Origin of Whales in Epicontinental Remnant Seas," p. 405.

xxiv   "Fossils contradict the notion": Gingerich, "Whales of Tethys," p. 88.

"Robert Bakker traced his conversion to . . . this 'hot-blooded dinosaur' paradigm": Bakker, *Dinosaur Heresies*, p. 15.

"Before my Peabody visit": Mary Dawson and Chris Beard, Carnegie Museum, Pittsburgh, personal comm., October 2000.

xxv   "[Dunbar] had always wanted the mammal mural": Grimes, "Interview with Rudolph F. Zallinger," p. 34.

"Dunbar did shed some light": Dunbar, "Recollection on the Renaissance of Peabody Museum Exhibits," p. 22.

xxvi "The Peabody hired him again in 1951 to do the mammal cartoon": ibid.

"A colleague of Zallinger's told me": Andrew Petryn, phone comm., April 1, 2001.

"Zallinger was essentially not allowed to 'invent' at all": Scully et al., *Age of Reptiles*, p. 11.

xxvii "I ultimately proposed a different convention, that of using the entire available wall . . . for a 'panorama of time'": Zallinger, "Making of the Age of Reptiles Mural."

"Osborn's last conversation with Cope": Osborn, *Cope, Master Naturalist*, p. 587.

xxviii "But mammal evolution has a special claim on our attention": Zhe-xi Luo, Carnegie Museum, Pittsburgh, personal comm., October 2000.

CHAPTER 1. PACHYDERMS IN THE CATACOMBS

3 "Some animals are viviparous": Aristotle, *On the Parts of Animals* 1.5, quoted in Gregory, *Orders of Mammals*, p. 11.

4 "[T]he bones, horns, claws, etc. of land animals are seldom found in a petrified state": Buffon, *Natural History*, 1: 252.

6 "It differed from the elephant the way 'the dog differs from the jackal and hyena'": Rudwick, *Georges Cuvier*, p. 22.

"The wooly mammoth is the classic of paleontology": Osborn, "Romance of the Wooly Mammoth," p. 227.

8 "I found myself as if placed in a charnel house": Eiseley, *Darwin's Century*, p. 84.

"The second genus, which he called *Anoplotherium*": Rudwick, *Georges Cuvier*, p. 65.

9 "So I sacrificed the remains of these vertebrae ": ibid., p. 71.

10 "Indeed, one of the species, *Palaeotherium crassum*": ibid., p. 67.

11 "Since 1725, scholars had regarded a human-sized fossil skull and ribcage from a German limestone quarry as a victim of Noah's flood": Schopf, *Cradle of Life*, p. 290.

"Indeed, he was the first to say such things": Rudwick, *Georges Cuvier*, p. 183.

12 "[S]uch reconstructions . . . were the most vivid expressions of his ambition": ibid., p. 267.

Cuvier's "eulogy" of Lamarck: Peattie, *Green Laurels*, p. 178.

13 "Have you ever plunged into the immensity of space and time": Balzac, *Wild Ass's Skin*, trans. Hunt, p. 40.

CHAPTER 2. DR. JEKYLL AND THE STONESFIELD JAWS

15 "Another young rock fancier, John Ruskin": Batchelor, *John Ruskin*, p. 38.

16 "So wonderfully alike were these bones": Cadbury, *Terrible Lizard*, p. 74.

17 "Cuvier did concede . . . that the fossil would be 'a remarkable exception' to his rule of mammal occurrence": Desmond, *Politics of Evolution*, p. 308.

Balzac's preference for Geoffroy's ideas: Gould, *Structure of Evolutionary Theory*, p. 311.

18 "Owen's youth had been troubled. His merchant father had died bankrupt": Rev. Richard Owen, *Life of Richard Owen*, 1: 10.

"Following the autopsy of a 'negro patient in the gaol hospital'": ibid., p. 23.

"As soon as I recovered": ibid., p. 25.

19 "His physiological experiments had included": Asma, *Stuffed Animals and Pickled Heads*, p. 60.

"Coveting the eight-foot skeleton": ibid., p. xi.

"Anatomy students often brag": Rev. Richard Owen, *Life of Richard Owen*, p. 22.

"Ambivalently mourning her father": ibid., p. 22.

"One acquaintance, the naturalist Edward Forbes": Desmond, *Archetypes and Ancestors*, p. 26.

20 "It is astonishing with what an intense feeling of hatred Owen is regarded": Rupke, *Richard Owen*, p. 6.

23 "From now on there was open hostility between Owen and the radicals": Desmond, *Politics of Evolution*, p. 320.

"Owen's wife Caroline commented in her diary": ibid., p. 319.

"It is an interesting circumstance": Chambers, *Vestiges of the Natural History of Creation*, ed. Secord, p. 112.

26 "Their variety was greater than that presented by their carnivorous successors": Cope, "Creodonta," p. 255.

"I have been studying the skeleton of a fossil carnivorous beast": Osborn, *Cope, Master Naturalist,* p. 297.

28 "*[Coryphodon]* older in time, earlier in date, than *Palaeotherium*": Owen, "On the Occurrence in North America of Rare Extinct Vertebrates," p. 217.

"It is to be presumed that no true researcher after truth can have a prejudiced dislike to conclusions based upon adequate evidence": Secord, *Victorian Sensation,* p. 422.

"In 1848, Owen led Chambers through the Hunterian Museum": ibid., p. 423.

29 "Although the bones of mammalia": Lyell, *Principles of Geology,* 3: 47.

30 "I confess to my shame": Desmond, *Huxley,* p. 13.

"To what natural laws or secondary causes": Desmond, *Archetypes and Ancestors,* p. 47.

31 "Dr. Jekyll recalls that his 'scientific studies . . . led wholly towards the mystic and transcendental'": Stevenson, *Strange Case of Dr. Jekyll and Mr. Hyde,* p. 53.

32 "In no other saurian": Owen, *Paleontology,* p. 268.

"Cynodracon major, for example": Desmond, *Archetypes and Ancestors,* p. 198.

33 "It is impossible to reflect": Darwin, *Voyage of the H.M.S. Beagle,* p. 157.

"Earlier in the evening": Moyal, *Platypus,* p. 105.

"During a morning's discussion at Owen's": Darwin, letter, July 30, 1837.

"Zallinger painted Darwin's 'great quadrupeds'": Barnett et al., *Wonders of Life on Earth,* pp. 65–67.

34 "Even the outlandish platypus had survived": Moyal, *Platypus,* p. 110.

"In 1856, he invited the thirty-one year-old Huxley": Desmond, *Huxley,* p. 223.

35 "Darwin's variety seemed to acknowledge little": Ruse, *Darwinian Revolution,* p. 248.

36 "[O]bservation of the actual change": Owen, *Paleontology,* p. 443.

"Owen wrote a spiteful review": Owen, "Darwin on the Origin of Species," in *Darwin,* ed. Appleman, p. 296.

"In 1848, an American visitor": Secord, *Victorian Sensation,* p. 425.

"It is consolatory": Darwin, "Historical Sketch," in *On the Origin of Species,* 6th rev. ed., Modern Library edition, p. 7.

"His enmity was mild": Browne, *Charles Darwin,* p. 538.

"Asked for a recommendation, Owen at first": Desmond, *Archetypes and Ancestors,* p. 26.

38 "[T]he monotremes now connect": Darwin, *Descent of Man,* p. 527.

"[T]hese facts appear to me to point out": Bowler, *Life's Splendid Drama,* p. 283.

"The English Cuvier was a notorious": Desmond, *Archetypes and Ancestors,* p. 198.

"Huxley let Owen snap away": ibid.

39 "They had 'well-developed limbs'": Bowler, *Life's Splendid Drama,* p. 302 (Croonian Lecture, p. 4).

40 "A few pages in the middle": Rev. Owen, *Life of Richard Owen,* 2: 312.

"According to his only recent biographer": Rupke, *Richard Owen,* p. 3.

"With age, his raw-boned looks": ibid., p. 4.

"He was wild": Stevenson, *Strange Case of Dr. Jekyll and Mr. Hyde,* p. 14.

CHAPTER 4. THE NOBLEST CONQUEST

41 "A paleontologist estimated in 1930": McDonald, "Hagerman Fossil Beds," p. 325.

42 "Darwin explained that the 'extremely imperfect" geological record was the reason'": Gould, *Panda's Thumb,* p. 181.

"As Lyell wrote Darwin": Eiseley, *Darwin's Century,* p. 162.

"The noblest conquest man has ever made": Peattie, *Green Laurels,* p. 74.

43 "Indeed, Zallinger evoked its half-prehistoric": Ley, *Worlds of the Past,* pp. 105, 109.

"Gaudry was not a Darwinian": Rudwick, *Meaning of Fossils,* p. 241.

45 "With his publicist's flair": Desmond, *Huxley*, p. 401

"Huxley's borrowed lineage": Desmond, *Archetypes and Ancestors*, p. 166.

"But he admitted in an 1870 talk": Bowler, *Life's Splendid Drama*, p. 331

46 "From the uniform monotonous prairie": Lanham, *Bone Hunters*, p. 33.

47 "You have no idea how much my mind": Foster, *Strange Genius*, p.40.

"Although sympathetic to Darwinism": Rainger, *Agenda for Antiquity*, p. 11.

Leidy "called his work 'a record of facts'": Osborn, *Cope, Master Naturalist*, p. 160.

48 "My belief is that he always lived apart": Schuchert and LeVene, *Othniel C. Marsh*, p. 345.

"When a student in Germany some twelve years ago": Marsh, "Introduction and Succession of Vertebrate Life in America," p. 358

49 "He tipped the conductor to hold the train": Schuchert and LeVene, *O. C. Marsh*, p. 98.

"Probably to Hull's surprise": ibid., p. 343.

50 "I am surprised that any scientific observers should not have at once detected the unmistakable evidence against its antiquity": *New York Herald*, December 1, 1869, p. 8.

"The line of descent appears": Marsh, "Fossil Horses of America," p. 294.

51 "If Huxley questioned a link": Leonard Huxley, *Life and Letters of Thomas Henry Huxley*, 1: 495.

"One of Huxley's lectures": Schuchert and LeVene, *O. C. Marsh*, p. 236.

"Huxley had swallowed Marsh's genealogy": *New York Herald*, September 23, 1876, p. 5.

52 "Seldom has a prophecy": Leonard Huxley, *Life and Letters of Thomas Henry Huxley*, p. 501.

"I had him 'corralled' in the basement": Desmond, *Huxley*, p. 485.

"Marsh failed to find the creature's 'grandfather'" Schuchert and LeVene, *O. C. Marsh*, p. 480.

"[N]o collection which has hitherto been formed approaches that made by Prof. Marsh": ibid., p. 238.

"When I was in America": Leonard Huxley, *Life and Letters of Thomas Henry Huxley*, p. 6.

"Darwin wrote Marsh that his work had 'afforded the best support to the theory of evolution which has appeared in the last 20 years'": Schuchert and LeVene, *O. C. Marsh*, p. 247.

CHAPTER 5. TERRIBLE HORNS AND HEAVY FEET

56 "Its author was Edward D. Cope": Simpson, "Resurrection of the Dawn Horse," p. 196.

"As a cause for many changes of structure in mammals": Marsh, "Vertebrate Life in America," p. 376.

"The real progress of mammalian life": ibid., p. 377.

57 "The origin of variation of animal structures is . . . the object of the doctrine of evolution to explain": Cope, "Relation of Animal Motion to Animal Evolution," p. 40.

"The ascending development of the bodily structure": ibid., p. 47.

58 "Under his expert guidance": Czerkas and Glut, *Dinosaurs, Mammoths, and Cavemen*, p. 16.

"Inside everything was unique . . . the floor was completely hidden by the massive bones of some vast creature": ibid., p. 14.

"Cope's brilliance spawned many stories": David Wake, personal comm., 1998.

59 "Charles Knight found his lab": Czerkas and Glut, *Dinosaurs, Mammoths, and Cavemen*, p. 14.

"If I know myself": Osborn, *Cope, Master Naturalist*, p. 136.

"Charles Knight told Osborn": Davidson, *Bone Sharp*, pp. 108–9.

"Knight perhaps exaggerated": Osborn, *Cope, Master Naturalist*, p. 545.

"He flirted with creationist ideas": Cope, "Fossil Reptiles of New Jersey," p. 114.

"In an 1871 pamphlet, he acknowledged that 'the law of natural selection . . . '": Cope, *On the Method of the Creation of Organic Types*.

60 "In a letter to General E. O. C. Ord": Osborn, *Cope, Master Naturalist*, p. 184.

"One of the most curious of the extinct mammals": Leidy, *Report of the United States Geological Survey on the Territories*, p. 72.

61 "These animals nearly equaled the elephant in size": Marsh, *Vertebrate Life in America*, p. 358.

"The utter desolation of the scene": Leidy, *Report of the United States Geological Survey on the Territories*, p. 19.

62 "My motive in going with Cope": Wallace, *Bonehunters' Revenge*, p. 81

"In a word, *Eobasileus*": ibid., p. 83.

63 "In the March issue": Cope, "Gigantic Mammals of the Genus *Eobasileus*," p. 159.

"Marsh called Cope's": Marsh, "Fossil Mammals of the Order Dinocerata," p. 152.

"Cope defended his trunked *Eobasileus*": Cope, "On Some of Professor Marsh's Criticisms," p. 295.

"He called Cope's work": Marsh, "Reply to Professor Cope's Explanations," p. ix.

"Cope responded": Cope, "On Professor Marsh's Criticisms," p. i–ii.

"In December, Cope wrote his father": Osborn, *Cope, Master Naturalist*, p. 182.

"He provided fine reconstructions": Schuchert and LeVene, *O. C. Marsh*, p. 475.

64 "The best he could do": Marsh, *Dinocerata*, p. 173.

"It is evident that the greater part": Cope, "Marsh on the Dinocerata," p. 703.

66 "It is the oldest Mammalian fauna of any extent": Osborn, *Cope, Master Naturalist*, p. 324.

68 "The function of the organism": Cope, "Theism of Evolution," p. 265.

CHAPTER 6. MR. MEGATHERIUM VERSUS
PROFESSOR MYLODON

70 "And although small, arboreal": Spencer G. Lucas, "Pantodonta," in Janis et al., eds., *Evolution of the Tertiary Mammals in North America*, p. 281.

"He took no pains": Osborn, *Cope, Master Naturalist*, p. 247.

"It has quite annoyed me": Davidson, *Bone Sharp*, p. 172.

71 "The paragraph and crude diagram that he devoted": Cope, *Vertebrata of the Tertiary Formations of the West*, 1: 617.

"Albert Gaudry, who nominated Cope to the French Geological Society": *New York Herald*, January 16, 1890; Osborn, *Cope, Master Naturalist*, p. 251.

"America's first great evolutionary theoretician": Gould, *Ontogeny and Phylogeny*, p. 85.

"Henry Fairfield Osborn recalled": Osborn, *Impressions of Great Naturalists*, p. 57.

72 "The generalizations were dictated by George Bauer": *New York Herald*, January 12, 1890.

"Marsh's 1869 debunking of the 'Cardiff giant' had especially stung Bennett": "The Onandaga Giant," *New York Herald*, November 18, 1869.

73 "The vertebra of the sea snakes in the Yale Museum": "The Rocky Mountains," *New York Herald*, December 24, 1870.

"It fanned the flames by goading Marsh": *New York Herald*, January 19, 1890.

"Prodded by a *Herald* reporter the next day": "Scientist Cope Fires Back at Marsh," *New York Herald*, January 20, 1890.

74 "Joseph Leidy told": *Philadelphia Inquirer*, January 13, 1890, p. 2.

"Cope tried to dismiss": Osborn, *Cope, Master Naturalist*, p. 411.

"Think of Mr. . . . Megatherium . . . out on the warpath for Professor . . . Mylodon": *New York Herald*, March 14, 1897, sec. 6, p. 3.

"I have not a blind faith": Beccari, *Wanderings in the Great Forests of Borneo*, p. 93.

76 "There are three hundred men": *New York Herald*, September 4, 1898, sec. 5, p. 6.

"According to Cope's will": Werkeley, "Professor Cope, Not Alive but Well," p. 73.

"In 1914, a Smith College biology professor": Stewart, "Where Is Cope's Face?" p. 76.

77 "Eiseley had lent Cope's skull to the Museum of Natural History": Christiansen, *Fox at the Wood's Edge*, p. 436.

"His book repeated the legends": Psihoyos and Knoebber, *Hunting Dinosaurs*, pp. 17–29.

"Sensationally calling Cope's Victorian prejudices 'Hitleresque,' it used a rumor": Bowden, "On the Trail of a Wayward Skull," *Philadelphia Inquirer*, October 6, 1994, A-1.

"The cranium that the photographer": Alan Mann, University of Pennsylvania Museum of Anthropology, letter of January 21, 1998.

"They may, if anything, be closer": Spencer G. Lucas and Robert M. Schock, "Dinocerata," ch. 19 of Janis et al., eds., *Evolution of the Tertiary Mammals in North America*, p. 258.

CHAPTER 7. FIRE BEASTS OF THE ANTIPODES

79 "His appearance made me gasp": Conan Doyle, *Lost World*, p. 16.

80 "Two creatures like large armadilloes": ibid, p. 119.

81 "If Buffon had known of the gigantic sloth": Darwin, *Voyage of the H.M.S. Beagle*, p. 157.

"Zallinger's art for Time/Life's book on the *Beagle* voyage": Barnett et al., *Wonders of Life on Earth*, pp. 70–75.

"He found 'many curious resemblances . . .' between edentates and tillodonts": Marsh, "Vertebrate Life in America," p. 372.

83 "These consisted mainly of volcanic rock": Simpson, *Discoverers of the Lost World*, p. 67.

84 "Then, in 1898, penetrating to the remote Lake Colhue-Huapi": ibid., p. 70.

"On February 15, 1899, Carlos wrote to Florentino": ibid., p. 71.

"Florentino had written Carlos earlier": ibid., p. 69.

"Three years later, he published a profusely illustrated 568-page book": Ameghino, *Formations sédimentaires du Crétacé supérieur et du Tertiaire de Patagonie.*

"When Albert Gaudry, recalling his Pikermi adventures": Simpson, *Discoverers of the Lost World*, p. 94.

85 "Evolution, we thought, had run its course through the ages without anything stopping it": ibid., p. 102.

86 "Reaching the antipodes in April": Hatcher, *Bone Hunters in Patagonia*, p. 19.

"Sea cliffs at a place called Corriguen Aike": ibid., p. 73.

87 "Not a day passed that I did not find remains of dinosaurs": ibid., p. 135.

"Hatcher had a prickly personality": Simpson, *Discoverers of the Lost World*, p. 85.

"In a later paper, he sarcastically doubted": ibid., p. 86.

88 "In 1903, he wrote Florentino": ibid., p. 120.

"He had left Princeton in 1899": ibid., p. 122.

"He too went to Argentina in 1900": ibid., p. 133.

"I was received with the utmost cordiality": Scott, *Some Memories of a Paleontologist*, p. 250.

"Scott was not afraid to criticize": Simpson, *Discoverers of the Lost World*, p. 142.

89 "As Scott observed, Florentino's vision": Scott, *Some Memories of a Paleontologist*, p. 251.

"Simpson called Carlos": Simpson, *Discoverers of the Lost World*, p. 74.

90 "Simpson found Florentino": ibid., p. 75.

"[T]he only 'Argentine fauna' are a 'sow harnessed in chains' and werewolves": Borges, *Book of Imaginary Beings*, p. 209.

"But the fire beast story is so 'Borgesian' that its very absence from his work seems suspect": Borges, "The Argentine Writer and Tradition," in *Labyrinths*, p. 181.

"I have said that the City was founded on a stone plateau": Borges, "The Immortal," in *Labyrinths*, p. 109.

CHAPTER 8. TITANS ON PARADE

92 "At one point, these bones": Marsh, "Return of Professor Marsh's Expedition," p. 117.

"One morning Marsh and I went from camp": Burt quoted in Mattes, *Indians, Infants, and Infantry*, pp. 196–97.

94 "His papers classed it": Marsh, "Principal Characters of the Brontotheriidae," p. 335.

"It has been well said that what one truly": Wallace, *Bonehunters' Revenge*, p. 298.

95 "Thus 901 young were produced": Weismann, *Essays upon Heredity*, p. 445.

"[W]hen examining an extensive series of fossils": Rainger, *Agenda for Antiquity*, p. 41.

96 "[H]ardly had the uintatheres gone to earth": Osborn, "Prehistoric Quadrupeds of the Rockies," p. 714.

97 "Professor Marsh seems to spend": Osborn, letter of February 13, 1898, Osborn Library, American Museum of Natural History.

"It contains a good many species": Osborn, *Cope, Master Naturalist,* p. 297.

"It was indeed a rare bit of one's education": Osborn, "J. L. Wortman," p. 652.

98 "Although allowing that Marsh had 'made the largest and most valuable contributions'": Osborn, *Titanotheres of Ancient Wyoming, Dakota, and Nebraska,* p. 145.

"There, modeled in clay, was Osborn's view": Rainger, *Agenda for Antiquity,* p. 166.

"The museum's titanothere work lasted twenty years": ibid., p. 92.

99 "One of the most fascinating of the many problems of paleontology": Osborn, "Hunting the Ancestral Elephant in the Fayum Desert," p. 815.

100 "In the year 1899, paleontology had advanced to such a point that the origin of many families was known": ibid., p. 816.

"I described this African 'Garden of Eden'": ibid., p. 817.

101 "It has proved to be epoch-making": ibid., p. 819.

"The reader can imagine": ibid.

"Weather and labor problems": ibid., p. 825.

102 "Their Egyptian workforce, which hitherto": ibid., p. 826.

"One must find coal, or oil, or gold": ibid., p. 823.

"The workers learned to go easy on the fragile bones": ibid., p. 827.

"After two weeks, however, 'one of the prospecting party' found a 'splendid skull of the *Palaeomastodon'*": ibid.

"The finding of the two heads": ibid., p. 835.

"[W]e cannot regard the little *Moeritherium*": ibid., p. 832.

"Osborn had no doubt": ibid., p. 833.

"Since the fore and hind limbs were elongate": ibid., p. 835.

103 "In the whole history of creation": ibid.

"Zallinger adapted it from a Knight illustration of *Palaeomastodon*": ibid., p. 834.

CHAPTER 9. FIVE-TOED HORSES AND MISSING LINKS

104 "You are off in the morning, stiffened by a frosty night": Osborn, "Prehistoric Quadrupeds of the Rockies," p. 707.

105 "The fossil hunter is predestined to his work": ibid., p. 708.

"In fact, the 'member of the prospecting party' who found the two skulls": Morgan and Lucas, *Notes from Diary — Fayum Trip, 1907*, p. 32.

"His old foe Marsh had considered Mesozoic mammals a *mysterium tremendum*": Schuchert and LeVene, *O. C. Marsh*, p. 403.

"Jacob Wortman's discovery in 1882 of the first known Cretaceous mammal fossils": ibid., p. 456.

106 "[T]eeth, fragmentary jaws, and even skeletal parts began to pour into Marsh's eager hands": ibid., p. 457.

"MOST COLOSSAL ANIMAL EVER": *New York Journal and Advertiser*, December 11, 1898.

"NEW YORK'S NEWEST, OLDEST, BIGGEST CITIZEN": *New York World*, December 26, 1898, p. 10.

"He sniffed after the most spectacular ones": Bird, *Bones for Barnum Brown*, p. 116.

107 "Bird recalled a trip to Big Bend, Texas": ibid., p. 194.

"In the dispersal center, during the close of the Age of Reptiles": Osborn, *Science*, April 13, 1900, p. 567.

"He believed that man had evolved, . . . from bipedal, tool-making creatures": Osborn, "Is the Ape Man a Myth?" p. 4.

108 "Conceivably, wrote the author of *The Invisible Man*": Wells, *Outline of History*, p. 49.

"Ever since 1912": Andrews, *On the Trail of Ancient Man*, p. 3.

"'My Gawd,' wrote one female admirer": Gallenkamp, *Dragon Hunter*, p. 299.

109 "When the plans for the expedition": Andrews, *On the Trail of Ancient Man*, p. 17.

"Ninety nine out of a hundred persons": ibid., p. 21.

"The newspapers . . . hoped": ibid., p. 19.

"He vacillated between": ibid., p. 20.

"At the base of a hill": ibid., p. 147.

110 "One of his ten 'narrow escapes'": ibid., p. 148.

"There was not a sign of human life": ibid., p. 317.

"In a spot": ibid., p. 132.

"Our efforts to discover fossils": ibid., p. 135.

111 "It required three months": Osborn, "Extinct Rhinoceros *Baluchitherium*," p. 224.

"Paraceratherium was one of the few mammals": Prothero, *Eocene-Oligocene Transition*, p. 199.

"To one who could read the language": Andrews, *New Conquest of Central Asia*, pp. 279–80.

112 "It was one of the greatest days": Andrews, *On the Trail of Ancient Man*, p. 245.

"This is the high point": Gallenkamp, *Dragon Hunter*, p. 178.

"A specimen in which he was greatly interested": Andrews, *On the Trail of Ancient Man*, p. 245.

"Discovery of the ancestral five-toed horses": Osborn, "Discovery of an Unknown Continent," pp. 145–7.

113 "Granger walked to the base of a formation": Andrews, *On the Trail of Ancient Man*, p. 328.

114 "They were 'among the earliest placental mammals'": ibid., p. 331.

CHAPTER 10. THE INVISIBLE DAWN MAN

115 "The earliest Tertiary fauna is . . . of Paleocene age, and of curious and rather unexpected character": Matthew and Granger, "Most Significant Finds of the Mongolian Expeditions," p. 533.

116 "[I]t is remarkable that we find": ibid., p. 534.

"The 'reference to the primates' was 'uncertain' he had concluded": Osborn, "New Fossil Mammals from the Fayum": p. 272.

"The animal is certainly a new genus": Gould, *Bully for Brontosaurus*, p. 438.

117 "It was at the dramatic moment": Osborn, "Why Central Asia?" p. 266.

"We prophecy that the Dawn Man": ibid., p. 269.

"If the family of mankind has *always* been superior": Gregory and McGregor, "A Dissenting Opinion as to Dawn Men and Ape Men," p. 271.

118 "The main trend of mammal evolution": Matthew, Lecture to the Linnean Society of London, quoted in Rainger, *Agenda for Antiquity*, p. 206.

119 "Matthew considered it 'utterly impossible'": Matthew, "Phylogeny of the Felidae," p. 307.

"His saber-toothed protagonist": Matthew, "Scourge of the Santa Monica Mountains," p. 469.

"This was his favorite prey": ibid., p. 470.

"Strive as he might": ibid., p. 471.

120 "He told Matthew": Rainger, *Agenda for Antiquity*, p. 76.

122 "Gregory called Osborn 'a terrific problem'": ibid., p. 77.

CHAPTER II. A BONAPARTE OF BEASTS

125 "The earliest ones shown, *Poebrotherium*": Ostrom et al., *Age of Mammals*, p. 11.

126 "My good friend the paleontologist": Rainger, *Agenda for Antiquity*, p. 135.

"I am sorry to hear": ibid., p. 136.

"The nature of these variations": ibid., p. 212.

"Yet, according to Ronald Rainger": ibid.

"One of Matthew's oddities": Simpson, *Concession to the Improbable*, p. 34.

"Our car was delivered to us": ibid.

127 "My horse somehow guessed": ibid., p. 28.

"Months later this block was received": Laporte, *George Gaylord Simpson*, p. 60.

"I was an unexpected last child": Simpson, *Concession to the Improbable*, p. 98.

128 "I went collecting in the basement": ibid., p. 17.

"Simpson became an authority": Simpson, "Mammals Were Humble When Dinosaurs Roamed," p. 11.

129 "The very thought of Mesozoic Mammals": Laporte, ed., *Simple Curiosity*, letter of April 9, 1927, p. 49.

"We quarrel rather continually": ibid., p. 48.

"Applying for the coveted Yale position": Simpson, *Concession to the Improbable*, p. 38.

130 "When Osborn presented him with . . . his titanothere monograph": ibid., p. 40. Edwin Colbert attributes a similar story to another museum employee in his autobiography, *Digging into the Past*.

"Simpson stayed with his wife until 1929": ibid., p. 181.

"[T]hey are generally considered the greatest paleontological discovery of the present century": Laporte, ed., *Simple Curiosity*, p. 49.

"Thus I was out of a specialty": Simpson, *Concession to the Improbable*, p. 41.

131 "The study of origins is always peculiarly difficult": Simpson, *Attending Marvels*, p. 68.

"So it fell to our expedition to take the next step": ibid., p. 69.

"I am quite sure he felt he could do anything": Burns, "A Memoir," in Simpson, *Dechronization of Sam Magruder*, p. 130.

"Simpson said he only regretted that he had but one liver to give for his museum": Laporte, *George Gaylord Simpson*, p. 7.

"[I]t was apparent that I had not yet found the proper atmosphere": Simpson, *Attending Marvels*, p. 10.

"I wish I could say": ibid., p. 17.

"[T]here were no trees": ibid., p. 25.

132 "The animals buried here": ibid., p. 74.

"They dug up the bones of 'one of the largest snakes'": ibid., p. 221.

"They were homalodontotheres": ibid., p. 74.

"Simpson later modified homalodontotheres": Simpson, *Splendid Isolation*, p. 127.

"Despite such minor difficulties": Simpson, *Attending Marvels*, xiv.

"Petroleum geologists had encountered": Simpson, *Concession to the Improbable*, p. 64.

133 "Only a few tantalizing scraps": ibid.

"Andrews had made a last attempt": Gallenkamp, *Dragon Hunter*, p. 288.

"I had long looked forward": Simpson, *Concession to the Improbable*, p. 76.

"As he wriggled in this double bind": ibid., p. 77.

"My retreat from Moscow": ibid., p. 79.

134 "I had long been intensely interested": ibid., p. 81.

CHAPTER 12. LOVE AND THEORY

135 "[S]he interests me tremendously": Laporte, *Simple Curiosity*, p. 48.

"My new wife is a lanky blonde hoodlum": ibid, p. 224.

136 "This complementary relationship": Simpson, *Concession to the Improbable*, p. 84.

"The highest possible scientific motive": Laporte, *Simple Curiosity*, p. 48.

"The most basic mammalian characteristic": Simpson, "Meek Inherit the Earth," p. 102.

"The level at which Dr. Simpson operated": Colbert, *Digging into the Past*, p. 144.

"Work at the Museum proceeds": Laporte, *Simple Curiosity*, p. 208.

137 "This led 'to what is now commonly called the synthetic theory'": Simpson, *Concession to the Improbable*, p. 115.

"[A]s I saw a possible synthesis beginning to take form, I also saw a serious gap in it": ibid., p. 84.

"During the Eocene, the record . . . does not show any net or average increase in size": Simpson, *Meaning of Evolution*, p. 136.

138 "Over time, this widespread *Eohippus* population": Simpson, *Tempo and Mode in Evolution*, p. 123.

"Argument that this trend was not oriented": Simpson, *Meaning of Evolution*, p. 157.

139 "Even today, writers not familiar with the actual fossils": Simpson, *Life of the Past*, p. 125.

"The little eohippus lived about 125,000,000 years after the mammals arose": Simpson, *Meaning of Evolution*, p. 79.

"The guanaco is a most improbable creature": Simpson, "Children of Patagonia," p. 135.

"If actions are a fair guide": Simpson, *Attending Marvels*, p. 191.

"Patagonia is the guanaco's country": ibid, p. 197.

140 "[T]he history of life . . . is consistent with the evolutionary processes": Simpson, *Concession to the Improbable*, p. 115.

"I am currently (when I can bear to work)": Laporte, *Simple Curiosity*, p. 232.

"Simpson's task": Bowler, *Evolution*, p. 317.

"A younger colleague of Simpson's, Stephen Jay Gould": Afterword to Simpson, *Dechronization of Sam Magruder*, p. 114.

142 "In the hospital": Laporte, *Simple Curiosity*, p. 291.

"My life seems to be turning in on itself": ibid., p. 48.

143 "The faculty and student attitude": ibid, p. 312.

"Incidentally, there has been no heckling": ibid, p. 314

"Stephen Jay Gould, one of Simpson's students": Afterword to Simpson, *Dechronization of Sam Magruder*, p. 115.

"When Magruder playfully tries": ibid, p. 101.

"Surely no one ever had such a trip": Laporte, *Simple Curiosity*, p. 185.

144 "Almost completely devoid of vegetation": Simpson, *Dechronization of Sam Magruder*, p. 26.

CHAPTER 13. SIMPSON'S CYNODONT-TO-SMILODON SYNTHESIS

145 "Brandishing his double-barreled 'sauropod gun'": de Camp, "A Gun for Dinosaur," in *Dinosaurs!* pp. 1–2.

146 "There are as many different ways of collecting": Simpson, "How Fossils Are Collected," p. 329.

"In one article about Paleocene mammals": Simpson, "Meek Inherit the Earth," p. 103.

"The World We Live In's text describes *Hoplophoneus"*; Barnett et al., *The World We Live In*, p. 115.

148 "The complicated reptilian jaws": Simpson, "Meek Inherit the Earth," p. 102.

149 "The beasts that thronged the Mesozoic": William A. Clemens, personal comm., September 20, 2002.

150 "They appeared in the Mesozoic fossil record": Lillegraven et al., eds., *Mesozoic Mammals*, p. 192.

"[I]t has been common practice for zoologists": Colbert, *Evolution of the Vertebrates*, p. 272.

"[T]hey were undergoing constant and fundamental evolutionary changes": Simpson, "Meek Inherit the Earth," p. 102.

"Simpson thought that it had allowed better shearing and grinding than before": Lillegraven et al., eds., *Mesozoic Mammals*, p. 174.

151 "[T]he main reason for that extinction": Simpson, "Meek Inherit the Earth," p. 102.

"I have studied the dinosaurs": Simpson, *Dechronization of Sam Magruder*, p. 78.

"All that can be said is that conditions changed": Colbert, *Evolution of the Vertebrates*, p. 232.

"That the Cretaceous fauna would be less resistant": Kurten, *Age of Reptiles*, p. 239.

152 "Simpson wrote that the mammals": Simpson, "Meek Inherit the Earth," p. 102.

"Our *Diatryma*": Matthew and Granger, "Giant Eocene Bird," p. 418.

"The surface of the earth was open to conquest": Romer, *Man and the Vertebrates*, p. 98.

"Towering over the mammals of its day": Kurten, *Age of Mammals*, p. 49.

153 "Flightless birds continued to increase": Desmond, *Hot-Blooded Dinosaurs*, p. 179.

"The next step—from theory to attested fact, would be the finding of the ancestry": Simpson, "Great Animal Invasion," p. 210.

155 "Following the major uplift": Dunbar, "Recollections on the Renaissance of Peabody Museum Exhibits," p. 22.

156 "Those extant in the Plio-Pleistocene were the ones": Simpson, *Geography of Evolution*, p. 197.

CHAPTER 14. SHIFTING GROUND

158 "The Royal Navy's *Challenger* expedition": Linklater, *Voyage of the Challenger*, p. 274.

"The probability of spread": Simpson, *Evolution and Geography*, p. 87.

159 "Armed with quantitative methods": McKenna, "Sweepstakes, Filters, Corridors," p. 298.

"I did not deny the possibility": Simpson, *Concession to the Improbable*, p. 273.

"Simpson was preeminent among American geologists": Laporte, *George Gaylord Simpson*, p. 196.

160 "Simpson gave himself less credit": Simpson, *Fossils and the History of Life*, p. 112.

"Students of mammal origins tell us that many of the familiar animals we know originated on the Asiatic mainland": Scheele, *First Mammals*, p. 22.

161 "It is absurd that these two principles . . . should be regarded as mutually exclusive": Simpson, *Splendid Isolation*, p. 252.

162 "Although I am still sometimes cited": Simpson, *Concession to the Improbable*, p. 273.

"[It was] highly probable that continental drift . . . had little effect on . . . land faunas during . . . the Cenozoic": ibid.

"In 1971, when many geologists": Kurten, *Age of Mammals*, p. 24.

"He speculated that earth's land masses": ibid., p. 24.

"[A] 'Noah's Ark' was a piece of landmass that broke from a parent continent": McKenna, "Sweepstakes, Filters, Corridors."

163 "At the U.C. Berkeley Museum of Paleontology . . . I encountered two casts of jaws from early Paleocene Patagonia": William Clemens, personal comm., September 20, 2002.

"In the mid 1990s, paleontologists discovered": Weil, "Relationships to Chew Over," p. 29.

"Zhe-xi Luo and two co-authors favored": Luo et al., "Dual Origin of Tribosphenic Mammals," p. 53.

CHAPTER 15. DISSOLVING ANCESTRIES

167 "Previous taxonomists had taken parallelisms of structure": Secord, *Victorian Sensations*, p. 389.

168 "A horse cladogram might show": Benton, *Rise of the Mammals*, p. 107.

"Less famous cladistic analyses": e.g., Ji et al., "Chinese Triconodont Mammal"; Hu et al., "New Symmetrodont Mammal."

169 "In order to . . . construct a phylogeny": MacFadden, *Fossil Horses*, p. 20.

"A remark made by a colleague of MacFadden's": David Webb, Florida State Natural History Museum, Gainesville, personal comm., March 1993.

"One zoologist called Simpson's synthesis 'useless' and claimed that cladistics had transformed paleontology": Gee, *In Search of Deep Time*, pp. 136, 137.

"As though anticipating such attitudes": Simpson, *Concession to the Improbable*, p. 271.

"There's a problem with judging just from what survives": David Webb, Florida State Natural History Museum, Gainesville, personal comm., March 1993.

170 "Whether we like it or not . . . we have inherited a basic approach to animal classification from our earliest forebears": Gingerich, "Paleontology, Phylogeny, and Classification," p. 451.

"Although [Gingerich] acknowledges that Simpson's way of classifying mammals needed revision": ibid, p. 463.

"The paper proposing the idea": Farina and Blanco, "*Megatherium*," p. 1725.

"The forearms of Megatherium": Gee, *In Search of Deep Time*, p. 80.

"It makes an odd picture": ibid, p. 81.

171 "In truth, what ideas must we form": Matthew, "Ground Sloth Group," p. 115.

"Darwin called Lund's idea 'preposterous'": Darwin, *Voyage of the H.M.S. Beagle*, p. 76.

"Owen very properly ridiculed these fanciful theories": Matthew, "Ground Sloth Group," pp. 117–18.

"With no need for speed": Webb, "Successful in Spite of Themselves," pp. 52–53.

172 "Cladistic analysis is based": Gingerich, "Paleontology, Phylogeny, and Classification," p. 452.

"Given such confusion, you can only long for the eighteenth century": Gee, *In Search of Deep Time*, p. 136.

"As we know, Deep Time is not a movie": ibid.

"Early Miocene droughts killed": Matthew, "Fossil Bones in the Rock," p. 359.

173 "These extinct animals are commonly called giant pigs": ibid, p. 364.

"The name 'clawed ungulate' sounds like a contradiction": ibid.

"Let's watch *Moropus*": MacDonald and MacDonald, "Landscape Rich with Life," pp. 29–30.

174 "The skeletons are grouped around a tree trunk": Matthew, "Ground Sloth Group," p. 115.

"They have a wonderful array of whole skeletons": David Webb, Florida State Natural History Museum, Gainesville, personal comm., March 1993.

"Few persons are able to form an adequate idea of an animal": Osborn, "Restorations and Models," pp. 85–86.

CHAPTER 16. EXPLODING FAUNAS

176 "As if to emphasize the mystery": Osborn, *Cope, Master Naturalist*, p. 298.

177 "Although sequences of strata": Archibald and Clemens, "Late Cretaceous Extinctions," p. 380.

"Western North America is one of the regions": Clemens, Archibald, and Hickey, "Out with a Whimper," p. 294.

178 "[t]he earliest and lowest of three newly discovered distinct Cretaceous mammal faunas": Robert Sloan and Leigh Van Valen, "Cretaceous Mammals of Montana," p. 220.

179 *"Protoungulatum donnae* is presently the oldest": ibid., p. 226.

"[A] definite taxonomic radiation of rapidly evolving placental mammals": Lillegraven, "Fossil Mammals of the Upper Part of the Edmonton Formation," p. 7.

"The two basal stocks appear to have been independently derived": ibid.

180 "How did the Cretaceous World end?": Clemens, Archibald, and Hickey, "Out with a Whimper," p. 297.

181 "[S]o profound is our ignorance": Darwin, *On the Origin of Species*, p. 73.

"The inhabitants of each successive period": ibid., p. 345.

"Now that I am living among dinosaurs": Simpson, *Dechronization of Sam Magruder*, p. 80.

182 "He dismissed them as summarily": Simpson, *Fossils and the History of Life*, p. 137.

"The remarkable aspect of the Alvarezes's work": Gould, *Hen's Teeth*, p. 328.

"The excitement was extraordinary": Fortey, *Life*, p. 98.

183 "The ingredients for a juicy controversy were in place": ibid.

"The pattern is difficult to explain": Clemens, Archibald, and Hickey, "Out With a Whimper," p. 295.

"The multiplicity of patterns of extinction": ibid., p. 297.

"[T]he Alverez theory would remain merely a scientific curiosity": Powell, *Night Comes to the Cretaceous*, p. 124.

184 "[O]ur preferred scenario is tied solidly": Alvarez, "Experimental Evidence," p. 639.

"Alverez ended with arguments . . . as to the 'precise synchronicity' of the various extinctions": ibid, p. 640.

"And finally, if you feel that I have . . . have given the impression that physicists always wear white hats": ibid, p. 641.

185 "I became a believer": Kerr, "Dinosaurs and Friends," p. 162.

"They concluded that the mammal sites at Bug Creek": Smit and Van Der Kaars, "Terminal Cretaceous Extinctions," p. 1179.

"Data supporting the beginning of the ungulate radiation well in advance of dinosaur extinction": Sloan et al., "Gradual Dinosaur Extinction," p. 632.

"[Archibald] accepted the conclusion that the Bug Creek localities were Cretaceous in age": Archibald, *Dinosaur Extinction,* p. 41.

"[A] Paleocene age for all the Bug Creek sites is the majority opinion": ibid., p. 42.

186 "Because there was no significant change": Sheehan and Fastovsky, "Sudden Extinction of Dinosaurs," p. 837.

"There doesn't seem to be any evidence for a gradual decline in their diversity": Fastovsky in *The Dinosaurs!* "Death of the Dinosaur" (PBS, 1992).

"Gradualists . . . got a plug in some 1989 *National Geographic* Dig-a-Dinosaur coverage": Gore, "Extinctions," p. 691.

"Even among scientists . . . bias toward the impact hypothesis is often strong": Malcolm W. Browne, "Dinosaur Experts Resist Meteor Extinction Idea."

187 "[Alvarez] called . . . paleontologists generally 'not very good scientists'": Malcolm W. Browne, "Debate over Dinosaur Extinction."
" 'There was a little more to evolution than Darwin realized,' Alvarez said": *Dinosaurs!* "Death of the Dinosaur" (PBS, 1992).

CHAPTER 17. THE REVENGE OF THE SHELL HUNTERS

188 "Cope's associate, Alpheus Hyatt was equally important in developing neo-Lamarckism": Gould, *Ontogeny and Phylogeny,* p. 91.

"Simpson, with typical self-confidence thought vertebrate paleontologists prevailed in theorizing": Simpson, *Geography of Evolution,* p. 61.

189 "Simpson estimated in 1960 that there were only about 130 professional vertebrate paleontologists": ibid., p. 44.

"[I]t was invertebrates that caught the eyes of most of us": Eldredge, *Triumph of Evolution,* p. 84.

"Gould perceived similar patterns in snail evolution": Eldredge and Gould, "Punctuated Equilibria."

"The new theory was set up in opposition to the notion of . . . a slow and more or less continuous change": Fortey, *Trilobite!* p. 165.

190 "Stability is the norm until . . . suddenly many species become extinct": Eldredge, *Triumph of Evolution*, p. 86.

"Simpson seemed not to grasp punctuated equilibria's radical implications": Simpson, *Concession to the Improbable*, p. 269.

"I used to argue that mass extinctions": Gould, *Hen's Teeth*, p. 328.

191 "Darwin imagined the living world": Darwin, *On the Origin of Species*, p. 67.

"[F]or geographically widespread species, extinction is likely only if the killing stress is one so rare as to be beyond the experience of the species": Raup, "Role of Extinction in Evolution," p. 6758.

192 "A statistical analysis of all the data": Eldredge, *Life Pulse*, p. 211.

"[I]f mass extinctions are so frequent . . . then life's history either has an irreducible randomness or operates by new and undiscovered rules": Gould, *Flamingo's Smile*, p. 445.

"Analyzing 255 specimens of *Pelycodus* jaws and teeth": Gingerich, "Paleontology and Phylogeny," p. 17.

193 "There are no horrors here": Scully et al., *Age of Reptiles*, p. 6.

194 "Mass extinctions change the rules": Gore, "Extinctions," p. 673.

"He acknowledged that studies like Gingerich's": Gould, *Structure of Evolutionary Theory*, pp. 830–31.

"If evolution were a ladder": Gould, *Bully for Brontosaurus*, p. 276.

195 "The solution to the paradox of such adequate intelligence in such a primitive mammal is stunningly simple": ibid., p. 292.

"While praising the great synthesizer for seeing a branching bush instead of an orthogenetic ladder": ibid., p. 178.

"[Simpson] had erred in emphasizing 'the supposedly gradual and continuous transformation' between *Mesohippus* . . . and . . . *Miohippus*": ibid.

"The 'enormous expansion of collections . . .' had allowed . . . Prothero and Neil Shubin to 'falsify Simpson's gradual and linear sequence'": ibid., p. 179.

"Such stasis is apparent in most Neogene . . . horses": Prothero and Shubin quoted in ibid., p. 180.

196 "[T]he old story of . . . waves of differential migration": Gould, *Hen's Teeth*, p. 351.

"The gigantic head and short, powerful neck": Gould, "Play It Again, Life," p. 25.

"Birds had a second and separate try": ibid., p. 26.

197 "Far from posing a threat to animals high in the Eocene food chain": Andors, "Reappraisal," p. 122.

CHAPTER 18. SIMPSON REDIVIVUS

198 "Indeed . . . the mammoth . . . was exemplary of stasis": Gould, *Structure of Evolutionary Theory*, p. 1320.

"One section is devoted to the persistence": ibid., p. 747.

199 "But Gould thought his conclusion: ibid., p. 748.

"Vrba's classic studies of African antelopes": ibid., p. 866.

"According to the classic Darwinian view": Vrba, "Pulse That Produced Us," p. 50.

"Most species are static for 2–4 million years": Prothero and Heaton, "Faunal Stability," p. 257.

"Thanks to improved land mammal chronologies": Webb and Opdyke, "Global Climatic Influence," p. 184.

200 "[A]n impressive array of examples in the fossil record, from snails to horses": Davidson, "Theory Still Rocks Scientists' Equilibrium."

"'What we found suggests that the dinosaurs were thriving' Sheehan told the Associated Press": "Dinosaurs Killed off in Their Prime, Study Says," *San Francisco Chronicle*, June 1, 2000.

"'You need to consider the whole fauna,' Clemens concluded": ibid.

"The data doesn't agree with Sheehan": David Archibald, telephone comm., June 15, 2001.

"Archibald's own studies": "A Study of Mammalia and Geology across the Cretaceous-Tertiary Boundary in Garfield County, Montana," p. 257.

201 "If this interpretation is correct . . . the survival rate of mammal species is approximately 10% and of dinosaurs 0%": Lofgren, *Bug Creek Problem*, p. 71.

"At a Cretaceous-Paleocene site in Wyoming": Lillegraven and Eberle, "Vertebrate Faunal Changes," p. 706.

"The 'seventeen varieties of non-avian dinosaurs' . . . were about the same number as at Hell Creek": ibid., p. 701.

"We have been witnessing . . . an . . . over-zealous jump into a bandwagon": Eldredge, *Life Pulse*, p. 213.

"Both authors accepted that impacts": Prothero, *Paradise Lost*, p. 245.

202 "Thus far, marine regression": Archibald, *Dinosaur Extinction*, p. 160.

"[T]he biggest extinctions of the Cenozoic seem to coincide with the initial formation of glaciers on Antarctica": Prothero, *Paradise Lost*, p. 245.

"Gould sees contingency—evolutionary history based on the luck of the draw—as the major lesson": Conway Morris, "Showdown on the Burgess Shale," p. 51.

"[T]he idea that evolution proceeds in fits and starts, . . . moving nowhere in particular": Wright, "Accidental Creationist," p. 60.

203 "[Wright] cited examples of evolutionary progress through Darwinian competition": ibid., p. 58.

"He suggested that Gould's dislike of some early evolutionists' 'social Darwinist' ideology had motivated his attacks": ibid., p. 60.

205 "The diary of Arthur Lakes": Lakes, *Discovering Dinosaurs in the Old West*, p. 136.

"Walter Granger's 1907 Fayum notes": Morgan and Lucas, *Notes from Diary — Fayum Trip*, p. 29.

"An adventuresome American Museum paleontologist": Novacek, *Time Traveler*, p. 115.

"Novacek recalled that his teachers . . . regarded microfauna as 'the new frontier of research,' offering unique insights": ibid., p. 116.

"'It was a little smart cookie with an extended brain,' said . . . Zhe-xi Luo": Paul Recer, "Fossil Hints at Mammal Origins," *Columbus Dispatch*, May 25, 2001.

"Could it be . . . that the spectacular adaptations . . . were of a . . . more 'superficial' nature . . . ?": Lillegraven et al., *Mesozoic Mammals*, p. 4.

206 "One living species, Kitti's hog-nosed bat": Luo, Crompton, and Sun, "New Mammaliaform," p. 1538.

CHAPTER 19. WIND THIEVES OF THE KYZYLKUM

207 "Crawling around with magnifying glasses": Kielan-Jaworowska, "Late Cretaceous Mammals and Dinosaurs from the Gobi Desert," p. 152.

"Indeed, they collected so many new fossils": Kielan-Jaworowska, *Hunting for Dinosaurs*, p.132.

208 "[T]he choicest spot for Cretaceous fossils in the Gobi, if not the entire world": Novacek, *Time Traveler*, p. 298.

"Another skeleton, of the famous *Zalambdalestes*": ibid., p. 305.

"Undoubtedly the best preserved mammal fossils": Archibald, *Dinosaur Extinction and the End of an Era*, p. 95.

209 "[Nessov] could be irascible . . . but at the same time could be brilliant": David Archibald, phone comm., June 15, 2001.

"Nessov was very, very widely read": David Archibald, phone comm., June 15, 2001.

210 "They called them 'zhelestids,' a Kazakh-Greek composite": Archibald, "Fossil Evidence for a Late Cretaceous Origin of 'Hoofed' Mammals," p. 1150.

"Archaic ungulates retain so many characters": Nessov et al., "Ungulate-like Mammals from the late Cretaceous of Uzbekistan," p. 73.

"In 'zhelestids' and late Cretaceous eutherians": ibid., p. 72.

211 "But they definitely had lived "cheek by jowl . . . " with western Asian dinosaurs": Archibald, *Dinosaur Extinction*, p. 207.

"At U.C. Berkeley . . . a jaw fragment cast from a little creature named *Gallolestes*": William Clemens, personal comm., September 20, 2002.

"Lev did mostly surface collection": David Archibald, phone comm., June 15, 2001.

"Nessov's finds at Dzharakuduk included a few teeth": Archibald et al., "Late Cretaceous Relatives."

"Malcolm McKenna and Michael Novacek had studied *Zalambdalestes* skulls": Novacek, "Pocketful of Fossils," p. 42.

"*Zalambdalestes* also had skeletal similarities to . . . a late Cretaceous Mongolian genus": McKenna, "Early Relatives of Flopsy, Mopsy, and Cottontail," pp. 57–58.

212 "Their descendants appeared": Archibald, *Dinosaur Extinction*, p. 207.

"[Z]helestids might simply have been blunt-molared zalambdalestids": Novacek et al., "New Eutherian Mammal from the Late Cretaceous of Mongolia," p. 61A.

"[Z]zhelestid skulls are quite different from zalambdalestid ones": David Archibald, phone comm., November 8, 2002.

"Dinosaurs . . . kept mammals 'fairly generalized, smallish, and undifferentiated physically' until": Eldredge, *Pattern of Evolution*, p. 153.

"[T]he idea that 'mammals would still be rat-size creatures living in the ecological interstices' of a dinosaur world if a Late Cretaceous bolide hadn't hit": Gould, *Structure of Evolutionary Theory*, p. 1320.

213 "[The] discovery of a eutherian that had lived in the early Cretaceous, about 125 million years ago": Ji et al., "Earliest Known Eutherian Mammal," p. 816.

"*Eomaia* may have reached food sources that were not available": Weil, "Upwards and Onwards," p. 799.

"For studying early evolution of mammals": Perlman, "Shrewlike Early Mammal Fossil Found," p. 2.

"Some estimates were startling": Gore, "Rise of Mammals," pp. 14–16.

215 "The origin of most mammalian orders seems not to be tied to . . . niches left vacant by dinosaurs": Kumar and Hedges, "Molecular Timescale," p. 917.

"Ironically, modern vertebrate paleontologists": Normile, "New Views of the Origins of Mammals," p. 774.

"[W]hat they write about higher relationships seems perfectly reasonable": Pennisi, "Placentals' Family Tree," p. 2267.

"[N]o support for a Late Cretaceous diversification of extant placental orders": Archibald et al., "Late Cretaceous Relatives," p. 62.

"[C]ompetition from dinosaurs had made Mesozoic mammalian evolution 'pretty ho-hum' compared to the Cenozoic": David Archibald, telephone comm., June 15, 2001.

CHAPTER 20. THE SERPENT'S OFFERING

216 "Vincent Scully noted a basic difference between the mammal mural and its famous neighbor": Scully, *Age of Reptiles*, p. 14.

218 "I had felt it unfortunate that the dinosaur mural portrayed these great beasts in a modern landscape": Dunbar, "Recollections," p. 22.

219 "Five years later, he wrote the man, Gaston de Saporta": Francis Darwin, *Life and Letters of Charles Darwin*, 2: 284.

"[T]he close gradation of such forms seems to me a fact of *paramount* importance": Conry, *Correspondence entre Charles Darwin et Gaston de Saporta*, p. 119.

"In 1877, Darwin got an unexpected letter": Bowler, *Life's Splendid Drama*, p. 362.

"Temporarily forgetting 'close gradation,' Saporta took up Darwin's fertilization ideas": Conry, *Correspondence entre Charles Darwin et Gaston de Saporta*, pp. 99–101.

220   "Darwin's reaction was enthusiastic": ibid., p. 109.

"Darwin reacted differently, however": ibid., pp. 99–101.

"Darwin's response . . . was guarded and puzzling": ibid., p. 109.

221   "Plants provided animals with richer foods only as they became more developed": Saporta, *Monde des plantes*, p. 36.

223   "Wortman, Holland complained": Rea, *Bone Wars*, p. 121.

"The facts of mammalian distribution": Wortman, "Studies of Eocene Mammalia," p. 187.

224   "The existence of the higher types": ibid., p. 177.

"He complained to Charles Schuchert": Rainger, "Collectors and Entrepreneurs," p. 18.

"Osborn glossed over this waste of talent": Osborn, "Jacob L. Wortman," p. 653.

"It is to be regretted": Holland, "Obituary: Dr. Jacob L. Wortman," p. 201.

"It occupies a paragraph": Bowler, *Life's Splendid Drama*, p. 362.

225   "The radical changes in faunas in the Mesozoic-Cenozoic turnover": Simpson, *Fossils and the History of Life*, p. 187.

"[P]lant-mammal coadaptations in the mid-to-late Cretaceous": Lillegraven et al., eds., *Mesozoic Mammals*, p. 302.

"[T]he Paleocene saw the radiation of many small mammals": Novacek, *Time Traveler*, p. 111.

"Reflecting the rise of the angiosperms": William A. Clemens, "Mammalian Evolution in the Cretaceous," p. 165.

"Clemens thought a mid-Cretaceous predominance": ibid., p. 171.

"Writing before Nessov discovered zhelestids": ibid.

"But he thought that the coevolution with angiosperms": ibid., p. 173.

226 "A 1995 study suggested that the insect families": John Noble Wilford, "Long before Flowering Plants, Insects Evolved Ways to Use Them," in *Science Times Book of Fossils and Evolution*, ed. Wade, p. 117.

"Dinosaurs also may have affected": Bakker, *Dinosaur Heresies*, p. 182.

"In their way dinosaurs invented flowers": ibid., p. 198.

227 "When I mentioned the idea to David Archibald": Archibald, telephone comm., June 2001.

"Still, it is suggestive that Lev Nessov's Kyzylkum fossils": Archibald et al., "Late Cretaceous Relatives," p. 62.

228 "The flora, preserved in volcanic ash at Big Cedar Ridge": Wing et al., "Implications of an Exceptional Fossil Flora."

"[T]he environmental conditions . . . led to a greater diversity of mammals": Nessov, Russell, and Russell, "Survey of Cretaceous Tribosphenic Mammals," p. 82.

229 "It may well be that the limited areas of open vegetation . . . favored the evolution of small, frugivorous terrestrial, arboreal, and volant types": Scott L. Wing, "Tertiary Vegetation of North America," in Janis et al., eds., *Evolution of Tertiary Mammals in North America*, p. 55.

"[T]he early architects of the savanna landscape . . . each with its own internal microbial garden": Wakeford, *Liaisons of Life*, p. 103

230 "It's probably going to be dreadfully disappointing": David Webb, Florida Museum of Natural History, personal comm., March 1993.

"[A] coevolving set of primary consumer species": David Webb, "The Rise and Fall of the Late Miocene Ungulate Fauna in North America," in M. H. Nitecki, ed., *Coevolution*, p. 269.

"As the North American climate grew cooler and drier": Webb and Opdycke, "Global Climatic Influence," p. 194.

"You can tell from the chemistry of fossil teeth": Bruce J. MacFadden, Florida Museum of Natural History, personal comm., March 1993.

231 "In southern California's Pleistocene, for example": Edwards, "Rancholabrean Age, Latest Pleistocene Bestiary for California Botanists," p. 9.

"Natural selection talks only": Robert Wright, "Accidental Creationist," p. 59.

232 "Certain viruses have the capacity to fuse mammalian cells": Ryan, *Darwin's Blind Spot*, p. 214.

CHAPTER 21. ANTHROPOID LEAPFROG

233 "It may not have been the only member of our order": Gould, *Hen's Teeth*, p. 329.

234 "He did so in the introduction to his monograph on Marsh's primate fossils": Wortman, "Studies of Eocene Mammalia in the Marsh Collection, Peabody Museum," pp. 194–96.

235 "The cartoon in Time/Life's book": Barnett et al., *The World We Live In*, p. 112.

"In 1870, Joseph Leidy acquired a Bridger Basin fossil that he named *Notharctus*": Leidy, *Report of the United State Geological Survey*, p. 86.

"In many respects the lower jaw of *Notharctus* resembles that of the existing American monkeys quite as much as it does that of any living pachyderm": ibid., p. 89. *Pelycodus* is identified as *Notharctus* in Barnett et al., *The World We Live In*, p. 112.

236 "When Wortman sent him a bulbous-skulled little fossil": Osborn, *Cope, Master Naturalist*, p. 531.

"[T]here is no doubt that the genus *Anaptomorphus* is the most simian lemur yet discovered": Cope, *Vertebrata of Western Tertiary Strata*, p. 247.

"The energy of Cope has brought to notice many strange new forms": Marsh, "Introduction and Succession," p. 378.

"The relations of the American Primates": ibid., p. 375.

237 "[E]ither the adapiforms or tarsiiforms 'almost certainly gave rise to the anthropoid primates'": Rose, "Evolution and Diversification of Mammals in the Eocene," p. 125.

"These [two jaws] were widely figured in books on paleontology and human evolution": Simons, "Hunting the 'Dawn Apes' of Africa," p. 19.

"The paucity of the material limited understanding": ibid., p. 20.

238 "Brown's three anthropoid ape jaws indicated": Matthew, "Fossil Animals of India," p. 213.

"Looking for bones in jungle is one of the mortal sins of paleontology": Lilian Brown, *Bring 'Em Back Petrified*, p. 18.

"[R]ainbow-chasing, the rainbow being an elusive ribbon of varicolored rock": Lilian Brown, *I Married a Dinosaur*, p. 163.

"I remember a white man and woman coming on horseback": Ciochon, "Fossil Ancestors," p. 34.

"Something petrified filled every inch of cart space": Lilian Brown, *I Married a Dinosaur*, p. 233.

239 "Brown modestly called his Pondaung Hills haul": Barnum Brown, "Notes," p. 521.

"It was associated with some other fossils": Colbert, "New Primate from the Upper Eocene," p. 1.

"It seemed to me that because of the form and development of the teeth": Colbert, *Digging into the Past*, p. 216.

"Elwyn Simons 'came across a small piece of forehead bone'" Simons, "Egypt's Simian Spring," p. 58.

"William K. Gregory had noted the bone's monkeylike character": Gregory, *Origin and Evolution of the Human Dentition*, p. 289, cited in Morgan and Lucas, *Notes from Diary — Fayum Trip*, n. 184.

240 "Of the four primate jaws they had discovered": Ciochon, "Fossil Ancestors," p. 35.

"Once in Africa, these early higher primates continued to evolve": ibid., p. 36.

241 "This arguably represents more taxonomic diversity of primates": Simons, "Diversity in the Early Tertiary," p. 10743.

"[A]nthropoids might have evolved . . . from 'a third Eocene ancestral line'": Simons, "Egypt's Simian Spring," p. 59.

"The most likely area of origin of higher primates": Philip Gingerich, "Eocene Adapidae, Paleobiogeography, and the Origin of South American Platyrrhini," in Ciochon and Chiarelli, eds., *Evolutionary Biology*, p. 134.

"A balance of cranial, dental, soft tissue, and molecular evidence": A. L. Rosenberger and F. S. Szalay, "On the Tarsiiform Origins of the Anthropoidea," Ciochon and Chiarelli, eds., *Evolutionary Biology*, p. 139.

"Some isolated teeth from the late Paleocene": Beard, "Historical Biogeography of Tarsiers," p. 272.

242 "Within Anthropoidea, *Eosimias* appears to occupy a very basal phylogenetic position": Beard et al., "A Diverse New Primate Fauna from Middle Eocene Fissure-Fillings in Southeastern China," p. 609.

"Beard and his colleagues thought . . . *Amphipithecus* and *Pondaungia* might be": ibid.

"*Eosimias* is smaller than all known adapiforms except . . . *Anchomomys*": ibid.

"They were furry little things": Gugliotta, "Tiny Fossils, Big Find," p. A17.

243 "Dental and cranial resemblances between pondaungines and anthropoids": Gunnell et al., "New Assessment," p. 337.

"But in a 2002 paper, Ciochon and a colleague echoed Simons's earlier speculation": Ciochon and Gunnell, "Eocene Primates from Myanmar."

"A sidebar speculated": ibid., p. 167.

"They have the anthropoid skull": Pat Holroyd, U.C. Berkeley Museum of Paleontology, personal comm., September 20, 2002.

244 "[A]pes may have evolved because of 'the Oligocene primates' arboreal way of life'": Simons, *Earliest Apes*, p. 34.

"The most formidable fairly recent development": Simpson, *Concession to the Improbable*, p. 269.

245 "During the past decade the predominance of this mode has been challenged": Gould, *Hen's Teeth*, p. 336.

"My colleague, Guy Bush . . . tells me": ibid., p. 337.

EPILOGUE. CENOZOIC PARKS

249 "Zallinger's images look 'as if they had always existed'": Mitchell, *Last Dinosaur Book*, p. 189.

"[T]he world's first 'rendition in art of the mythological realm'": Campbell, *Masks of God*, vol. 1: *Primitive Mythology*, p. 398.

250 "[I]f you look at the foundations of human intelligence": Wright, "Accidental Creationist," p. 63.

251 "[W]e may owe our evolution, in large part, to the great Cretaceous dying": Gould, *Hen's Teeth*, p. 329.

"[H]umans are here by the luck of the draw": Wright, "Accidental Creationist," p. 64.

"Many classic 'trends' of evolution are stories of . . . unsuccessful groups": Gould, *Full House*, p. 64.

252 "[O]ne other prominent . . . mammalian lineage has an equally long and extensive history": ibid., p. 73.

"The 'main line of progression . . . has been an increase of perception'": Barnett et al., *The World We Live In*, p. 122.

"[A]n upstart on the planet Earth, man nevertheless stands alone": ibid.

253 "[T]he test of a first-rate intelligence": F. Scott Fitzgerald, *The Crack-Up* (1936; reprint, New York: New Directions, 1956), p. 69.

254 "[T]he Messel fossils had 'close relationships . . . most of all to South-east Asia'": Gerhard Storch and Friedemann Scharschmitt, "The Messel Fauna and Flora: A Biogeographical Puzzle," in Schall and Ziegler, eds., *Messel*, p. 297.

256 "[R]esurrecting *[Mammuthus]* 'would rediscover the circular image of eternal time'": Cohen, *Fate of the Mammoth*, p. 228.

257 "They are still primitive": Simpson, "Great Animal Invasion," p. 211.

258 "[N]ot only flying lizards, but flying squirrels, flying foxes, flying frogs, and . . . flying snakes": Beccari, *Wanderings in the Great Forests of Borneo*, pp. 35–36.

"[Beccari's] 'formative epochs' might apply to mass extinctions like the late Cretaceous one": ibid., p. 93.

SELECT BIBLIOGRAPHY

Alvarez, Luis. "Experimental Evidence That an Asteroid Impact Led to the Extinction of Many Species 65 Million Years Ago." *Proceedings of the National Academy of Sciences* 80, no. 2 (1983): 627–42.

Ameghino, Florentino. *Les Formations sédimentaires du Crétacé supérieur et du Tertiaire de Patagonie.* Annales del Museo Nacional de Buenos Aires, 15. Buenos Aires: Juan A. Alsina, 1906.

Andors, Allison Victor. "*Diatryma* among the Dinosaurs." *Natural History* 104, no. 6 (June 1995): 68.

———. "Reappraisal of the Eocene Groundbird *Diatryma*." Natural History Museum of Los Angeles County *Papers in Avian Paleontology,* Science ser., ed. Kenneth E. Campbell Jr., no. 36 (1992).

Andrews, Roy Chapman. *The New Conquest of Central Asia.* New York: American Museum of Natural History, 1932.

———. *On the Trail of Ancient Man.* New York: Putnam's Sons, 1926.

Archer, Michael. "Mammals Eggstraordinaire." *Natural History* 103, no. 4 (April 1994): 48.

———. "World Furry Weight Champions." *Natural History* 103, no. 4 (April 1994): 44.

Archibald, J. David. *Dinosaur Extinction and the End of an Era: What the Fossils Say.* New York: Columbia University Press, 1996.

———. "Fossil Evidence of a Late Cretaceous Origin of 'Hoofed' Mammals." *Science* 272, no. 5256 (May 24, 1996): 1150–53.

————. *A Study of Mammalia and Geology across the Cretaceous-Tertiary Boundary in Garfield County, Montana.* University of California Publications in Geological Sciences, vol. 122. Berkeley: University of California Press, 1982.

Archibald, J. David, Alexander O. Averianov, and Eric G. Ekdale. "Late Cretaceous Relatives of Rabbits, Rodents, and Other Extant Eutherian Mammals." *Nature* 414, no. 6859 (2001): 62–65.

Archibald, J. David, and Laurie J. Bryant. "Limitations on K-T Mass Extinction Theories Based upon the Vertebrate Record." *Global Catastrophes in Earth History.* Snowbird, Utah: Lunar and Planetary Institute and National Academy of Sciences, 1988.

Archibald, J. David, and W. A. Clemens. "Late Cretaceous Extinctions." *American Scientist* 70, no. 44 (1982): 377–85.

Asma, Stephen T. *Stuffed Animals and Pickled Heads: The Culture and Evolution of Natural History Museums.* New York: Oxford University Press, 2001.

Bailly, André. *Défricheurs d'inconnu: Peiresc, Tournefort, Adanson, Saporta.* Aix en Provence: Edisud, 1992.

Bakker, Robert T. *The Dinosaur Heresies.* New York: Morrow, 1986.

Balzac, Honoré de. *The Wild Ass's Skin.* Translated by Herbert J. Hunt. New York: Penguin Books, 1977. Originally published as *La Peau de chagrin* (Paris, 1830).

Barnett, Lincoln, et al. *The Wonders of Life on Earth.* New York: Time, Inc., 1960.

————, et al. *The World We Live In.* New York: Time, Inc., 1955.

Batchelor, John. *John Ruskin: A Life.* New York: Carroll & Graf, 2000.

Beard, K. Christopher. "Historical Biogeography of Tarsiers." In *Dawn of the Age of Mammals in Asia,* ed. id. and Mary R. Dawson, Carnegie Museum of Natural History, Pittsburgh, *Bulletin,* no. 34 (1998): 260–77.

Beard, K. Christopher, Tao Qui, Mary R. Dawson, Banyue Wang, and Chuankei Li. "A Diverse New Primate Fauna from Middle Eocene Fissure-Fillings in Southeastern China." *Nature* 368, no. 6472 (1994): 604–9.

Beccari, Odoardo. *Wanderings in the Great Forests of Borneo.* Translated by Enrico H. Giglioli. 1904. Reprint. Singapore: Oxford University Press, 1989. Originally published as *Nelle foreste di Borneo, viaggi e ricerche di un naturalista* (Florence, 1902).

Benton, Michael. *The Rise of the Mammals.* New York: Crescent Books, 1991.

Bird, Roland T. *Bones for Barnum Brown: Adventures of a Dinosaur Hunter.* Fort Worth: Texas Christian University Press, 1985.

Borges, Jorge Luis. *The Book of Imaginary Beings.* New York: Dutton, 1969.

————. *Labyrinths.* New York: Modern Library, 1983.

Bowden, Mark. "On the Trail of a Wayward Skull." *Philadelphia Inquirer,* October 6, 1994, A-1.

Bowler, Peter J. *Evolution: The History of an Idea.* Berkeley: University of California Press, 1983. 2d rev. ed., 1989. 3d rev. ed., 2003.

———. *Fossils and Progress: Paleontology and the Idea of Progressive Evolution in the Nineteenth Century.* New York: Science History Publications, 1976.

———. *Life's Splendid Drama: Evolutionary Biology and the Reconstruction of Life's Ancestry.* Chicago: University of Chicago Press, 1996.

Brown, Barnum. "Notes." *Natural History* 23, no. 5 (1923): 521.

Brown, Lilian. *Bring 'Em Back Petrified.* New York: Dodd, Mead, 1956.

———. *I Married a Dinosaur.* New York: Dodd, Mead, 1950.

Browne, Janet. *Charles Darwin: Voyaging.* New York: Knopf, 1995.

Browne, Malcolm W. "Dinosaur Experts Resist Meteor Extinction Idea." *New York Times,* October 29, 1985.

———. "The Debate over Dinosaur Extinction Takes a Rancorous Turn." *New York Times,* January 19, 1988.

Buffon, Georges. *Natural History: General and Particular,* vol. 1: *Theory of the Earth.* London: William Smellie, 1781. Originally published as *Histoire naturelle, générale et particulière* (1749).

Byrne, S. "The Distribution and Ecology of the Non-Native Tree Squirrels *Sciurus carolensis* and *Sciurus niger* in Northern California." Ph.D. diss., University of California, Berkeley, 1979.

Cadbury, Deborah. *Terrible Lizard: The First Dinosaur Hunters and the Birth of a New Science.* New York: Holt, 2000.

Campbell, Joseph, *The Masks of God.* Vol. 1: *Primitive Mythology.* New York: Viking, 1971.

Chambers, Robert. *Vestiges of the Natural History of Creation and Other Evolutionary Writings.* 1844. Edited by James A. Secord. Chicago: University of Chicago Press, 1994.

Chiappe, Luis M. "A Diversity of Early Birds." *Natural History* 104, no. 6 (1995): 52.

Christiansen, Gale E. *Fox at the Wood's Edge: A Biography of Loren Eiseley.* New York: Holt, 1990.

Ciochon, Russell L. "Fossil Ancestors of Burma." *Natural History* 94, no. 10 (1985): 26–36.

Ciochon, Russell L., and A. Brunetto Chiarelli, eds. *Evolutionary Biology of the New World Monkeys and Continental Drift.* New York: Plenum Press, 1980.

Ciochon, Russell L., and Gregg F. Gunnell. "Eocene Primates from Myanmar: Historical Perspectives on the Origin of Anthropoidea." *Evolutionary Anthropology* 11 (2002): 156–68.

Clemens, Elisabeth S. "Of Asteroids and Dinosaurs: The Role of the Press in Shaping Scientific Debate. *Social Studies of Science* 16 (1986): 421–56.

———. "The Impact Hypothesis and Popular Science: Conditions and Consequences of Interdisciplinary Debate." In *The Mass-Extinction Debates: How Science Works in a Crisis,* ed. William Glen, pp. 92–120. Stanford: Stanford University Press, 1994.

Clemens, William A. "Evolution of the Terrestrial Vertebrate Fauna during the Cretaceous-Tertiary Transition." In *Dynamics of Extinction*, ed. David K. Elliot, pp. 63–85. New York: Wiley, 1986.

———. "Mammalian Evolution in the Cretaceous." *Zoological Journal of the Linnaean Society* 50 (1971), suppl. 1, *Early Mammals*, ed. D. M. Kermack and K. A. Kermack, pp. 165–80.

———. "Patterns of Evolution across the Cretaceous-Tertiary Boundary." *Mitt. Mus. Nat.kd. Berl., Zool.*, 77, no. 2 (2001): 175–91.

Clemens, William A., J. David Archibald, and Leo J. Hickey. "Out with a Whimper, not a Bang." *Paleobiology* 7, no. 3 (1981): 293–98.

Cohen, Claudine. *The Fate of the Mammoth: Fossils, Myths, and History*. Translated by William Rodarmor. Chicago: University of Chicago Press, 2002. Originally published as *Le Destin du mammouth* (Paris: Seuil, 1994).

Colbert, Edwin H. *Digging into the Past: An Autobiography*. New York: Dembner Books, 1989.

———. *Evolution of the Vertebrates*. New York: Wiley, 1955.

———. "A New Primate from the Upper Eocene Pondaung Formation of Burma." *American Museum Novitates,* no. 951 (October 1, 1937): 1–18.

———. "W. D. Matthew's Early Western Fossil Trips." *Earth Sciences History* 9, no. 1 (1990): 41–44.

Coleman, William. *Georges Cuvier: A Study of the History of Evolutionary Thought*. Cambridge, Mass.: Harvard University Press, 1964.

Conry, Yvette. *Correspondance entre Charles Darwin et Gaston de Saporta, précédée d'Histoire de la paléobotanique en France au XIX siècle*. Paris: Presses Universitaires de France, 1972.

Conway Morris, Simon. "Showdown on the Burgess Shale: The Challenge." *Natural History* 107, no. 10 (1999): 48–55.

Cope, Edward D. "The Creodonta." *American Naturalist* 18, no. 3 (1884): 255–67.

———. "The Gigantic Mammals of the Genus *Eobasileus*." *American Naturalist* 7, no. 3 (1873): 157–60.

———. "Marsh on the Dinocerata." *American Naturalist* 19, no. 255 (1885): 703–5.

———. *On the Method of the Creation of Organic Types*. Pamphlet. Philadelphia: M. Calla & Stavely, 1871.

———. "On Professor Marsh's Criticisms." *American Naturalist* 7, no. 7 (1873): i–ii.

———. "On Some of Professor Marsh's Criticisms." *American Naturalist* 7, no. 7 (1873): 290–99.

———. "The Relation of Animal Motion to Animal Evolution." *American Naturalist* 12, no. 1 (1878): 40.

———. "Synopsis of the Vertebrate Fauna of the Puerco Series." *Transactions of the American Philosophical Society* 16 (1888).

———. "The Tertiary Formations of the Central Region of the United States." *American Naturalist* 16, no. 3 (1882): 13.

———. "The Theism of Evolution." *American Naturalist* 22, no. 255 (1888): 265.

———. *The Vertebrata of the Tertiary Formations of the West: Book One.* Washington, D.C.: Government Printing Office, 1883.

———. "The Wheeler Geological Survey of New Mexico for 1874." *American Naturalist* 9, no. 1 (1875): 49.

Crepet, William. "Early Bloomers: A Clay Pit in New Jersey Holds the World's Richest Trove of Fossil Flowers." *Natural History,* May 1999, p. 60.

Cuvier, Georges. *Essay on the Theory of the Earth.* Edinburgh, 1817. Facsimile. New York: Arno Press, 1978.

———. *Recherches sur les ossemens fossiles: Où l'on rétablit les caractéres de plusieurs animaux dont les révolutions du globe ont détruit les espèces.* 2d rev. ed. 5 vols. Paris: G. Dufour & E. d'Ocagne, 1821–24. 3d ed. 1825.

Czerkas, Sylvia Massey, and Donald F. Glut. *Dinosaurs, Mammoths, and Cavemen: The Art of Charles R. Knight.* New York: Dutton, 1982.

Darwin, Charles. *The Descent of Man, and Selection in Relation to Sex.* 2 vols. London: John Murray, 1871.

———. *On the Origin of Species by Means of Natural Selection; or, The Preservation of Favoured Races in the Struggle for Life.* London: John Murray, 1859. 6th rev. ed., 1872.

———. *The Origin of Species and The Descent of Man.* New York: Modern Library, 1936.

———. *The Variation of Animals and Plants under Domestication.* 2 vols. London: J. Murray, 1868.

Davidson, Jane Pierce. *The Bone Sharp: The Life of Edward Drinker Cope.* Philadelphia: Academy of Natural Sciences, 1997.

Davidson, Keay. "Theory Still Rocks Scientists' Equilibrium." *San Francisco Chronicle,* May 27, 2002, p. A6.

Dawson, Mary R., and K. Christopher Beard, eds. *Dawn of the Age of Mammals in Asia.* Carnegie Museum of Natural History, Pittsburgh, *Bulletin,* no. 34 (1998).

De Camp, L. Sprague. "A Gun for Dinosaur." In *Dinosaurs!* ed. Jack Dann and Gardner Dozois, pp. 1–28. New York: Ace Books, 1990.

Desmond, Adrian. *Archetypes and Ancestors: Paleontology in Victorian London, 1850–1875.* London: Blond & Briggs, 1982.

———. *The Hot-Blooded Dinosaurs.* London: Blond & Briggs, 1975.

———. *Huxley: From Devil's Disciple to Evolution's High Priest.* Reading, Mass.: Addison-Wesley, 1997.

———. *The Politics of Evolution.* Chicago: University of Chicago Press, 1984.

Dodson, Peter. "Dodson on Dinosaurs." *American Paleontologist* 7, no. 1 (1999): 5–7.

Dorf, Erling. *Petrified Forests of Yellowstone.* Washington, D.C.: U.S. National Park Service, 1980.

Doyle, Arthur Conan. *The Lost World.* 1912. Reprint. San Francisco: Chronicle Books, 1989.

Dunbar, Carl O. "Recollections on the Renaissance of Peabody Museum Exhibits, 1939–1959." *Discovery: Magazine of the Peabody Museum* 12, no. 1 (1976): 17–27.

Edwards, Stephen W. "A Rancholabrean-Age Latest Pleistocene Bestiary for California Botanists." *The Four Seasons: Journal of the Regional Parks Botanic Garden* 10, no. 2 (1996): 5–34.

Eiseley, Loren. *Darwin's Century: Evolution and the Men Who Discovered It.* Garden City, N.Y.: Doubleday, 1958.

Eldredge, Niles. *Life Pulse: Episodes from the Story of the Fossil Record.* New York: Penguin Books, 1987.

———. *The Miner's Canary: Unraveling the Mysteries of Extinction.* New York: Prentice-Hall, 1991.

———. *The Pattern of Evolution.* New York: W. H. Freeman, 1999.

———. *The Triumph of Evolution.* New York: W. H. Freeman, 2000.

Eldredge, Niles, and Stephen Jay Gould. "Punctuated Equilibria: An Alternative to Phyletic Gradualism." In *Models in Paleontology,* ed. T. J. Schopf, pp. 82–115. San Francisco: Freeman Cooper, 1977.

Farina, R. A., and R. E. Blanco. "*Megatherium,* the Stabber." *Proceedings of the Royal Society of London,* ser. B, 263 (1996): 1725–29.

Fortey, Richard. *Life: A Natural History of the First Four Billion Years of Life on Earth.* New York: Knopf, 1998.

———. *Trilobite! Eyewitness to Evolution.* New York: Knopf, 2000.

Gallencamp, Charles. *Dragon Hunter: Roy Chapman Andrews and the Central Asian Expeditions.* New York: Viking, 2001.

Gaudry, Albert. *Fossiles de Patagonie: Dentition de quelques mammifères.* Paris: Société géologique de France, 1904.

Gee, Henry. *In Search of Deep Time: Beyond the Fossil Record to a New History of Life.* New York: Free Press, 1999.

Gingerich, Philip D. "Origin of Whales in Epicontinental Remnant Seas: New Evidence from the Early Eocene of Pakistan." *Science* 220, no. 4595 (April 22, 1983): 403–6.

———. "Paleontology, Phylogeny, and Classification: An Example from the Fossil Record." *Systematic Zoology* 28 (1979): 451–63.

———. "Paleontology and Phylogeny: Patterns of Evolution at the Species

Level in Early Tertiary Mammals." *American Journal of Science* 276 (January 1976): 1–28.

———. "The Whales of Tethys." *Natural History* 103, no. 4 (1994): 86.

Gingerich, Philip D., S. Mahmood Raza, Muhammad Arif, Mohmmad Anwar, and Xiaoyuan Zhou. "New Whale from the Eocene of Pakistan and the Origin of Cetacean Swimming." *Nature* 368, no. 6474 (1994): 844–47.

Gore, Rick. "Extinctions." *National Geographic,* June 1989, pp. 662–700.

———. "The Rise of Mammals." *National Geographic,* April 2003, pp. 2–37.

Gould, Stephen Jay. *Bully for Brontosaurus,* New York: Norton, 1991.

———. *The Flamingo's Smile.* New York: Norton, 1985.

———. *Full House: The Spread of Excellence from Plato to Darwin.* New York: Harmony Books, 1996.

———. *Hen's Teeth and Horses' Toes.* New York: Norton, 1983.

———. *Ontogeny and Phylogeny.* Cambridge, Mass.: Harvard University Press, 1977.

———. "Play It Again, Life." *Natural History* 95, no. 2 (1986): 20–26.

———. *The Structure of Evolutionary Theory.* Cambridge, Mass.: Harvard University Press, 2002.

Gould, Stephen Jay, and Niles Eldredge. "Punctuated Equilibria: An Alternative to Phyletic Gradualism." In *Models in Paleontology,* ed. T. J. Schopf, pp. 82–115. San Francisco: Freeman Cooper, 1977.

Granger, Walter, and Charles P. Berkey. "Discovery of Cretaceous and Older Tertiary Strata in Mongolia." *American Museum Novitates,* no. 42 (August 7, 1922).

Gray, Denis, Collin Piprell, and Mark Graham. *National Parks of Thailand.* Bangkok: Industrial Finance Corporation of Thailand, 1994.

Green, Harry. "Agonistic Behavior of Three-Toed Sloths *(Bradypus variegatus).*" *Biotropica* 21, no. 4 (1989): 369–72.

Gregory, William K. *The Orders of Mammals.* American Museum of Natural History *Bulletin* 27 (February 1910).

———. *The Origin and Evolution of the Human Dentition.* Baltimore: Williams & Wilkins, 1922.

Gregory, William K., and J. Howard McGregor. "A Dissenting Opinion as to Dawn Men and Ape Men." *Natural History* 26, no. 3 (1926): 270–71.

Grimes, Lee. "An Interview with Rudolph F. Zallinger." *Discovery: Magazine of the Peabody Museum* 11, no. 1 (1975): 33–35.

Gugliotta, Guy. "Tiny Fossils, Big Find: Small Primates in China Give Clues to Evolution." *San Francisco Chronicle,* March 16, 2000, pp. A1, 17. Reprinted from the *Washington Post.*

Gunnell, Greg F., Russell Ciochon, Philip Gingerich, and Patricia Holroyd. "New Assessment of *Pondaungia* and *Amphipithecus* (Primates) from the Late

Middle Eocene of Myanmar, with a Comment on 'Amphipithecidae.'" *Contributions from the Museum of Paleontology, University of Michigan* 30, no. 13 (2002): 337–72.

Haines, Tim. *Walking with Prehistoric Beasts.* New York: DK Publishing, 2001.

Hatcher, John Bell. *Bone Hunters in Patagonia.* Woodbridge, Conn.: Ox Bow Press, 1985.

Hellman, Geoffrey T. "Go On, Investigators! Scrutinize!" *New Yorker,* November 3, 1962, pp. 142–75.

Hoffmann, Hillel J. "Messel: Window on an Ancient World." *National Geographic,* February 2000, p. 34.

Holland, William J. "Obituary: Dr. Jacob L. Wortman." *Annals of the Carnegie Museum* 1926: 199–201.

Hu, Yaoming, Yuanqing Wang, Zhe-xi Luo, and Chuankui Li. "A New Symmetrodont Mammal from China and Its Implications for Mammalian Evolution." *Nature* 390, no. 6656 (1997): 137–42.

Hutton, James. *Theory of the Earth, or, An Investigation of the Laws Observable in the Composition, Dissolution and Restoration of Land upon the Globe.* Edinburgh: Royal Society of Edinburgh, 1788.

Huxley, Leonard. *Life and Letters of Thomas Henry Huxley.* 1900. 2 vols. New York: Appleton, 1901.

Irvine, William. *Apes, Angels, and Victorians: Darwin, Huxley, and Evolution.* New York: Time, Inc., 1963.

Jaffe, Mark. *The Gilded Dinosaur: The Fossil War between E. D. Cope and O. C. Marsh and the Rise of American Science.* New York: Crown, 2000.

Janis, Christine M. "The Sabertooth's Repeat Performance." *Natural History* 103, no. 4 (1994): 78–83.

Janis, Christine M., Kathleen M. Scott, and Louis Jacobs, eds. *Evolution of the Tertiary Mammals in North America.* Cambridge: Cambridge University Press, 1998.

Ji, Quiang, Zhe-xi Luo, and Ji Shu-an. "A Chinese Triconodont Mammal and Mosaic Evolution of the Mammalian Skeleton." *Nature* 398, no. 6725 (1999): 226–30.

Ji, Quiang, Zhe-xi Luo, Chong-Xi Yuan, John R. Wible, Jian-Ping Zhang, and Justin A. Georgi. "The Earliest Known Eutherian Mammal." *Nature* 416, no. 6883 (2002): 816–22.

Jordi, Augustí, and Mauricio Antón. *Memoria de la tierra: Vertebrados fósiles de la Península Ibérica.* Barcelona: Ediciones del Serbal, 1997.

Jullian, Philippe. *Montmartre.* Translated by Anne Carter. New York: Dutton, 1977.

Kerr, R. W. "Dinosaurs and Friends Snuffed Out?" *Science* 251, no. 4990 (1991): 160–62.

Kielan-Jaworowska, Zofia. *Hunting for Dinosaurs.* Cambridge, Mass.: MIT Press, 1969.

———. "Late Cretaceous Mammals and Dinosaurs from the Gobi Desert." *American Scientist* 63 (March 1975): 150–59.

Klesius, Michael. "The Big Bloom." *National Geographic,* July 2002, pp. 104–21.

Krause, David W. "Mammalian Evolution in the Paleocene: Beginning of an Era." In *Mammals: Notes for a Short Course Organized by P. D. Gingerich and C. E. Badgley,* ed. T. W. Broadhead. University of Tennessee Studies in Geology, no. 8. Knoxville: University of Tennessee, Dept. of Geological Sciences, 1984.

Kumar, Sudhir, and S. Blair Hedges, "A Molecular Timescale for Vertebrate Evolution." *Nature* 392, no. 6679 (1998): 917–20.

Kurten, Bjorn, *The Age of Mammals.* New York: Columbia University Press, 1971.

———. *The Age of Reptiles.* New York: McGraw-Hill, 1968.

Lakes, Arthur. *Discovering Dinosaurs in the Old West: The Field Journals of Arthur Lakes.* Michael F. Kohl and John S. McIntosh, eds. Washington, D.C.: Smithsonian Institution Press, 1997.

Lanham, Url. *The Bone Hunters.* New York: Columbia University Press, 1973.

Laporte, Leo F. *George Gaylord Simpson: Paleontologist and Evolutionist.* New York: Columbia University Press, 2000.

———, ed. *Simple Curiosity: Letters from George Gaylord Simpson to His Family, 1921–70.* Berkeley: University of California Press, 1987.

Leed, Eric J. *The Mind of the Traveler: From Gilgamesh to Global Tourism.* New York: Basic Books, 1991.

Leidy, Joseph. *Report of the United States Geological Survey on the Territories.* vol. 1, pt. 1. Washington, D.C.: Government Printing House, 1873.

Ley, Willy. *Worlds of the Past.* Illustrated by Rudolph Zallinger. New York: Golden Press, 1971.

Lillegraven, Jason A. "Latest Cretaceous Mammals of Upper Part of Edmonton Formation of Alberta, Canada." *University of Kansas Paleontological Contributions,* art. 50 (Vertebrata 12), March 7, 1969.

Lillegraven, Jason A., and Jaclyn J. Eberle. "Vertebrate Faunal Changes through Lancian and Puercan Time in Southern Wyoming." *Journal of Paleontology* 73, no. 4 (1999): 691–710.

Lillegraven, Jason A., Zofia Kielan-Jaworowska, and William A. Clemens, eds. *Mesozoic Mammals: The First Two-Thirds of Mammalian History.* Berkeley: University of California Press, 1979.

Linklater, Eric. *The Voyage of the Challenger.* Garden City, N.Y.: Doubleday, 1972.

Lockley, Martin. *The Eternal Trail: A Tracker Looks at Evolution.* Reading, Mass.: Perseus Books, 1999.

Lofgren, Donald L. *The Bug Creek Problem and the Cretaceous-Tertiary Transition at McGuire Creek, Montana.* Berkeley: University of California Press, 1995.

Lucas, Spencer G., J. Keith Rigby Jr., and Barry S. Kues, eds. *Advances in San Juan Basin Paleontology.* Albuquerque: University of New Mexico Press, 1981.

Luo, Zhe-xi, Richard L. Cifelli, and Zofia Kielan-Jaworowska. "Dual Origin of the Tribosphenic Molar." *Nature* 409, no. 6816 (2001): 53–57.

Luo, Zhe-xi, Alfred W. Crompton, and Ai-Lin Sun, "A New Mammaliaform from the Early Jurassic and Evolution of the Mammalian Characteristics." *Science* 292, 5521 (May 25, 2001): 1535–40.

Lyell, Charles, Sir. *Principles of Geology.* 3 vols. London: J. Murray, 1830–33. Facs. ed. Chicago: University of Chicago Press, 1990–91.

MacDonald, James R., and Laurie J. MacDonald. "A Landscape Rich with Life." In *Agate Fossil Beds.* National Park Service Handbook no. 107. Washington, D.C.: Government Printing Office, 1980.

MacFadden, Bruce J. *Fossil Horses: Systematics, Paleobiology, and Evolution of the Family Equidae.* New York: Cambridge University Press, 1992.

———. "The Heyday of Horses." *Natural History* 103, no. 4 (1994): 61–62.

Mader, Bryn J. "Distant Thunder." *Natural History* 103, no. 4 (1994): 61–62.

Marsh, O. C. *Dinocerata: A Monograph of an Extinct Order of Gigantic Mammals.* U.S. Geological Survey Monograph 10. Washington, D.C.: Government Printing Office, 1884. Misc. doc., U.S. House of Representatives, 49th Cong., 1st sess., no. 305. Washington, D.C.: Government Printing Office, 1886.

———. "Fossil Horses of America." *American Naturalist* 8, no. 5 (1874): 288–94.

———. "The Fossil Mammals of the Order Dinocerata." *American Naturalist* 7, no. 3 (1873): 146–53.

———. "Introduction and Succession of Vertebrate Life in North America." *American Journal of Science,* 3d ser., 14, nos. 79–84 (1877).

———. "Notice of New Equine Mammals from the Tertiary Formation." *American Journal of Science,* 3d ser., 7 (1874): 247–58.

———. "Polydactyl Horses: Recent and Extinct." *American Journal of Science,* 3d ser., 17 (1879): 499–505.

———. "Principal Characters of the Brontotheriidae." *American Journal of Science,* 3d ser., 11, no. 64 (1876): 335–40.

———. "Reply to Professor Cope's Explanation." *American Naturalist* 7, no. 6 (1873): i–x.

———. "Return of Professor Marsh's Expedition." *American Naturalist* 9, no. 2 (1875): 117–18.

Mattes, Merrill J. *Indians, Infants, and Infantry: Andrew and Elizabeth Burt on the Frontier.* Denver: Old West Publishing Co., 1960.

Matthew, William Diller. "The Asphalt Group of Fossil Skeletons." *American Museum Journal* 13, no. 7 (1913): 291–97.

———. "Fossil Animals of India." *Natural History* 24, no. 2 (1924): 208–14.

———. "Fossil Bones in the Rock." *Natural History* 23, no. 4 (1923): 358–69.

———. "The Ground Sloth Group." *American Museum Journal* 11, no. 4 (1911): 113–19.

———. "Illustrations of Evolution among Fossil Mammals." *American Museum Journal* 3, no. 1 (1903): 3–30.

———. "New Carnivora from the Tertiary of Mongolia." *American Museum Novitates,* no. 104 (January 15, 1924).

———. "New Insectivores and Ruminants from the Tertiary of Mongolia, with Remarks on the Correlations." *American Museum Novitates,* no. 105 (January 18, 1924).

——— "Phylogeny of the Felidae." *Bulletin of the Museum of Natural History* 28 (1910): 307.

——— "Scourge of the Santa Monica Mountains." *American Museum Journal* 16, no. 7 (1916): 469–72.

———. "Two New Perissodactyls from the Arshanto Eocene Formation of Mongolia." *American Museum Novitates,* no. 208 (February 16, 1926).

Matthew, William Diller, and Walter Granger. "The Most Significant Fossil Finds of the Mongolian Expeditions." *Natural History* 26, no. 5 (1926): 532–34.

McDonald, Gregory H. "Hagerman Fossil Beds." *Rocks and Minerals* 68 (September–October 1993): 322–26.

McGowan, Christopher. *The Dragon Seekers: How an Extraordinary Circle of Fossilists Discovered the Dinosaurs and Paved the Way for Darwin.* Cambridge, Mass.: Perseus Publishing, 2001.

McKenna, Malcolm. "Early Relatives of Flopsy, Mopsy, and Cottontail." *Natural History* 103, no. 4 (1994): 56.

———. "Sweepstakes, Filters, Corridors, Noah's Arks, and Beached Viking Funeral Ships in Palaeogeography." In NATO Advanced Study Institute, *Implications of Continental Drift to the Earth Sciences,* ed. D. H. Tarling and S. K. Runcorn, 1: 295–308. London: Academic Press, 1973.

McLoughlin, John C. *Synapsida: A New Look into the Origin of Mammals.* New York: Viking, 1980.

Mitchell, W. J. T. *The Last Dinosaur Book.* Chicago: University of Chicago Press, 1998.

Morgan, Vincent L., and Spencer G. Lucas, eds. *Notes from Diary–Fayum Trip, 1907.* New Mexico Museum of Natural History and Science *Bulletin,* no. 22. Albuquerque: New Mexico Museum of Natural History and Science, 2002.

———. *Walter Granger, 1872–1941, Paleontologist.* New Mexico Museum of Natural History and Science *Bulletin,* no. 19. Albuquerque: New Mexico Museum of Natural History and Science, 2002.

Mouhot, Henri. *Travels in Siam, Cambodia, and Laos, 1858–1860.* [1864]. Singapore: Oxford University Press, 1989.

Moyal, Ann. *Platypus: The Extraordinary Story of a How a Curious Creature Baffled the World.* Washington, D.C.: Smithsonian Institution Press, 2001.

Nessov, Lev, J. David Archibald, and Zofia Kielan-Jaworowska. "Ungulate-like Mammals from the Late Cretaceous of Uzbekistan and a Phylogenetic Analysis of Ungulatomorpha." In *Dawn of the Age of Mammals in Asia,* ed. K. Christopher Beard and Mary R. Dawson, Carnegie Museum of Natural History, Pittsburgh, *Bulletin,* no. 34 (1998), pp. 40–88.

Nessov, Lev, Denise Sigogneau Russell, and Donald Russell. "A Survey of Cretaceous Tribosphenic Mammals from Middle Asia (Uzbekistan, Kazakhstan, and Tajikstan), of their Geological Setting, Age, and Faunal Environment." *Palaeovertebrata* 23 (May 1994): 51–92.

Nitecki, M. H., ed. *Coevolution.* Chicago: University of Chicago Press, 1983.

Normile, Dennis. "New Views of the Origins of Mammals." *Science* 281, no. 5378 (1998): 774.

———. "Whale-Ungulate Link Strengthens." *Science* 281, no. 5378 (1998): 775.

Novacek, Michael J. "A Pocketful of Fossils." *Natural History* 103, no. 4 (1994): 41.

———. *Time Traveler.* New York: Farrar, Straus & Giroux, 2002.

Novacek, Michael J., G. W. Rougier, D. Dashzeveg, and M. C. McKenna. "New Eutherian Mammal from the Late Cretaceous of Mongolia and Its Bearing on the Origins of Modern Placental Radiation." *Journal of Vertebrate Paleontology* 20 (suppl. to no. 3) (September 2000): 61A.

O'Connor, Richard. *The Scandalous Mr. Bennett.* Garden City, N.Y.: Doubleday, 1962.

Osborn, Henry Fairfield. *The Age of Mammals in Europe, Asia, and North America.* New York: Macmillan, 1910.

———. "*Andrewsarchus,* Giant Mesonychid of Mongolia." *American Museum Novitates,* no. 146 (November 11, 1924).

———. *Cope, Master Naturalist: The Life and Letters of Edward Drinker Cope, with a Bibliography of His Writings Classified by Subject. A Study of the Pioneer and Foundation Periods of Vertebrate Palaeontology in America.* Princeton, N.J.: Princeton University Press, 1931.

———. "The Discovery of an Unknown Continent." *Natural History* 24, no. 2 (1924): 133–49.

——— "*Eudinoceras,* Upper Eocene Amblypod of Mongolia." *American Museum Novitates,* no. 145 (November 10, 1924).

———. "The Extinct Giant Rhinoceros *Baluchitherium* of Western and Central Asia." *Natural History* 23, no. 3 (1923): 209–28.

———. "Hunting the Ancestral Elephant in the Fayum Desert." *Century Magazine* 74, no. 6 (1907): 815–35.

————. *Impressions of Great Naturalists.* New York: Scribner's, 1924.

————. "Is the Ape Man a Myth?" *Human Biology* 1 (January 1929): 4–9.

————. "J. L. Wortman—A Biographical Sketch." *Natural History* 26 (1926): 652–53.

————. "New Fossil Mammals from the Fayum." American Museum of Natural History *Bulletin* 24 (1908): 265–72.

————. "Prehistoric Quadrupeds of the Rockies." *Century Magazine* 52 (1896): 705–15.

————. "Restorations and Models of the Extinct North American Mammals." *American Museum Journal* 1, no. 6 (1901): 85–87.

————. "The Romance of the Wooly Mammoth." *Natural History* 30, no. 3 (1930).

———— "Titanotheres and Lophiodonts in Mongolia." *American Museum Novitates,* no. 91 (October 17, 1923).

————. *The Titanotheres of Ancient Wyoming, Dakota, and Nebraska.* U.S. Geological Survey Monograph 55. Washington, D.C.: Government Printing Office, 1929.

————. "Why Central Asia?" *Natural History* 26, no. 3 (1926): 263–69.

Ostrom, John H., Leo J. Hickey, David M. Schankler, Bruce H. Tiffney, and Scott L. Wing. *The Age of Mammals: A Guide to the Rudolph F. Zallinger Mural in the Peabody Museum of Natural History, Yale University.* Rev. ed. New Haven, Conn.: Peabody Museum, 1993. Originally published in 1973.

Owen, Richard [1804–92]. "Darwin on the Origin of Species." In *Darwin: A Norton Critical Edition,* ed. Philip Appleman. New York: Norton, 1970.

————. "On the Occurrence in North America of Rare Extinct Vertebrates Found Fragmentarily in England." *Annals and Magazine of Natural History,* 5th ser., 2, no. 9 (1878).

————. *Paleontology, or A Systematic Summary of Extinct Animals and Their Geological Relations.* 1860. 2d ed. Edinburgh: Adam & Charles Black, 1861.

Owen, Reverend Richard. *The Life of Richard Owen.* 2 vols. London: John Murray, 1894.

Peattie, Donald Culross. *Green Laurels: The Lives and Achievements of the Great Naturalists.* New York: Simon & Schuster, 1936.

Pennisi, Elizabeth. "Placentals' Family Tree Drawn and Quartered." *Science* 294, no. 5550 (2001): 266–68.

Perlman, David. "Shrewlike Early Mammal Fossil Found." *San Francisco Chronicle,* April 25, 2002, p. 2.

Powell, James Lawrence. *Night Comes to the Cretaceous: Dinosaur Extinction and the Transformations of Modern Geology.* New York: W. H. Freeman, 1998.

Prothero, Donald R. *The Eocene-Oligocene Transition: Paradise Lost.* New York: Columbia University Press, 1994.

Prothero, Donald R., and T. H. Heaton. "Faunal Stability during the Early Oligocene Climatic Crash." *Palaeogeography, Palaeoclimatology, Palaeoecology* 127 (1996): 257–83.

Prothero, Donald R., and Neil Shubin. "The Evolution of Oligocene Horses." In *The Evolution of Perissodactyls*, ed. D. R. Prothero and R. M. Schoch, 142–75. Oxford: Oxford University Press, 1989.

Psihoyos, Louie, and John Knoebber, *Hunting Dinosaurs.* New York: Random House, 1994.

Rainger, Ronald. *An Agenda for Antiquity: Henry Fairfield Osborn and Vertebrate Paleontology at the American Museum of Natural History, 1890–1935.* Tuscaloosa: University of Alabama Press, 1991.

———. "Collectors and Entrepreneurs: Hatcher, Wortman, and the Structure of American Vertebrate Paleontology circa 1900." *Earth Sciences History* 9, no. 1 (1990): 14–21.

———. "W. D. Matthew, Fossil Vertebrates, and Geological Time." *Earth Sciences History* 8, no. 2 (1989): 159–66.

Raup, David. "The Role of Extinction in Evolution." *Proceedings of the National Academy of Sciences* 91, no. 15 (1994): 6758–63.

Raup, David, and J. J. Sepkoski Jr. "Periodicity of Extinctions in the Geologic Past." *Proceedings of the National Academy of Sciences* 81, no. 3 (1984): 801–5.

Rea, Tom. *Bone Wars: The Excavation and Celebrity of Andrew Carnegie's Dinosaurs.* Pittsburgh: University of Pittsburgh Press, 2001.

Romer, Alfred S. *Man and the Vertebrates.* Chicago: University of Chicago Press, 1933.

Rose, Kenneth D. "Evolution and Radiation of Mammals in the Eocene, and the Diversification of Modern Orders." In *Mammals: Notes for a Short Course Organized by P. D. Gingerich and C. E. Badgley*, ed. T. W. Broadhead. University of Tennessee Studies in Geology, no. 8. Knoxville: University of Tennessee, Dept. of Geological Sciences, 1984.

Rudwick, Martin J. S. *Georges Cuvier: Fossil Bones and Geological Catastrophes.* Chicago: University of Chicago Press, 1997.

———. *The Meaning of Fossils.* Chicago: University of Chicago Press, 1985.

———. *Scenes from Deep Time: Early Pictorial Representations of the Prehistoric World.* Chicago: University of Chicago Press. 1992.

Rupke, Nicolaas A. *Richard Owen: Victorian Naturalist.* New Haven, Conn.: Yale University Press, 1994.

Ruse, Michael. *The Darwinian Revolution.* Chicago: University of Chicago Press, 1999.

Ruspoli, Mario. *The Cave at Lascaux: The Final Photographs.* New York: Abrams, 1986.

Ryan, Frank. *Darwin's Blind Spot: Evolution beyond Natural Selection.* Boston: Houghton Mifflin, 2002.

Saporta, Gaston, Marquis de. *Le Monde des plantes avant l'apparition de l'homme.* Paris: G. Masson, 1879.

Schall, Stephen, and Willi Ziegler, eds. *Messel: An Insight into the History of Life and of the Earth.* Translated by Monika Shaffer-Fehre. Oxford: Clarendon Press, 1992.

Scheele, William E. *The First Mammals.* Cleveland: World Publishing Co., 1955.

Schiebout, Judith A. "The Paleocene / Eocene Transition on Tornillo Flat in Big Bend National Park, Texas." In *National Park Service Paleontological Research,* Vincent L. Santucci and Lindsay McClelland, eds. Denver: U.S. Department of the Interior, 1995.

Schopf, J. William. *Cradle of Life: The Discovery of Earth's Earliest Fossils.* Princeton, N.J.: Princeton University Press, 1999.

Schuchert, Charles, and Clara Mae LeVene. *O. C. Marsh: Pioneer in Paleontology.* New Haven, Conn.: Yale University Press, 1940.

Scott, William Berryman. *Some Memories of a Paleontologist.* Princeton, N.J.: Princeton University Press, 1939.

Scully, Vincent, Rudolph Zallinger, Leo J. Hickey, and John H. Ostrom. *The Age of Reptiles: The Great Dinosaur Mural at Yale.* New York: Abrams, 1990.

Secord, James A. *Victorian Sensation: The Extraordinary Publication, Reception, and Secret Authorship of "Vestiges of the Natural History of Creation."* Chicago: University of Chicago Press, 2000.

Sheehan, P. M., and D. E. Fastovsky. "Sudden Extinction of Dinosaurs: Latest Cretaceous, Upper Great Plains, U.S.A." *Science* 254, no. 5033 (1991): 835–39.

Shipman, Pat. *Taking Wing: Archaeopteryx and the Evolution of Bird Flight.* New York: Simon & Schuster, 1998.

Shor, Elizabeth Noble. *The Fossil Feud between E. D. Cope and O. C. Marsh.* Hicksville, N.Y.: Exposition Press, 1974.

Simons, Elwyn. "Dawn Ape of the Fayum." *Natural History* 93, no. 5 (1984): 18–20.

———. "Diversity in the Early Tertiary Anthropoidean Radiation in Africa." *Proceedings of the National Academy of Sciences* 89, no. 22 (1992): 10743–47.

———. *Early Cenozoic Mammalian Faunas: Fayum Province, Egypt.* New Haven, Conn.: Peabody Museum of Natural History *Bulletin* 28 (1968).

———. "Egypt's Simian Spring." *Natural History* 102, no. 4 (1993): 58–59.

———. "Hunting the 'Dawn Apes' of Africa." *Discovery* 4, no. 1 (1968): 19–32.

———. *Primate Evolution: An Introduction to Man's Place in Nature.* New York: Macmillan, 1972.

Simpson, George Gaylord. *American Mesozoic Mammalia.* New Haven, Conn.: Yale University Press, 1929.

———. *Attending Marvels: A Patagonian Journal.* 1934. Chicago: University of Chicago Press, 1982.

———. "Children of Patagonia." *Natural History* 32, no. 1 (1932).

———. *Concession to the Improbable: An Unconventional Autobiography.* New Haven, Conn.: Yale University Press, 1978.

———. *The Dechronization of Sam Magruder.* New York: St. Martin's Press, 1996.

———. *Discoverers of the Lost World.* New Haven, Conn.: Yale University Press, 1984.

———. *Fossils and the History of Life.* New York: W. H. Freeman, 1982.

———. *The Geography of Evolution.* Philadelphia: Chilton Books, 1965.

———. "The Great Animal Invasion." *Natural History* 49, no. 4 (1942): 206–211.

———. "Hayden, Cope, and the Eocene of New Mexico." *Proceedings of the Academy of Natural Sciences of Philadelphia* 103 (1951): 1.

———. *Horses: The Story of the Horse Family in the Modern World and through Sixty Million Years of History.* New York: Oxford University Press, 1951.

———. "How Fossils Are Collected." *Natural History* 39, no. 5 (1937).

———. *Life of the Past: An Introduction to Paleontology.* New Haven, Conn.: Yale University Press, 1953.

———. "Mammals Were Humble When Dinosaurs Roamed." *New York Times,* October 18, 1925, sect. 10, p. 11.

———. *The Meaning of Evolution.* New Haven, Conn.: Yale University Press, 1967.

———. "The Meek Inherit the Earth." *Natural History* 59, no. 2 (1942): 98–103.

———. "A Mesozoic Mammal Skull from Mongolia." *American Museum Novitates,* no. 201 (November 24, 1925).

———. "Resurrection of the Dawn Horse." *Natural History* 66, no. 4 (1940): 194–99.

———. *Splendid Isolation: The Curious History of South American Mammals.* New Haven, Conn.: Yale University Press, 1980.

———. *Tempo and Mode in Evolution.* New York: Columbia University Press, 1944.

Sloan, Christopher P. "Feathers for T. Rex? New Birdlike Fossils Are Missing Links in Dinosaur Evolution." *National Geographic,* November 1999.

Sloan, Robert E., and Leigh Van Valen. "Cretaceous Mammals from Montana." *Science* 148, no. 3667 (April 9, 1965): 220–27.

Sloan, Robert E., J. Keith Rigby, Leigh Van Valen, and Diane Gabriel. "Gradual Dinosaur Extinction and Simultaneous Ungulate Radiation in the Hell Creek Formation." *Science* 232, no. 4750 (May 2, 1986): 629–33.

Smit, Jan, and S. Van Der Kaars. "Terminal Cretaceous Extinctions in the Hell Creek Area, Montana: Compatible with Catastrophic Extinction." *Science* 223, no. 4641 (1984): 1177–80.

Spitzka, Edward A. "A Study of the Brains of Six Eminent Scientists and Scholars Belonging to the American Anthropometric Society." *Transactions of the American Philosophical Society* 21, no. 4 (1907): 175–308.

Standhardt, Barbara R. "Early Paleocene (Puercan) Vertebrates of the Dogie Locality, Big Bend National Park, Texas." In *National Park Service Paleontological Research,* Vincent L. Santucci and Lindsay McClelland, eds. Denver: U.S. Department of the Interior, 1995.

Stegner, Wallace. *Beyond the Hundredth Meridian: John Wesley Powell and the Second Opening of the West.* Boston: Houghton Mifflin, 1954.

Stevenson, Robert Louis. *The Strange Case of Dr. Jekyll and Mr. Hyde.* New York: Dutton, 1924.

Stewart, T. D. "Where is Cope's Face?" *Earth Science History* 2, no. 1 (1983): 76–77.

Vrba, Elizabeth. "The Pulse That Produced Us." *Natural History* 102, no. 5 (1993): 47–52.

Wade, Nicholas, ed. *The Science Times Book of Fossils and Evolution.* New York: Lyons Press, 1998.

Wakeford, Tom. *Liaisons of Life: From Hornworts to Hippo, How the Unassuming Microbe has Driven Evolution.* New York: Wiley, 2001.

Wallace, Alfred Russel. *The Malay Archipelago, the Land of the Orang-utan and the Bird of Paradise. A Narrative of Travel, with Studies of Man and Nature.* London: Macmillan, 1869. 2 vols. Reprint of last rev. ed. New York: Dover, 1962.

Wallace, David Rains. *The Bonehunters' Revenge: Dinosaurs, Greed, and the Greatest Scientific Feud of the Gilded Age.* Boston: Houghton Mifflin, 1999.

Warren, Leonard. *Joseph Leidy: The Last Man Who Knew Everything.* New Haven, Conn.: Yale University Press, 1998.

Webb, S. David. "Successful in Spite of Themselves." *Natural History* 103, no. 4 (1994): 50–55.

Webb, S. David, and Neil D. Opdyke. "Global Climatic Influence on Cenozoic Mammal Faunas." In *Effects of Past Global Change on Life,* pp. 184–208. Washington, D.C.: National Academy Press.

Wegener, Alfred. *Die Entstehung der Kontinente und Ozeane.* 1915. 4th rev. ed. Braunschweig, Germany: Vieweg, 1929. Translated by John Biram as *The Origin of Continents and Oceans* (New York: Dover, 1966).

Weil, Anne. "Relationships to Chew Over." *Nature* 409, no. 6816 (2001): 28–30.

———. "Upwards and Onwards." *Nature* 416, no. 6883 (2002): 798–99.

Weismann, August. *Essays upon Heredity and Kindred Biological Problems.* 1889. 2d ed. 2 vols. Oxford: Clarendon Press, 1891–92.

———. *Über die Vererbung.* Jena: Gustav Fischer, 1883.

Wells, H. G. *The Outline of History.* New York: Macmillan, 1921.

Werkeley, Caroline. "Professor Cope, Not Alive but Well." *Smithsonian* 6, no. 5 (1975): 72–75.

Wheeler, E. A. "Systematic and Ecologic Significance of Fossil Hardwoods." In *National Park Service Paleontological Research,* Vincent L. Santucci and Lindsay McClelland, eds. Denver: U.S. Department of the Interior, 1995.

Wheeler, Walter H. "The Uintatheres and the Cope-Marsh War." *Science* 131 (1960): 1171–76.

Wing, Scott L., Leo J. Hickey, and Carl C. Swisher. "Implications of an Exceptional Fossil Flora for Late Cretaceous Vegetation." *Nature* 363, no. 6427 (1993): 342–44.

Wong, Kate. "The Mammals That Conquered the Seas." *Scientific American* 286, no. 5 (2002): 70–79.

Wortman, J. L. "Studies of Eocene Mammalia in the Marsh Collection, Peabody Museum." *American Journal of Science,* 4th ser., 15, no. 90 (1903).

Wright, Robert. "The Accidental Creationist: Why Stephen Jay Gould Is Bad for Evolution." *New Yorker,* December 13, 1999, pp. 56–65.

Wyss, Andre, John Flynn, and Reynaldo Charrier. "Fire, Ice, and Fossils." *Natural History* 108, no. 5 (1999): 38–41.

Zallinger, Rudolph. "The Making of the Age of Reptiles Mural." http://www.peabody.yale.edu/mural/Rudy.html.

Zimmer, Carl. *At the Water's Edge: Macroevolution and the Transformation of Life.* New York: Free Press, 1998.

INDEX

Badlands National Park, 46, 93; author's visit to, 46
Bahinia, 242, 243
Bain, Andrew, 31
Baird, Spencer, 47
Baja California, 211
Bakker, Robert: on asteroid theory, 186; on dinosaur-angiosperm coevolution, 226, 227; on dinosaur extinction, 164, 186, 187; on Zallinger's *Age of Reptiles* mural, xvi, xxiv
Baldwin, David, 65, 66
Ballou, William Hosea, 72, 97, 106
Balzac, Honore de, 8, 12, 13, 17, 28, 58
Bancroft, George, 11
Bancroft Library, xiv
Bangkok, 254
Barbour, Erwin, 173
Bartram, John and William, 46
Barunlestes, 211
Barylambda, xvii, 14, 132, 152, 153, 204
basal Z coal, 184
basilisks, 88
Basilosaurus, 21–23, 46
bathmism, 68
Bathmodon, 65
bats, xxviii, 209, 251; evolution of, 153, 156, 194, 214; Kitti's hog-nosed, 206; flying foxes, 258; fossil, 255; wrinkle-lipped, 255
Bauer, George, 72
BBC, 39, 170, 197, 202
Beadnell, H. L., 100, 101
Beagle, H. M. S., 32, 33, 81
Beard, Christopher, xii; on anthropoid evolution, 242; and *Eosimias*, 242; on *Planetetherium*, xxv
bears, 6, 26, 43, 204, 226
Beccari, Odoardo: on Darwinism, 74; on hybridization, 258
bees, 220, 226
beetles, 41, 220, 226, 254
Beijing Academy, 242
Bennett, James Gordon Jr.: and Marsh, 72, 73; and *New York Herald,* 72; and newspaper sensationalism, 72
Bering Strait, 159
Berkeley, city of, 257, 259
Big Bend, 107

Big Cedar Ridge, 228
binturong, 255, 256
birches, 218
Bird, Roland T., 106, 107
birds, 24, 181, 196, 197, 254, 255; and dinosaurs, xxiv, 206; in Zallinger's *Age of Mammals* mural, 1, 152; and Zallinger's *Age of Reptiles* mural, 206
bison, 119, 173, 193, 249; in Zallinger's *Age of Mammals* mural, 121, 159, 193, 252
Bissekty Formation, 209
Blainville, Henri de, 21, 22
Bonaparte, Napoleon, 10, 11, 123, 134
Bone Cabin, 106, 107
bone sharps, 50
Book of Imaginary Beings (Borges), 90
Borges, Jorge Luis, 90
borhyaenids, 83, 84, 89
Bosch, Hieronymus, 203
Botheratiotherium, 21
Bothriodon, 47
Botticelli, xxvi
boulevard Raspail, 3
Bowler, Peter, xiv; on Saporta, 224; on Simpson, 140; on Wortman 224
brain size, 56, 57, 76
Brazil, 171
Breithaupt, Brent, xiii
Br'er Bear, 204
Br'er Fox, 119
Br'er Wolf, 204
British Empire, 31, 100
British Museum, 23, 31, 101
broadbills, 255
Brontops, 92; in Zallinger's *Age of Mammals* mural, 91, 93, 98, 118, 146, 156, 194, 204
brontotheres, xxii; and Cope, 92, 94; and Leidy, 92; and Marsh, 92–94, 98; and Osborn, 94, 96–99; and Simpson, 138. *See also* titanotheres
Brontotherium, 92, 98, 118
Bronx Zoo, 20
Broom, Robert, 34
Brown, Barnum: Burma expedition of, 238–40; and dinosaurs, 106, 122; as fossil collector, 106, 108; and Granger, 106; and mammal fossils, 87, 88, 106, 107, 131, 142; and Osborn, 106, 107, 122; and

Simpson, 129, 136; and Siwalik Hills, 237, 238; and *Tyrannosaurus*, 106; and Wortman, 106, 107, 223
Brown, Lilian, 238, 239
Brown, Roland, xxv
Brownsville, Texas, 224
Buckland, William: as cleric, 15, 16, 34; and Cuvier, 16; eccentricities of, 15; as geologist, 15, 16, 18, 23, 34; and Grant, 21, 22; and Kirkdale Cave, 16; and Owen, 21, 22; and Stonesfield jaws, 16, 17, 21, 22
Buenos Aires, 82, 87, 90, 131
Buffon, Comte de: career of, 4, 5; Cuvier on, 6; on fossils, 4; on horses, 42; on mammoths, 5, 12; and South American mammals, 12; theories of, 4, 7
Bug Creek, 178, 179, 185, 200, 202, 233
Burma, 238, 240, 242
Burns, Joan, xiv
Burt, Andrew S.,
Bush, Guy, 242
bushbabies, 235, 247, 248

Caen, 5
Cairo, 100, 104
Caldwell, W. H., 39
California, 204, 227
Cambridge University, 32
Camelops, 123, 125, 147
Camelot, 122
camels, xxiii, 116, 158, 229; evolution of, 125, 139; Matthew on, 119, 122; and neo-Darwinism, 139, 147; in Zallinger's *Age of Mammals* mural, 123–25, 147
Campbell, Joseph, 249
Canada, 73, 118, 151, 179
capuchin monkeys, 248
capybaras, 33, 81, 83
carbon isotopes, 230, 231
Carboniferous Period, 29
Cardiff Giant, 49, 50, 72, 73
Caribbean, 197
Carlyle, Thomas, 20
Carnegie, Andrew, 223
Carnegie Institute, Pittsburgh, xiii, xxviii
carnivores, 26, 179, 221, 224, 226, 256
Carroll, Lewis, 41
Carson Valley, 74

Carter, Dr. James, 60, 61
Casamayor Formation, 132
Catacombs, Paris, 1–3, 12
Catarrhini, 42, 235, 240, 242, 244, 246
catastrophism, 6, 59, 181, 182
Catholics, 11
cats, 43, 89, 256
cattle, xxiii, 10
Cenozoic Era, 129, 155, 156, 159, 162, 196, 249, 256, 257; and birds, 152, 153, 196, 197; and mammals, 89, 110, 142, 150, 158, 196, 199, 202, 203, 206, 207–32, 243; origin of term, 29; and Zallinger's *Age of Mammals* mural, 176, 177, 193, 216, 218
Central America, 89, 156, 246
Century Magazine, The: and Ballou's dinosaur article, 106; and Osborn's Fayum expedition article, 99–103, 105, 116, 120; and Osborn's prehistoric quadrupeds article, xx, xxi, 96, 98, 104, 106
cetaceans, xxiii, 22
chalk, 10
Challenger Expedition, 158
Challenger, Professor (in Conan Doyle's *Lost World*), 79, 80
Chambers, Robert: and evolutionary diagrams 167; on marsupials, 23, 24; and Owen, 28; as *Vestiges* author, 28
Champsosaurus, 197
Chasmosaurus, xix
chestnut trees, 229
Chicago, 128
chickens, xvii, 255
Chickering Hall, 51
chimpanzees, 234, 248
China, 108, 122, 133, 205, 212, 213, 242
chipmunks, 204
Christchurch College, 15
Christianity, 181
Christiansen, Gale, 76
Christman, Erwin, 98, 99, 118
chronofaunas, 199, 230
Cimolestes, 178; in Zallinger's *Age of Reptiles* mural, xvii, xxvii, 152, 176, 177, 204
Cinderella, xxvii, 153, 248
Ciochon, Russell, 240–42, 246
civets, 255

dinosaurs *(continued)*
and Osborn, 105–7; and Owen, 23, 61;
Saporta on, 221; and Simpson, 128, 129,
136, 143, 145, 151, 181; and Wortman,
106, 222, 223; in Zallinger's *Age of Rep-
tiles* mural, xv–xvii, 25, 79, 176, 193, 203,
259
Diplacodon, 64, 93
*Discours sur les revolutions de la surface de
la globe* (Cuvier), 11
DNA analysis, 213–15, 225, 231, 235, 256
Dobzhansky, Theodosius, 137, 141
docodonts, 149
Dodson, Peter, xvi, xxv
dogs, 6, 84, 89, 109, 110, 174, 256, 258
dolphins, xxvii, 3
doves: domestic, 34; and Darwinism, 34,
35; rock, 34
Downe House, 34
Draco, 255, 258
Dromocyon, xix–xxi
Dubois, Eugene, 108
Duchy of Wurtemburg, 5
Dunbar, Carl O., 219; Zallinger on, xxv;
on Zallinger, xxv; on Zallinger's *Age of
Mammals* cartoon, xxv, 218; on Zal-
linger's *Age of Reptiles* mural, 218
Dwight, Timothy, 48
Dzharakuduk, 209, 211

eastern gray squirrel, 257, 258
Eberle, J. J., 201
echidna, xxvii, 39, 151, 163; Gould on, 195;
intelligence of, 195
Eden, xvii, 153, 212
edentates, 83; fossils of, 83, 253, 254; and
Marsh, 81; and Wortman, 107. *See also*
xenarthra
Edinburgh, 27, 28
Edmonton Formation, 179
Edvardcopeia, 82
Egypt, 100–105, 107, 237, 240
Eiseley, Loren, 76, 77
Eisenhower, General Dwight D., 141
Eldredge, Niles: on dinosaur extinction,
192, 201; on invertebrate paleontology,
189; on mammals, 212; on punctuated
equilibria, 189, 190, 198

elephants, xxviii, 4, 6, 100–103, 105, 214,
215, 256
Eliot, T. S., 183
elk, 25
embranchements, Cuvier's, 10
embryology, 17, 68, 147, 148, 231, 232
Emerson, Ralph Waldo, 36
England, 15, 17, 30, 33, 52, 70
Enlightenment, the, 3, 4, 7
entelodonts, 145, 147
Eobasileus: and Cope, 62, 63; and Marsh,
63; in Zallinger's *Age of Mammals* mural,
55, 62, 63
Eocene Epoch, xx, 47, 81, 83, 107, 238;
and mammal evolution, 34, 60, 64,
65, 83, 84, 89, 91, 93, 100, 137, 235,
237–44, 257; and mass extinction,
197, 201, 202; origin of term, 29;
in Zallinger's *Age of Mammals* mu-
ral, xx, 26, 52, 91, 197, 235, 252,
254–57
Eohippus, 56, 169, 255; Marsh on, 51,
52, 74; Simpson on, 138, 139, 152
Eohomo, 51–53
Eomaia scansoria, 213–15
Eomanis, 254
Eosimias, 242
Eotitanops, 98
Epigaulis, 204
Equus, 44, 51, 71, 95, 96, 138, 167, 168, 245;
parvulus, 49, 50; *simplicidens,* 41, 127; in
Zallinger's *Age of Mammals* mural, 53,
175, 250
Essay on the Principle of Population
(Malthus), 34
Essex, 26
Etna, Mount, 28
Euarchontoglires, 214
eucaryotic cells, 231
Europe, 4, 6, 44, 107, 129, 135; and anthro-
poid evolution, 235, 243; and horse evo-
lution, 42–45, 138; paleontological po-
tential of, 45
Eurotamandua, 254
eutherians, 149, 150, 208, 209, 223,
226–28. *See also* placentals
Evans, John, 46
Everest, Mount, 176

Central Asian expeditions, 113, 114, 122; on *Dinohyus*, 172, 173; evolutionary ideas of, 118–20, 126; on ground sloths, 171, 174; on horses, 120, 127; on mesonychids, xxi, xxiii; on *Moropus*, 172, 173 and Osborn, 118–20, 122, 130; and Simpson, 126, 127, 130; on Siwalik ape fossils, 238; on *Smilodon*, 119, 120, 146

Mayr, Ernst, vii, 141, 167

McDonald, Greg, xiii, 42

McKenna, Malcolm: on biogeography, 159, 162, 164; and Mongolia, 208, 211; and zalambdalestids, 212

Mead, Margaret, 130

Megaceratops, 92

Megalosaurus, 16, 17

Megatherium: Cuvier on, 7, 81; Darwin and, 33; Gee on, 170–72; Lund on, 171; Matthew on, 119; Owen on, 33, 37; in *Walking with Prehistoric Beasts*, 170; in *Wonders of Life on Earth*, 81; in Zallinger's *Age of Mammals* mural, 1–3, 166, 175, 250, 253

Mendel, Gregor, 71, 126

Merychippus, 53, 175

Mesohippus, 50, 53, 123, 175, 195

mesonychids, 197; nomenclature of, xix–xxii; and whales, xxii–xxiv, xxvii; in Zallinger's *Age of Mammals* cartoon, xxi; in Zallinger's *Age of Mammals* mural, xix–xxi, 154, 253

Mesonyx, xxi, xxii, 53, 54, 154, 253

Mesozoic Era, 31, 88, 151, 157, 225–28; mammals of, 105, 106, 108, 128, 129, 133, 149, 156, 168, 212, 213, 215, 223, 225–28, 233, 243; origin of term, 29; and Zallinger's *Age of Reptiles* mural, 176, 177, 193

Messel Shale, 254, 255, 257

Metacheiromys, 253, 254

metatherians, 150. *See also* marsupials

Mexico, 196

mice, 94, 95, 204, 208

Michelangelo, xv, xvi, xviii, xxvi, 249

Milwaukee Public Museum, 185

mimotonids, 212

Miocene epoch, 53, 89, 127, 197, 238, 255; and Agate Fossil Beds, 172–74; origin of

term, 29; and Pikermi fauna, 43, 44; in Zallinger's *Age of Mammals* mural, 123, 124, 147, 155, 159

Miohippus, 50, 74, 195

missing link, 109

Mitchell, W. J. T.: on dinosaurs, xvii; on Zallinger's *Age of Reptiles* mural, xvi, xxvi; on Zallinger, xvii, xviii, xv

modernism, xv–xvii

Moeritherium, 102, 105

Mogaung, 238, 240

mollusks, 7, 10, 188–90

Mongolia, 108–22, 130, 133, 149, 207, 208, 211, 237

monkeys, New World, 235, 242, 242, 266, 247; capuchin, 248; howler, 248

monkeys, Old World, 42, 235, 240, 242, 244, 246; baboon, 42, 244; langur, 256; macaque, 244, 256

monocotyledons, 220

monotremes, 66; Cuvier on, 20; Darwin on, 33, 34; definition of, xxvii, 20; evolution of, 20, 31, 151, 163; Gould on, 194–95; Lamarck on, 20; Owen on, 20, 31, 39

Montana, 106, 177, 186

Montbeliard, 5

Montmartre, 4, 13

Montpellier, 19

Mordred, 115

Moreno, F. P., 84

Morgan, J. P., 86

Morgan, Thomas H., 126

Morgan, Vincent, xiv

Morganucodontids, 148

Moriarty, Dr., 72

Morocco, 215

Moropus, 173, 174

Mosasaurus, xix, 10

Moscow, 133, 134

Moscow University, 45

Mount Blanco, 127

multituberculates: at Como Bluff, 149; Cope on, 66; definition of, 66, 149; evolution of, 66, 149, 225, 225; extinction of, 197, 244; at Hell Creek, 178, 179; in Mongolia, 149, 207, 208, 228; in San Juan Basin, 66; teeth of, 149; in

pyrotheres, 83, 84, 89, 90
Pyrotherium, 85, 87, 88, 89
pythons, 254, 255

Quakers, 58, 59
Queen Victoria, 23
Qui, Tao, 242

rabbits, xxviii, 9, 52, 235; evolution of, 211, 212, 214, 227; in Zallinger's *Age of Mammals* mural, 204
Rabinowitz, Alan, xiii
raccoons, 169, 209, 259
racial senescence, 151
rainforest, 218
Rainger, Ronald, xiv, 98, 126
Rand, Ayn, 129, 141
Raphael, xxvi
Raton Basin, 184
rats, 195
Raup, David: on asteroid theory, 182, 191, 196, 197; on punctuated equilibria, 191
Ray, John, 3
recapitulation, 57, 68
Recherches sur les ossemens fossiles (Cuvier), 9, 17
red squirrels, 257
redwoods, 259
reindeer, 249
Renaissance, xv
reptiles, 11, 21–24, 31, 57, 147, 148, 181; in Thailand, 255; in Zallinger's *Age of Mammals* mural, xviii, 216–19
retroviruses, 231, 232
rhinos, xxiii, xxviii, 4, 5, 6, 43, 57, 60, 64, 94, 108, 111, 114, 147, 171, 229, 230; in Agate Fossil Beds diorama, 173; in Zallinger's *Age of Mammals* mural, 91, 152, 158, 173
Rigby, Keith, 185
Rio Chico, 87, 132
Rocky Mountains, xxv, 46, 104, 155, 184
rodents, 153, 251; evolution of, 211, 212, 214, 227; in Zallinger's *Age of Mammals* mural, 204, 257
Roemer, Ferdinand, 45, 48
Romans, 4

Romer, Alfred S., 152, 196
Rose, Kenneth D., 237
Rousseau, Jean Jacques, 6
Royal College of Surgeons, 16
Royal Institute, 16
Rudwick, Martin J. S., xiv, 11, 12
Rupke, Nicolaas, 40
Ruskin, John, 15
Russia, 44, 109, 133, 134, 208, 209
Ryan, Frank, 232

saber-tooths, 43, 81, 170; Matthew on, 118–20; and neo-Darwinism, 119, 146, 147; and orthogenesis, 119, 120; Scott on, 119; Simpson on, 138, 139; in Zallinger's *Age of Mammals* cartoon, 146, 147; in Zallinger's *Age of Mammals* mural, 121, 146
Sahara, 124
Saint Denis, 4
Saint Louis, 46
salamanders, 11, 58, 71, 177, 181; as mammal ancestors, 38
sambur, 256
San Diego State University, xiii
San Juan Basin: author's visit to, 65; and Baldwin, 65, 66; and Cope, 65–69; fossils from, 65–69; and Lucas, 65; and Marsh, 65, 66; and Simpson, 144
Santa Cruz (Argentina), 83, 87,
Santa Cruz Formation, 83, 84, 87, 89
Saporta, Gaston de: career of, 219; and Darwin, 219–21; and insect-angiosperm coevolution, 219, 220, 226; and mammal-angiosperm coevolution, 220–22, 224, 226, 227; and paleobotany, 219, 225; and Wortman, 223, 224
sauropods, 106, 223, 225
Sauropsida, 38
Savage, Donald, 240
savanna: and mammal evolution, 44, 155, 199, 229, 230; in Zallinger's *Age of Mammals* mural, 147, 193, 218
Saxony, 5
Scarritt, H. S., 131
Schlosser, Max, 108, 116, 237
Schuchert, Charles, 106, 188, 224
Science, 178, 180, 185, 186

South America, 12, 155, 156, 158, 159, 161, 169, 196, 203, 252; and Ameghinos 79–90; and anthropoid evolution, 241, 246; and Darwin, 33, 34, 36; and edentates, 6, 7, 33, 34; and Simpson, 131–34,

South Dakota, 46

South Pacific, 30

South Pole, 161, 202

Soviet Union, 133, 134, 207, 209, 210

Spencer, Herbert, 45, 68

spiders, 226

Spitzka, Edward, 76, 77

Sporomiella, 231

Springer, Mark, 213–15, 231, 253

squirrels, ground, 204

squirrels, tree, 226, 232; in Berkeley, 257; eastern gray, 257–58; and evolution, 257–59; flying, 204, 257, 258; fox, 257–59; giant, 256; and hybridization, 258; origin of, 257; red, 257; Simpson on, 257; in Thailand, 256, 257; western gray, 257

Stalin, Joseph, 133

Stanley, Henry M., 72

Stebbins, G. Ledyard, 141

stegosaurs, 80, 203

Stevenson, Robert Louis, 19, 40

Stonesfield jaws, 29, 30, 39; Blainville on, 21, 22; Buckland on, 16, 21, 22; Chambers on, 23, 24; Cuvier on, 16, 17; Owen on, 17, 21, 22, 31; Simpson on, 149

Stonesfield slate, 15–17, 65

Structure of Evolutionary Theory, The (Gould), 198, 199, 212, 213

Stuttgart, 5

Subhyracodon, 91, 147

superorders, 214, 253

swamps, 81, 248; and mammal evolution, 44; in Zallinger's *Age of Mammals* mural, 176, 254; in Zallinger's *Age of Reptiles* mural, 176, 193

sweepstakes distribution, 158, 159, 164

sycamores, 198, 209, 218

symbiology, 231, 232

symbiosis, 227, 232

symmetrodonts, 149, 168

synapsids, 148, 157, 168

Synoplotherium, xviii, xix, xxi

Systema Naturae (Linnaeus), 3

taeniolabids, 225

tamandua, 253, 254

tapirs, xxiii, 8, 10, 14, 81, 113, 116, 158, 256

tarsiers, 235–37, 241, 242, 246, 248

tarsiiforms, 235, 237, 241–43

Tasmanian wolf, xx

tayras, 66, 67, 226

teeth, of *Basilosaurus,* 22; of Cope, 76; of *Coryphodon,* 26, 28, 67; of crocodiles, xxviii; of dinosaurs, 15, 226; as fossils, 4, 230; of hippos, 5; of horses, 42, 43, 44, 52, 67, 137, 138; of humans, xxviii; of mammals, xxviii, 3, 4, 60, 81, 83, 117, 128, 148–53, 163, 164, 225, 230; of mammoths, 103, 198, 256; of mastodons, 103, 256; of multituberculates, 66, 225; of other vertebrates, xxviii; of pantodonts, 67, 83; of platypus, 66, 163; of primates; 116, 117, 233, 234–36, 238, 239; of reptiles, 21, 148; of saber-tooths, 118–20, 139, 146–48; of sloths, 171; of uintatheres, 67; of whales, 21, 22; of zhelestids, 209, 210, 212. *See also* tusks, tribosphenic molar

Teleoceras, 103

Tempo and Mode in Evolution (Simpson), 140

Tennessee, 30

Terror, the, 5, 7

Tertiary Period, 11, 28, 29, 36, 42, 74, 97; and biogeography, 158, 159, 164; and Hell Creek, 178, 180, 184–87; and Marsh's Darwinism 56, 57; origin of term, 5; and Saporta, 219, 220; and Stonesfield jaws, 16, 17, 23,

Tetonius, 235

Tetraclaenodon, 146, 152

Texas, 45, 126, 127

Thailand, xiii, 254–57

"theory of the earth," Buffon's, 6

Theory of the Earth (Hutton), 28

theriodonts, 32, 38, 39, 41, 147, 148, 157

therion, 3

Compositor:	G&S Typesetters, Inc.
Text:	11.25/13.5 Adobe Garamond
Display:	Adobe Garamond
Printer and binder:	Edwards Brothers, Inc.